Topics in
Current Physics

41

Topics in Current Physics   Founded by Helmut K. V. Lotsch

Volume 39   **Nonequilibrium Vibrational Kinetics**
            Editor: M. Capitelli

Volume 40   **Microscopic Methods in Metals**
            Editor: U. Gonser

Volume 41   **Structure and Dynamics of Surfaces I**
            Editors: W. Schommers and P. von Blanckenhagen

Volume 42   **Metallic Magnetism I**
            Editor: H. Capellmann

Volumes 1–38 are listed on the back inside cover

Structure and Dynamics of Surfaces I

Edited by W. Schommers
and P. von Blanckenhagen

With Contributions by
J. E. Black P. von Blanckenhagen H. Ibach
S. Lehwald L. D. Marks T. S. Rahman
K. H. Rieder M. Rocca W. Schommers
C. Varelas

With 175 Figures

Springer-Verlag Berlin Heidelberg New York
London Paris Tokyo

Dr. Wolfram Schommers
Dr. Peter von Blanckenhagen

Kernforschungszentrum Karlsruhe, Institut für Nukleare Festkörperphysik, Postfach 3640, D-7500 Karlsruhe, Fed. Rep. of Germany

ISBN-13:978-3-642-46576-5   e-ISBN-13:978-3-642-46574-1
DOI: 10.1007/978-3-642-46574-1

Library of Congress Cataloging-in-Publication Data. Structure and dynamics of surfaces. (Topics in current physics ; 41) Bibliography: p. 1. Surfaces (Physics) 2. Surface chemistry. I. Schommers, W. (Wolfram), 1941–. II. Blanckenhagen, P. (Peter), 1936–. III. Series. QC173.4.S94S76 1986   530.4'1   86-6654

Offset printing and bookbinding: Konrad Triltsch, Graphischer Betrieb, Würzburg.
2153/3150-543210

Preface

During the last decade, surface research has clearly shifted its interest from the macroscopic to the microscopic scale; a wealth of novel experimental techniques and theoretical methods have been applied and developed successfully. The Topics volume at hand gives an account of this tendency.

For the understanding of surface phenomena and their exploitation in technical applications, the theoretical and experimental analysis at the *microscopic level* is of particular interest. In heterogeneous catalysis, for example, a chemical reaction takes place at the interface of two phases, and the process occurring at the surface is composed of a sequence of individual *microscopic* steps. These individual steps include adsorption, desorption, surface diffusion, and reaction on the surface. These elementary steps are greatly influenced by the *structure* and the *dynamics* of the surface region. Especially the catalytic activity may strongly depend on the structure of the catalyst's surface. The necessity of performing surface investigations on a *microscopic scale* is also reflected clearly in research work relating to *metal-semiconductor interfaces* which determine essentially the properties of electronic device materials. The experimental probe on the atomic scale, coupled with parallel theoretical calculations, showed that the electronic properties of a metal-semiconductor interface strongly depend on the crystallographic *structure* of the semiconductor; in particular, it is important to know in this context the modification of the atomic arrangement in the surface region caused by the termination of the crystal by the surface. Also, processes which occur at *solid-electrolyte* interfaces are of considerable importance; the technological significance of corrosion, fuel cells, electrocatalysis and photoelectrochemical solar energy conversion are examples. In the analysis of such solid-electrolyte interfaces, the *structure* and the *dynamics* in the interfacial region will be essential parameters, too.

The individual chapters of this book have been prepared by experts who have made their own contributions to the subjects discussed in the book. However, a quasi-monograph written by a number of authors can have drawbacks as compared with a monograph written by a single author: The contents cannot achieve the same degree of homogeneity and a complete coverage of the subject can be realized only with difficulty. In the present book these deficiencies could not be avoided completely.

The treatment is divided into two parts, with each chapter including the *background* of the subject, *typical results*, and in most cases also trends of *future developments*. In the present volume (TCP 41), first some typical examples of the influence of structure and dynamics on surface phenomena are discussed and then six up-to-date reviews are presented of the structural and dynamical properties of surfaces: experimental methods for determining surface structures and corrugations (atomic-beam scattering, scanning tunneling microscopy, etc.); high-resolution electron microscopy; surface channeling: its application in surface-structure analysis, the determination of surface vibrational amplitudes; dynamical surface properties in the *harmonic* approximation (the theory of vibrations for bare surfaces and adsorbate layers); molecular dynamics and the study of *anharmonic* surface effects; experimental study of surface phonon dispersion curves of surface and adsorbate layers by means of electron energy-loss spectroscopy. In assembling a program such as this, it was necessary to bring together a team of leading experts from several countries: K.H. Rieder from IBM Zürich, L.D. Marks from Northwestern University, USA, C. Varelas from Munich, J.E. Black from Brock University, Canada, M. Rocca from Genova, H. Ibach and S. Lehwald from KfA Jülich, T.S. Rahman from Kansas State University. We would like to thank each of the authors for their excellent cooperation.

The second volume (*Structure and Dynamics of Surfaces II*), which is in preparation, will deal with the following topics: study of surface phonons by means of the Green's function method; layer growth and surface diffusion; phase transitions on single crystal surfaces and in chemisorbed layers; solid and liquid surfaces studied by synchrotron x-rays; statistical mechanics of the liquid surface and the effect of premelting; the roughening transition; structural and dynamical aspects of adsorption and desorption; many-body description of surface elementary excitations and its application to semiconductors.

Karlsruhe, October 1985   *W. Schommers · P. von Blanckenhagen*

Table of Contents

1. **Introduction: The Relevance of the Structure and Dynamics of Surfaces**
   By W. Schommers and P. von Blanckenhagen (With 12 Figures) 1
   1.1 Structure of Surfaces 1
       1.1.1 Effect of Relaxation on Electronic Surface Properties ... 1
       1.1.2 Heterogeneous Catalysis 4
   1.2 Dynamics of Surfaces 9
       1.2.1 Surface Thermodynamic Functions in the Harmonic
             Approximation 9
       1.2.2 Electron-Phonon Coupling 13
   1.3 Conclusion and Outlook 14
   References 15

2. **Experimental Methods for Determining Surface Structures
   and Surface Corrugations.** By K.H. Rieder (With 30 Figures) 17
   2.1 Basic Physical Principles 17
   2.2 Notation and Conventions 18
   2.3 Surface Diffraction from Regular Gratings 20
       2.3.1 Kinematical Interference Conditions for Surface Scattering 20
       2.3.2 Experimental Methods Employing Diffraction from
             Gratings 22
   2.4 Other Methods Based on Diffraction Effects 38
       2.4.1 Surface Extended X-Ray Absorption Fine Structure
             (SEXAFS) 38
       2.4.2 Surface Extended Energy-Loss Fine Structure (SEELFS) 40
       2.4.3 Photoelectron Diffraction 42
       2.4.4 Near-Edge X-Ray Absorption Fine Structure (NEXAFS) 43
   2.5 Microscopy Techniques with Atomic Resolution 46
       2.5.1 Field Ion Microscopy 46
       2.5.2 Scanning Tunneling Microscopy 49
       2.5.3 High Resolution Transmission Electron Microscopy
             (TEM) 55
   2.6 Ion Scattering Spectroscopy (ISS) 56
   2.7 Other Methods 62
       2.7.1 Stimulated Desorption 62
       2.7.2 Vibrational Spectroscopies 62

      2.7.3 Surface Electronic Properties and Optical Methods 63
2.8 Outlook 64
References 65

3. High-Resolution Electron Microscopy of Surfaces
  By L.D. Marks (With 29 Figures) 71
3.1 Background 71
3.2 Image Formation 74
      3.2.1 Basics 74
      3.2.2 Electron Scattering 76
      3.2.3 Lens Aberrations 80
      3.2.4 Image Localisation 82
      3.2.5 Image Simulation 84
3.3 Applications 87
      3.3.1 Qualitative Imaging 88
      3.3.2 Quantitative Analysis 90
3.4 A Note of Warning 95
3.5 Gold(110) 97
3.6 Other High-Resolution Techniques 103
      3.6.1 Transmission Geometry 103
      3.6.2 Reflection Geometry 106
3.7 Conclusions and the Future 106
References 107

4. Surface Channeling and Its Application to Surface Structures and Location of Adsorbates
  By C. Varelas (With 31 Figures) 111
4.1 Overview 111
4.2 Principles 111
      4.2.1 Continuum Potential 111
      4.2.2 Transverse Energy 115
      4.2.3 Flux Distribution 116
      4.2.4 Dechanneling 119
      4.2.5 Classical Scattering Versus Quantum Diffraction 119
4.3 Surface Channeling Angular Yield Profiles 120
      4.3.1 General Considerations 121
      4.3.2 Computer Simulation 125
4.4 Applications 128
      4.4.1 Experimental 129
      4.4.2 Surface Structure and Reconstruction 130
      4.4.3 Surface Reconstruction 132
      4.4.4 Location of Adsorbed Atoms 135
4.5 Applicability of Surface Channeling to Surface Studies 137
      4.5.1 Surface Sensitivity 137
      4.5.2 Selection of the Auger Transition 139

      4.5.3   Selection of the Angle of Incidence to the Surface 142
  4.6   Factors of Influence on the Angular Yield Profiles 145
      4.6.1   Radiation Damage 145
      4.6.2   Thermal Lattice Vibrations 146
      4.6.3   Mean-Free Path of Excited Electrons 148
      4.6.4   Surface Steps 148
  4.7   Concluding Remarks 149
  References 151

5. Dynamical Surface Properties in the Harmonic
Approximation. By J.E. Black (With 16 Figures) 153
  5.1   Introductory Remarks 154
  5.2   The Theory of Vibrations in the Harmonic Approximation 157
      5.2.1   The Case of No Periodicity 157
      5.2.2   The Case of a Bare Periodic Surface 161
      5.2.3   The Case of a Periodic Overlayer 163
  5.3   Bare Metals 164
      5.3.1   Methods of Mode-Frequency Calculation 164
      5.3.2   General Theoretical Calculations 171
      5.3.3   Experimental Results and Theoretical Interpretation 173
  5.4 Metals with Adsorbates 179
      5.4.1   The Calculation of Adsorbate Mode Frequencies 180
      5.4.2   Experimental and Theoretical Results — Commensurate
           Overlayers 183
      5.4.3   Experimental and Theoretical Results —
           Incommensurate Overlayers 190
  5.5   Concluding Remarks 193
      5.5.1   Bare Surfaces 193
      5.5.2   Adsorbates 194
  References 195

6. Molecular Dynamics and the Study of Anharmonic Surface
Effects. By W. Schommers (With 35 Figures) 199
  6.1   Introductory Remarks 199
  6.2   Analysis of Many-Particle Systems 200
      6.2.1   Pair Potential Approximation 200
      6.2.2   "Simple Models" 202
      6.2.3   Molecular Dynamics 203
  6.3   Average Values 205
      6.3.1   Density in Phase Space 205
      6.3.2   Statistical-Mechanical Ensembles 206
      6.3.3   Measureable Quantities 209
  6.4   Molecular Dynamics Systems 211
      6.4.1   Models for the Bulk 211
      6.4.2   Models with Surfaces and Interfaces 213

      6.4.3   Interaction Potentials.................................... 213

      6.4.4   Time Evolution of Molecular Dynamics Systems 219

  6.5   Applications of the Molecular Dynamics Method 224

      6.5.1   Phase Transitions in Two-Dimensional Systems 224

      6.5.2   Diatomic Molecules Adsorbed on Surfaces.............. 227

      6.5.3   Melting Transition of Near Monolayer Xenon Films
              on Graphite 228

      6.5.4   Microscopic Behavior of Krypton Atoms at the Surface.. 230

  6.6   Final Remarks 240

  References.. 241

7. **Surface Phonon Dispersion of Surface and Adsorbate Layers**
  By M. Rocca, H. Ibach, S. Lehwald, and T.S. Rahman
  (With 22 Figures) 245

  7.1   Background and Overview..................................... 245

  7.2   Description of the Experiment 246

      7.2.1   General Features of Electron-Surface Interaction 246

      7.2.2   Experimental Details 249

      7.2.3   Multiple-Scattering Effects............................. 252

  7.3   Phonon Dispersion Curves.................................... 258

      7.3.1   The Clean Ni Surface.................................. 258

      7.3.2 (2×2) Adsorbate Phases on Ni(100) 262

      7.3.3   Effect of Random Adsorption of Oxygen on the
              Rayleigh Wave 266

  7.4   Internal Strains, Phonon Anomalies and Reconstruction........ 268

      7.4.1   A Lattice-Dynamical Model with Internal Strains 268

      7.4.2   The Stress-Model and Other Experimental Observations 271

  References.. 274

Subject Index 277

List of Contributors

Black, J.E.
Physics Department, Brock University, St. Catharines,
Ontario L2S 3A1, Canada

Blanckenhagen, von P.
Kernforschungszentrum Karlsruhe, Institut für Nukleare Festkörperphysik,
Postfach 3640, D-7500 Karlsruhe, FRG

Ibach, H.
Institut für Grenzflächenforschung und Vakuumphysik,
Kernforschungsanlage Jülich, Postfach 1913, D-5170 Jülich, FRG

Lehwald, S.
Institut für Grenzflächenforschung und Vakuumphysik,
Kernforschungsanlage Jülich, Postfach 1913, D-5170 Jülich, FRG

Marks, L.D.
Department of Materials Sciences, The Technological Institute,
Northwestern University, Evanston, IL 60201, USA

Rahman, T.S.
Cardwell Hall, Department of Physics, Kansas State University,
Manhattan, KS 66506, USA

Rieder, K.H.
IBM-Zürich Research Laboratory, CH-8803 Rüschlikon, Switzerland

Rocca, M.
Dipartimento di Fisica dell'Università di Genova and GNSM Unità di Genova,
Via Dodecaneso 33, I-16146 Genova, Italy

Schommers, W.
Kernforschungszentrum Karlsruhe, Institut für Nukleare Festkörperphysik,
Postfach 3640, D-7500 Karlsruhe, FRG

Varelas, C.
Sektion Physik, Universität München, D-8000 München, FRG
Present address: Stiebel Eltron GmbH u. Co. KG, Dr.-Stiebel-Strasse,
D-3450 Holzminden, FRG

1. Introduction: The Relevance of the Structure and Dynamics of Surfaces

W. Schommers and P. von Blanckenhagen

With 12 Figures

Structural and dynamical properties of surfaces are of considerable importance for the elucidation of surface phenomena. For example, the adsorption of particles and chemical reactions at surfaces are influenced by these properties. In this introductory chapter we shall discuss some typical examples of the influence of structure and dynamics on surface phenomena.

1.1 Structure of Surfaces

The termination of the crystal by a surface can modify the atomic arrangement in the surface region. Two effects are of particular interest:

1. The distances between atomic planes parallel to the surface deviate from their bulk values; this phenomena is called *relaxation*.
2. Lateral structural modifications may occur, and this effect is called *surface reconstruction*. In this case the shape and size of the unit cell are modified.

In general, several atomic layers participate in these two structural effects.

As the surface atomic structure, as described by relaxation and reconstruction, is crucial for the understanding of many surface phenomena, much effort has been spent on the development of theoretical and experimental methods for their determination.

1.1.1 Effect of Relaxation on Electronic Surface Properties

As an example, let us discuss the effect of structural relaxation on the electronic surface states of the n-type semiconductor InP(110). Indium phosphide is an important electronic device material [1.1]. These devices (e.g., Schottky barrier diodes and metal-insulator-semiconductor (MIS) devices) depend critically on the electronic properties of the interface formed between the InP crystal and the metal and insulator, respectively. In the study of such interfaces it is essential to understand the electronic properties of the *free* InP surface.

The structure of the InP(110) surfaces has been obtained by LEED studies [1.2] and is schematically shown in Fig. 1.1. The P atoms in the outermost layer move outward (0.06 Å) and the In atoms inward (0.63 Å), giving a vertical shear of 0.69 Å. In the second layer the In atoms move outward (0.03 Å) and the

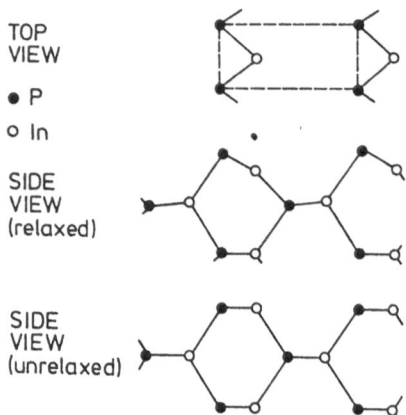

TOP
VIEW

● P

○ In

SIDE
VIEW
(relaxed)

SIDE
VIEW
(unrelaxed)

Fig. 1.1. Schematic representation of the surface atomic geometry for the (110) surface of InP. More details were given in [1.2]

P atoms inward (0.03 Å). The electronic surface states of the system strongly depend on the atomic arrangement at the surface: While in the *relaxed* geometry (Fig. 1.1) the electronic surface states are driven out of the gap between the valence and conduction band, the *unrelaxed* geometry (Fig. 1.1) admits surface states in the gap. This is demonstrated in Fig. 1.2 by model calculations. Let us briefly discuss these results.

Occupied and unoccupied electronic surface states can exist in the following three regions (Fig. 1.2):

1. in the lower part of the bulk valence-band gap (~9 eV);
2. in the central region of the bulk valence-band gap (~5 eV); and
3. in and around the fundamental gap, i.e. the gap between the valence and conduction bands.

There are also surface resonances existing in the shaded regions and mix with the bulk states.

The electronic states of InP(110) (Fig. 1.2) have been investigated theoretically by self-consistent pseudopotential and tight-binding methods, and experimentally by angle-resolved photoelectron spectroscopy [1.3]. The results of the self-consistent pseudopotential calculation for the relaxed and the unrelaxed geometries (Fig. 1.1) are given by the solid and broken lines in Fig. 1.2a and b. The surface features are labeled A_i and C_i, respectively, according to whether the feature is localized on the anion or cation. The surface states in the gap between the valence and conduction bands are of most interest; it can be seen from Fig. 1.2 that due to the *relaxation* the surface states are driven out of the gap, and this is supported by the tight-binding calculation (Fig. 1.2c). In particular, both calculations show that surface states are absent in the thermal gap (which exists at Γ). The occupied dangling-bond band (branch A_5 in Fig. 1.2b) is localized on the P atom and is p-like in character (Fig. 1.3a). The unoccupied dangling-bond band (branch C_3 in Fig. 1.2b) is localized on the In atom and is more s-like and a little p-like in character [1.3] (Fig. 1.3b).

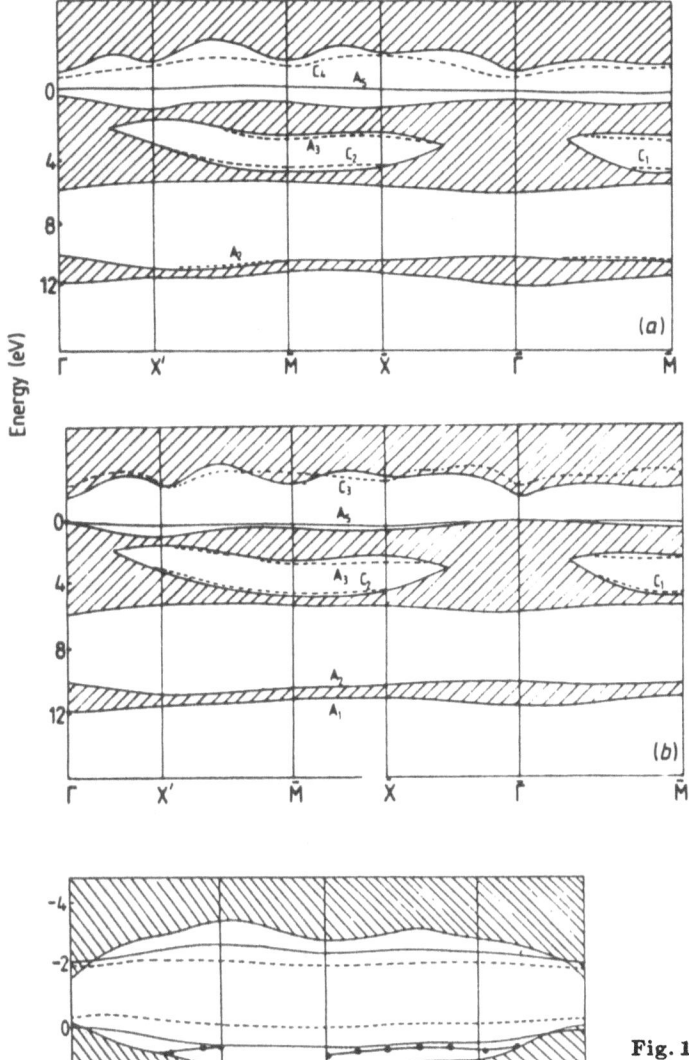

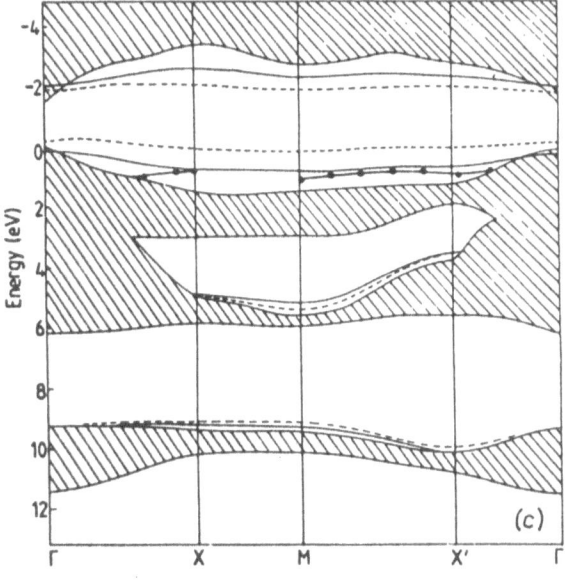

Fig. 1.2a–c. Dispersion curves for electronic surfaces states at the (110) surface of InP [1.3]. Self-consistent pseudopotential calculation: (a) unrelaxed geometry; (b) relaxed geometry. Tight-binding calculation (c): broken curves: unrelaxed geometry; solid curves: relaxed geometry. The hatched regions in (a, b and c) correspond to projected bulk states. In (c) also the dispersion of the uppermost occupied surface state determined by angle-resolved photoelectron spectroscopy [1.3] is plotted

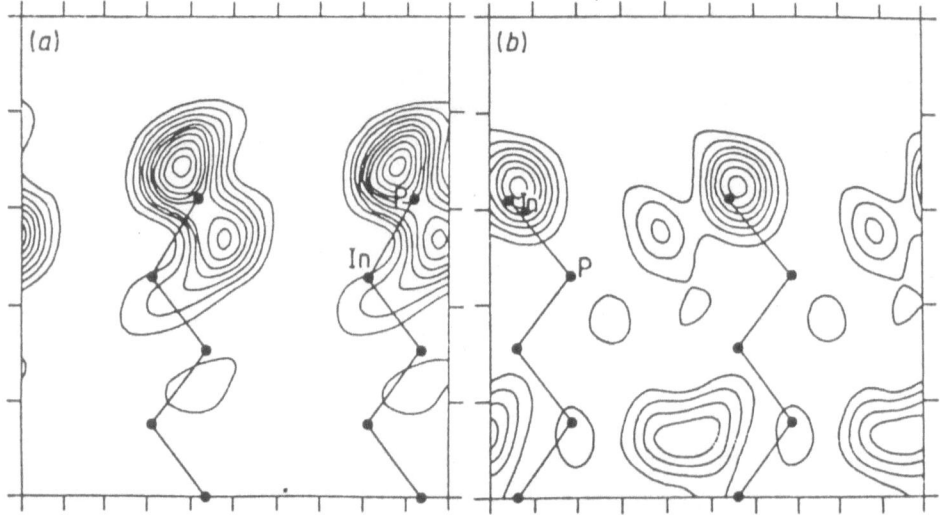

Fig. 1.3a,b. Charge-density contours for the relaxed (110) surface of InP at point $\bar{x}$ [1.3]: **(a)** occupied dangling-bond surface state localized on the P atom and **(b)** unoccupied dangling-bond surface state localized on the In atom. The charge density contours for the unrelaxed case are shown in [1.3] and differ from the relaxed case

The absence of surface states in the gap between the valence and conduction bands means that no surface *Fermi level pinning* can occur. If, however, *relaxation* is fully or partly removed (e.g., by adsorption of a surface metal layer) electronic surface states may again lie in the fundamental gap and Fermi level pinning becomes possible.

1.1.2 Heterogeneous Catalysis

Heterogeneous catalysis is a chemical reaction at the *interface* of two phases (for example, at the gas-solid interface). In a heterogeneously catalyzed reaction the catalyst (the solid) remains unchanged and the reaction may be influenced by the composition and by the atomic and electronic structures of the solid surface. Therefore, this phenomenon is a typical problem of surface science.

The chemical reaction at the catalyst proceeds at a higher rate than in the homogeneous phase, i.e. in the gas phase without a catalyst. The activation energy in the heterogeneous catalysis is much lower than in the homogeneous reaction, and since the reaction rate depends exponentially on the activation

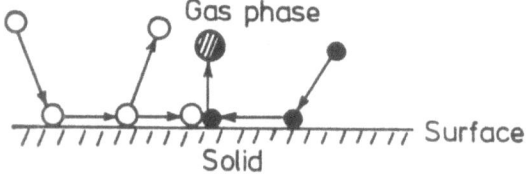

Fig. 1.4. Individual steps in heterogeneous catalysis (reproduced from [1.7])

energy the reaction rate in the heterogeneous reaction is higher than in the corresponding homogeneous process. In the reaction mechanism several *individual steps* of gas-surface interaction are involved (see also Fig. 1.4):

- – adsorption,
- – migration of adsorbed particles,
- – chemical reactions,
- – desorption.

The mechanism of these steps will be dependent on the surface properties; in particular, the *surface structure* will be involved. Thus, *kinetic measurements* alone (the dependence of the reaction rates on the concentration of the particles present in the gaseous phase is studied) are not qualified to analyze the microscopic reaction mechanism. More direct information is needed and this can be obtained by surface sensitive *diffraction experiments* and by application of *spectroscopic techniques* [1.4,5]. Instead of real catalysts, single crystals with well-defined surface structures and composition, prepared and investigated under ultrahigh-vacuum conditions, are used for these investigations. In this "surface science approach" the *individual steps* in a heterogeneously catalyzed reaction can, in principle, be analyzed. As examples for heterogeneous reactions let us briefly discuss the *ammonia synthesis* (Haber-Bosch process) and the oxidation of carbon monoxide; in particular, the influence of the surface structure on these reactions will be considered.

a) Ammonia Synthesis

The potential energy diagram for the synthesis from the elements

$$N_2 + 3H_2 \rightarrow 2NH_3 \tag{1.1}$$

is shown in Fig. 1.5. For a homogeneous reaction in the gas phase the activation energy is so large that the industrial synthesis can be performed only via catalysts (in an N_2/H_2 mixture of about 100 atm at 400°C). Studies [1.6] of real catalysts (iron mixed with oxidized K, Al or Si) have shown that nitrogen adsorption is the rate-limiting step, but it has not been clear whether the ammonia synthesis proceeds via molecular nitrogen

$$N_{2,ad} + 6H_{ad} \rightarrow 2NH_3, \tag{1.2}$$

or via atomic nitrogen

$$N_{ad} + 3H_{ad} \rightarrow NH_3. \tag{1.3}$$

Single-crystal-surface (Fe) studies [1.7] helped to resolve this problem: the catalyst produces *atomic* nitrogen and then adds atomic hydrogen. The reaction mechanism in the synthesis of ammonia is given by the following steps [1.7]:

$$H_2 \rightleftharpoons 2H_{ad},$$

$$N_2 \rightleftharpoons N_{2,ad} \rightleftharpoons 2N_{ad},$$

$$N_{ad} + H_{ad} \rightleftharpoons NH_{ad},$$

$$NH_{ad} + H_{ad} \rightleftharpoons NH_{2,ad},$$

$$NH_{2,ad} + H_{ad} \rightleftharpoons NH_{3,ad} \rightleftharpoons NH_3. \qquad (1.4)$$

The *reaction rate* in the ammonia synthesis is strongly dependent on the *surface structure* of the catalyst (iron), i.e. on the crystallographic orientation of the iron surface. The sequence in activity is

$$(111)>(100)>(110),$$

with differences of up to about *two orders* of magnitude. Theoretical aspects of the heterogeneous catalysis were discussed in [1.8] and the literature therein.

Also *defects at the surface* (e.g., monoatomic steps) can influence the reactivity at the surface. Furthermore, the reactivity of *reconstructed* surfaces can be different from that of an *unreconstructed* one. In this connection let us briefly discuss now the *kinetic oscillations* of the steady-state rate of catalytic CO oxidation on clean Pt surfaces.

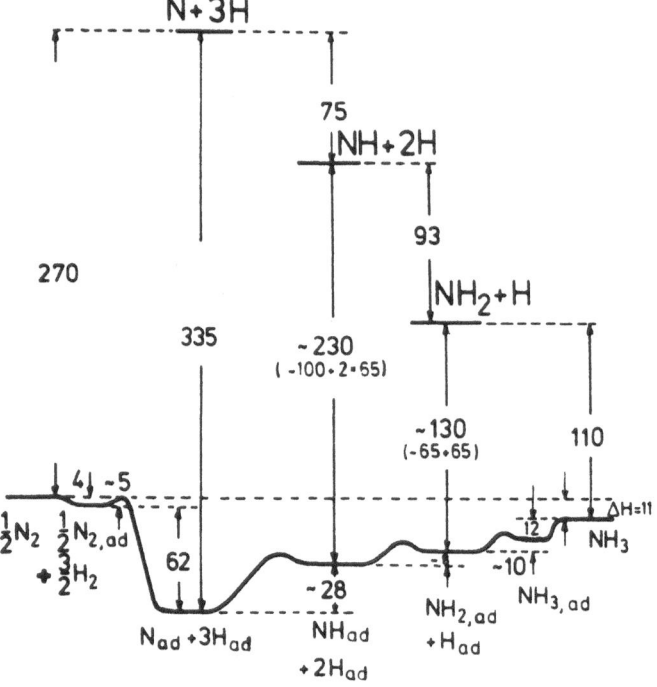

Fig. 1.5. Potential energy diagram for ammonia synthesis catalyzed by an iron surface (energy values in kcal/mole) (reproduced from [1.7])

6

b) Oxidation of CO on Pt Surfaces and Periodic Variations of the Surface Structure

Recent experiments [1.9,10] have demonstrated that there is a connection between a structural phase transformation and temporal oscillations in the steady-state rate of catalytic reactions. In most of these experiments the catalytic carbon monoxide on the platinum surface were studied. The mechanism of this reaction is given by [1:11,12]

$$CO + {}^* \rightleftharpoons CO_{ad},$$

$$O_2 + {}^* \rightarrow 2O_{ad},$$

$$O_{ad} + CO_{ad} \rightarrow CO_2, \tag{1.5}$$

where the asterisk * denotes a free adsorption site. The kinetic oscillations have been observed on polycrystalline Pt and Pt(100) single-crystal samples [1.9].

If one works at constant O_2 pressure and constant temperature, the steady-state rate r of CO_2 formation behaves as a function of p_{CO} (p_{CO} is the CO par-

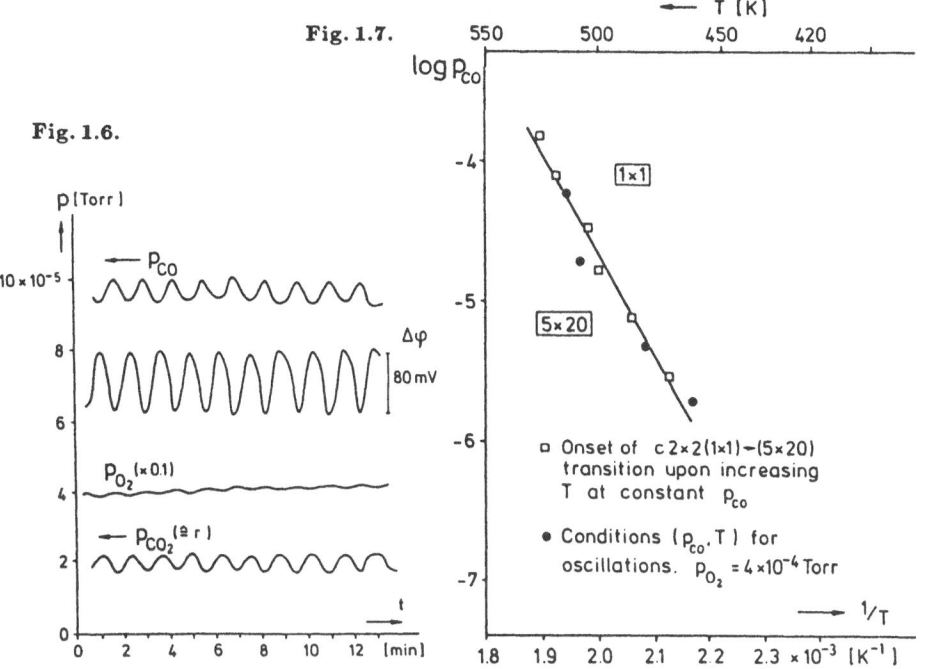

Fig. 1.6. Kinetic oscillations on polycrystalline Pt: p_{CO}, p_{O_2}, p_{CO_2} (= r of CO_2 formation), and $\Delta\phi$ as a function of time [1.9]. T = 502 K, p_{O_2} = 4×10^{-4}Torr and $p_{CO,av}$ = 1×10^{-4}Torr

Fig. 1.7. Conditions for kinetic oscillations (circles) at p_{O_2} = 4×10^{-4}Torr. Squares mark the onset of the c(2×2) − (1×1) → (5×20) phase transition induced by thermal CO desorption [1.9,15]

tial pressure) as follows: r increases continuously with increasing p_{CO}, reaches a plateau and decreases again. The kinetic oscillations occur between 450 and 530 K and at a steady-state rate r which is close to the plateau mentioned above.

Results for the partial pressures p_{O_2}, p_{CO} and p_{CO_2} as a function of time, obtained with the polycrystalline Pt-catalyst, are shown in Fig. 1.6. In Fig. 1.6, too, the work-function $\Delta\phi$ is plotted and $\Delta\phi$ exhibits the same oscillations, as observed for the partial pressures. This effect already indicates a possible change with time of the surface structure. Thus, single-crystal studies including structural analysis should elucidate this phenomenon.

For a fixed O_2 pressure and a given temperature the kinetic oscillations are restricted to a small p_{CO} range; the conditions for the oscillations of the Pt(100) sample are shown by the circles in Fig. 1.7. These conditions are correlated with the borderline for the *structural phase transition* of the Pt(100) surface: the *clean* Pt(100) forms a stable reconstructed (5×20) structure [1.13]. If the

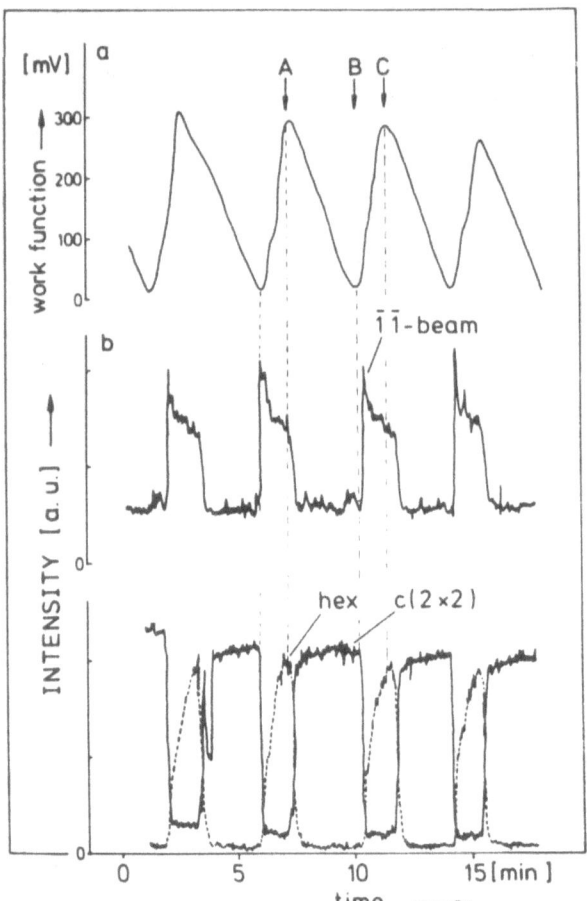

Pt (100)
T = 500 K
$p_{O_2} = 4.0 \times 10^{-4}$ Torr
$p_{CO} = 4.0 \times 10^{-5}$ Torr

Fig. 1.8a,b. Oscillatory kinetics on Pt(100) at T = 500 K, $p_{O_2} = 4 \times 10^{-4}$ Torr, $p_{CO} = 4 \times 10^{-5}$ Torr [1.15]: (a) work-function as a function of time; and (b) periodic variation of the intensities of various LEED spots. ($\bar{1}\bar{1}$): substrate lattice, $(1/2\,1/2) = c(2\times2) - (1\times1)$ structure, $(3/5\,\bar{1})$: hexagonal structure

coverage of CO exceeds a critical value [1.14], a phase transition takes place and the (5×20) structure changes into the unreconstructed (1×1) structure. The adsorbed CO overlayer forms a c(2×2) structure with a local critical coverage $\theta_{CO,loc} = 0.5$. If θ_{CO} is lowered again, the surface switches back into the (5×20) structure. In Fig. 1.7 the conditions are indicated (squares) for which the (5×20) structure just starts to develop [1.15]. It can be seen from Fig. 1.7 that there is obviously a strong correlation between these and the conditions for the occurrence of the kinetic oscillations. Continuous LEED observations [1.10] during the kinetic oscillations showed that these are indeed well correlated with variations of the surface structure (Fig. 1.8). Recently, it has been observed that structural transformation propagates over the surface in the form of waves [1.16].

1.2 Dynamics of Surfaces

The knowledge of the dynamics (vibrations, diffusion, etc.) of the particles at the surface is of relevance in connection with the analysis of certain processes taking place at the surface (e.g., adsorption of particles) and also for the determination of surface properties (e.g., thermodynamic functions).

At sufficiently low temperatures the surface dynamics can be described by the *harmonic* approximation (Chap. 5); for example, in the case of noble-gas crystals the harmonic approximation should be valid below about one sixth of the melting temperature T_m (as compared to about $1/3\,T_m$ for the bulk). In a more general treatment of surface dynamics *anharmonic* effects have to be taken into consideration (Chap. 6).

1.2.1 Surface Thermodynamic Functions in the Harmonic Approximation

On the basis of the *phonon frequencies* (Chap. 5) it is straightforward to calculate thermodynamic properties. Assuming that the surface does not undergo a change of structure we obtain, for example, for monoatomic systems for the entropy S, the vibrational free energy F, and the specific heat at constant volume C_v, the following expressions [1.17,18]:

$$S = k_B \sum_{p,q} \left[-\ln\left(1 - e^{-\alpha}\right) + \frac{\alpha}{e^\alpha - 1} \right], \tag{1.6}$$

$$F = k_B T \sum_{p,q} \left[\frac{1}{2}\alpha + \ln\left(1 - e^{-\alpha}\right) \right], \tag{1.7}$$

$$C_v = k_B \sum_{p,q} \left(\frac{\alpha^2 e^\alpha}{\left(e^\alpha - 1\right)^2} \right) \quad \text{with} \tag{1.8}$$

$$\alpha = \frac{\hbar \omega_p(q)}{k_B T}, \tag{1.9}$$

where $\omega_p(q)$ are the phonon frequencies. Here, q is the wave vector, and p labels the phonon branches. Methods for the determination of the phonon frequencies are given in Chap. 5.

Equations (1.6–8) can be used for the determinination of S, F and C_v in the bulk as well as for crystals with surface. From this information we can calculate the *surface* thermodynamic functions by [1.17]

$$D^s = \frac{N}{A}(d - d^b), \tag{1.10}$$

where D stands for S, F or C_v; d is the value of the thermodynamic function D per particle for the crystal with surface, i.e. d = D/N, N being the number of particles of the system; and d^b is the same quantity in an infinite crystal without surface. A is the surface area of the crystal with surface.

In analogy to (1.10) we can define the *phonon density of states* (PDS) $f^s(\omega)$ at the surface [1.17]

$$f^s(\omega) = \frac{N}{n}[f(\omega) - f^b(\omega)], \tag{1.11}$$

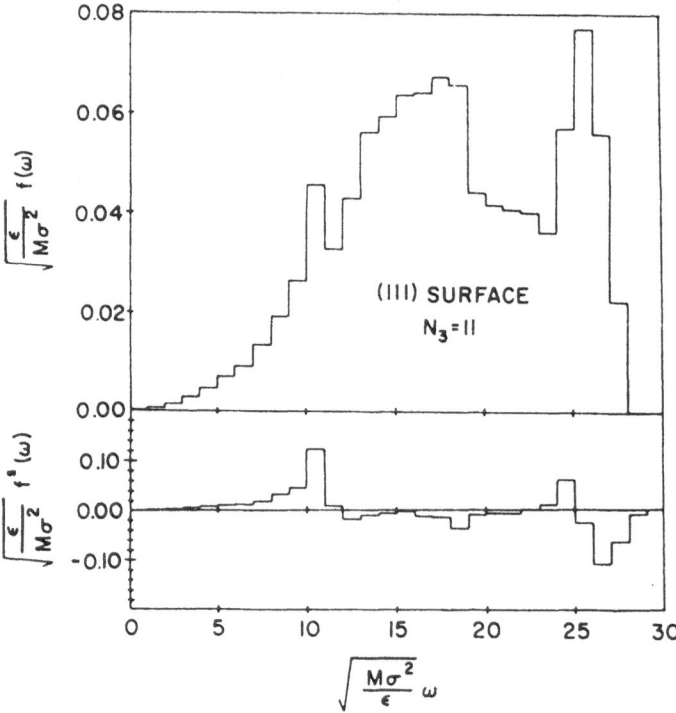

Fig. 1.9. Phonon density of states $f(\omega)$ for the crystal with surface. $f^s(\omega)$ is the surface phonon density of states. M is the atomic mass, and σ and ϵ are Lennard-Jones potential parameters. The results are based on a model calculation for a slab-shaped crystal with $N_3 = 11$ layers and (111) surfaces (more details of the model were given in [1.17])

where $f^b(\omega)$ is the PDS in the bulk, and $f(\omega)$ is the PDS for the crystal with surface; n is the number of surface particles. The quantity $f^s(\omega)$, which is often called the surface-excess phonon density of states, can be measured, for example, by comparing the bulk PDS with PDS of a powder of the same material [1.19].

$f(\omega)$ and $f^s(\omega)$ for a slab-shaped Lennard-Jones crystal consisting of 11 layers are shown in Fig. 1.9. In these calculations it was assumed that the surface does not undergo a change of structure below the melting temperature, and that the thermal expansion at the surface is the same as in the bulk of the crystal. More details concerning the model are given in [1.17]. In contrast to $f(\omega)$, the PDS $f^s(\omega)$ can be positive and negative. With $f^s(\omega)$ the *surface* thermodynamic functions S^s, F^s and C_v^s can be expressed by [1.17]

$$S^s = \frac{3k_B T}{A_0} \int_0^\infty d\omega\, f^s(\omega) \left[-\ln(1 - e^{-\alpha}) + \frac{\alpha}{e^\alpha - 1} \right], \tag{1.12}$$

$$F^s = \frac{3k_B T}{A_0} \int_0^\infty d\omega\, f^s(\omega) \left[\frac{1}{2}\alpha + \ln(1 - e^{-\alpha}) \right], \tag{1.13}$$

$$C_v^s = \frac{3k_B}{A_0} \int_0^\infty d\omega\, f^s(\omega) \frac{\alpha^2 e^\alpha}{(e^\alpha - 1)^2} \tag{1.14}$$

with $A_0 = A/n$. The numerical results (Figs. 1.10–12) for S^s, F^s and C_v^s are based on the same Lennard-Jones model which has been used to calculate $f^s(\omega)$ (Fig. 1.9). In the numerical determination of S^s, F^s and C_v^s it is, however, more accurate [1.17] to carry out the summation of (1.6–8) for the crystals *with* and *without* surface and then use (1.10), than to calculate them from (1.12–14) on the basis of $f^s(\omega)$; the results given in Figs. 1.10–12 have been

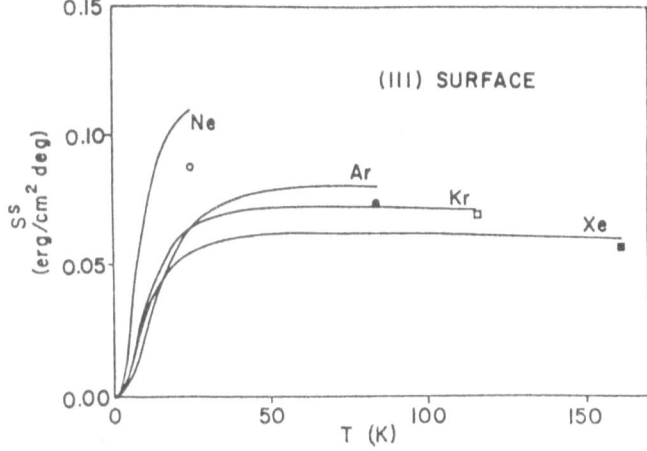

Fig. 1.10. Surface entropy S^s for Ne, Ar, Kr and Xe for the (111) surface [1.17,20]

11

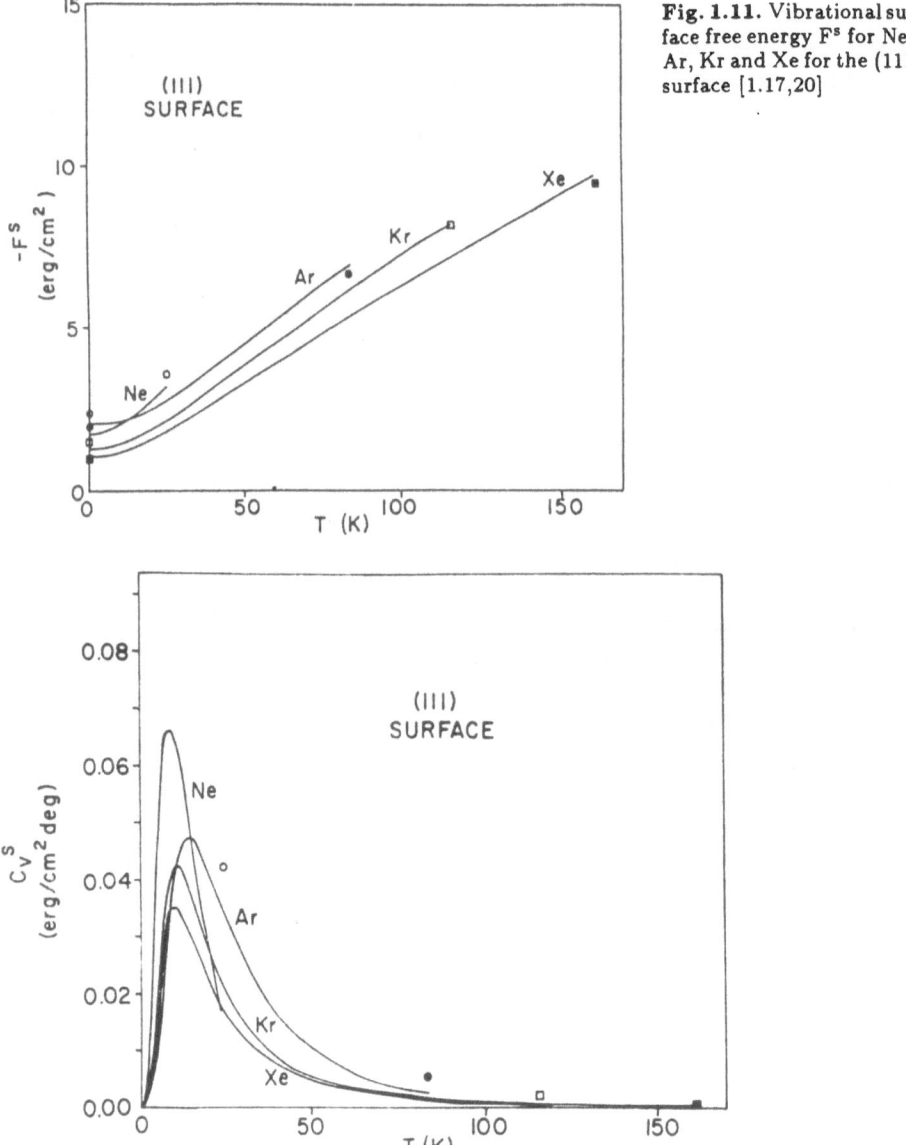

Fig. 1.11. Vibrational surface free energy F^s for Ne, Ar, Kr and Xe for the (111) surface [1.17,20]

Fig. 1.12. Surface specific heat at constant volume C_v^s for Ne, Ar, Kr and Ne for the (111) surface [1.17,20]

obtained by application of the more accurate procedure. The circles and squares in Figs. 1.10–12 are the results for the thermodynamic functions at the surface in the *Debye approximation* at $T = 0\,K$ and the melting temperature [1.20].

   The curves given in Figs. 1.10–12 were discussed in [1.17]. Let us briefly comment here only the behavior of C_v^s. At $T = 0\,K$ we have

$$C_v = 0, \quad C_v^b = 0. \tag{1.15}$$

Consequently,

$$\lim_{T \to 0} C_v^s = 0. \tag{1.16}$$

On the other hand, at high temperatures $(T \to \infty)$ we have

$$C_v = 3k_B, \quad C_v^b = 3k_B \tag{1.17}$$

and, therefore,

$$\lim_{T \to \infty} C_v^s = 0. \tag{1.18}$$

The narrow peak (Fig. 1.12) at intermediate temperatures reflects the presence of the surface modes. For low temperatures the main contribution to C_v^s comes from the lower frequencies where $f^s(\omega)$ is *positive*. With increasing temperatures those frequencies are getting effective at which $f^s(\omega)$ is *negative*, and this leads to a decrease of C_v^s.

The functions S^s, F^S, C_v^s, etc. will be dependent on the *surface structure*, and this can be understood in terms of the number of bonds cut by the surface. In the case of the (110) surface each surface particle is missing six nearest-neighbor interactions, four in the case of the (100) surface, and three in the case of the (111) surface. Thus, the effect of the surface on the functions S^s, F^s, C_v^s, etc. should be greatest for the (110) surface and smallest for the (111) surface.

The results given in Figs. 1.10–12 are only valid for sufficiently low temperatures, the reason being that at high temperatures anharmonic effects have to be considered (Chap. 6).

1.2.2 Electron-Phonon Coupling

The behavior of $f^s(\omega)$ in Fig. 1.9 (positive at low and negative at high frequencies) means that the PDS $f(\omega)$ of the crystal *with* surface is shifted to lower frequencies compared to the PDS $f^b(\omega)$ of the crystal *without* surface. The modification of the phonon spectrum at the surface has an influence on the electron-phonon coupling constant λ, which is given within the approximate *McMillan* theory [1.21] by

$$\lambda = N(0) \langle g^2 \rangle / M \langle \omega^2 \rangle, \tag{1.19}$$

where $N(0)$ is the electron density of states at the Fermi level, M is the effective atomic mass, $\langle g^2 \rangle$ is the average of the squared electron-phonon matrix element, and $\langle \omega^2 \rangle$ is expressed by

$$\langle \omega^2 \rangle = \frac{\displaystyle\int_0^\infty \omega f(\omega) d\omega}{\displaystyle\int_0^\infty \frac{1}{\omega} f(\omega) d\omega}. \tag{1.20}$$

In [1.22–24] the phonon spectra have been calculated for thin crystalline films and used to estimate electron-phonon coupling constants and also superconducting transition temperatures. The surface-excess phonon density of states should enhance the transition temperature for thin films in comparison to the bulk transition temperature. However, λ (and therefore the transition temperature) is also influenced by N(0), [1.19], which can be substantially modified at the surface [1.25].

1.3 Conclusion and Outlook

The purpose of the preceeding discussion was to demonstrate by a few examples that there is a strong influence of surface structure and dynamics on other important surface properties. Thus, it is of considerable importance to determine both theoretically and experimentally the structure and the dynamics of surfaces, see also [1.26,27].

In this Topics volume the structure and dynamics are discussed at the *microscopic* level. Novel experimental techniques and theoretical developments, which have been applied successfully during the last decade, are integrated. Each contribution includes the *background* of the subject, *typical results,* and in most cases also trends of *future developments.*

In Chap. 2, K.H. Rieder gives an up-to-date review to the present developments in surface-structural research using diffraction effects, microscopy techniques and ion-scattering methods. In particular, atomic-beam scattering and scanning tunneling microscopy are discussed. Rieder presents selected examples of results for clean unreconstructed and reconstructed surfaces as well as for adsorbate overlayers and thin films to illustrate the current possibilities of the different methods discussed and the various research areas in which they are applied.

The current status of *high-resolution electron microscopy* of surfaces is reviewed by L.D. Marks in Chap. 3. The emphasis is on the *profile method.* The background to the technique is described, in particular the electron-scattering process, the effects of the microscope imaging system, the importance of one-to-one mapping between the object and the image and the role of image simulations in obtaining quantitative data. Furthermore, different applications are discussed, illustrated by a detailed description of results on the gold(110) surface. Two other techniques, namely *plan-view imaging* and *reflection imaging* are briefly described.

In Chap. 4, C. Varelas gives an introduction to the *surface channeling* technique. Since its first experimental realization only few years ago, surface channeling has proved to be a very versatile method for surface analysis. In particular, it is sensitive to surface structures, reconstructions, the location of adsorbated atoms, and to the vibrational amplitudes at the surface. These applications are discussed in Chap. 4 in connection with selected up-to-date experimental results.

Dynamical surface properties in the *harmonic* approximation are discussed in Chap. 5 by J.E. Black. The objective of this contribution is to focus attention on the theory needed for the study of vibration frequencies, the theory of lattice dynamics of discrete systems. Where possible Black also compares the theoretical results with those obtained in experiments. He treats cases with non-periodical surface structures, cases with bare but periodic surface, and finally cases in which a periodic layer of adatoms is deposited on a bare periodic substrate.

The mean square amplitudes of the particles are significantly larger at the surface of the crystal than in the bulk and, therefore, the *harmonic approximation* is limited to low temperatures; for example, in the case of noble-gas crystals below one sixth of the melting temperature T_m (as compared to about $1/3\,T_m$ for the bulk). Thus, in a more general treatment of surface phenomena *anharmonic* effects have to be taken into consideration. *Molecular dynamics* calculations are important in studying classical many-particle systems with strong anharmonicities, since anharmonicity is treated without approximation. In Chap. 6 the molecular dynamics method is introduced by W. Schommers and the role of this method in many-particle physics is discussed. Applications of the molecular dynamics method to problems in solid state physics with particular emphasis on surface science are given. The following problems are discussed: Phase transitions in two-dimensional systems, diatomic molecules adsorbed on a surface, melting transition of near monolayer xenon films on graphite, microscopic behavior of krypton atoms at the surface.

The surface vibrations can be investigated by *high-resolution electron energy loss spectroscopy* in the impact scattering regime. In Chap. 7, M. Rocca, H. Ibach, S. Lehwald and T.S. Rahman discuss the general features of this experimental method. Also a brief overview of the electron phonon scattering is given placing particular emphasis on multiple scattering effects as they play a major role in the measurements of phonon frequencies and dispersions. As examples Rocca et al. present measurements performed for Ni(100) surfaces, both clean and covered with (2×2) overlayers of oxygen and sulfur, and after disordered adsorption of oxygen. The experimental results are analyzed in terms of lattice dynamical calculations.

References

1.1   H.H. Wieder: J. Vac. Technol. A**2** (2), 97 (1984)
1.2   R.J. Meyer, C.B. Duke, A. Paton, J.C. Tsang, J.L. Yeh, A. Kahn, P. Mark: Phys. Rev. Rev. B**22,** 6171 (1980)
1.3   G.P. Srivastava, I. Singh, V. Montgomery, R.H. Williams: J. Phys. C**16,** 3627 (1983)
1.4   G. Ertl: In *Electron and Ion Spectroscopy of Solids*, ed. by L. Fiermans, J. Vennik, and W. DeKeyser (Plenum, New York 1978) p. 144;
      G. Ertl, J. Küppers: *Low Energy Electrons and Surface Chemistry* (VCH, Weinheim 1985);
      G.A. Somorjai: *Chemistry in Two Dimensions Surfaces* (Cornell Univ. Press, Ithaca 1981)

1.5   R.L. Park, M.G. Lagally (eds.): *Solid State Physics: Surfaces* (Academic, New York 1985)

1.6   See, for example, P.H. Emmet: In *The Physical Basis for Heterogeneous Catalysis*, ed. by E. Drauglis, R.I. Jaffee (Plenum, New York 1975)

1.7   G. Ertl: Catal. Rev.-Sci. Engl. **21,** 201 (1980)

1.8   B.I. Lundquist: Vacuum **33,** 639 (1983)

1.9   G. Ertl, P.R. Norton, J. Rüstig: Phys. Rev. Lett. **49,** 177 (1982)

1.10   M.P. Cox, G. Ertl, R. Imbihl, J. Rüstig: Surf. Sci. **134,** L517 (1983)

1.11   C.T. Campbell, G. Ertl, H. Kuipers, J. Segners: J. Chem. Phys. **73,** 5862 (1980)

1.12   M.A. Barteau, E.I. Ko, R.J. Madix: Surf. Sci. **104,** 161 (1981)

1.13   M.A. van Hove, R.J. Joestner, P.C. Stair, J.P. Biberian, L.L. Kesmodel, I. Bartos, G.A. Somorjai: Surf. Sci. **103,** 218 (1981)

1.14   P.R. Norton, J.A. Davies, D.K. Creber, C.W. Sitter, T.E. Jackman: Surf. Sci. **108,** 205 (1981)

1.15   R.J. Behm, A.P. Thiel, P.R. Norton, G. Ertl: J. Chem. Phys. **78,** 7437 (1983)

1.16   G. Ertl: Surf. Sci. **152/153,** 328 (1985)

1.17   R.E. Allen, F.W. de Wette: J. Chem. Phys. **51,** 4820 (1969)

1.18   A.A. Maradudin, E.W. Montroll, G.H. Weiss, I.P. Ipatova: "Theory of Lattice Dynamics in the Harmonic Approximation", *Solid Stadt Physics* (Suppl. 3) (Academic, New York 1971)

1.19   K.H. Rieder, W. Drexel: Phys. Rev. Lett. **34,** 148 (1975)

1.20   G.C. Benson, K.S. Yun: In *The Solid-Gas Interface*, ed. by E.A. Flood (Dekker, New York 1967)

1.21   W.L. McMillan: Phys. Rev. **167,** 331 (1968)

1.22   R.E. Allen, F.W. de Wette: Phys. Rev. **187,** 883 (1969)

1.23   R.E. Allen, G.P. Alldredge, F.W. de Wette: Phys. Rev. **B2,** 2570 (1970)

1.24   G. Benedek, M. Miura, W. Kress, H. Bilz: Surf. Sci. **148,** 107 (1984)

1.25   N. Pessall, R.E. Gold, H.A. Johansen: J. Phys. Chem. Solids **29,** 19 (1968)

1.26   M.A. van Hove, S.Y. Tong (eds.): *The Structure of Surfaces*, Springer Ser. Surf. Sci., Vol. 2 (Springer, Berlin, Heidelberg 1985)

1.27   F. Nizzoli, K.-H. Rieder, R.F. Willis (eds.): *Dynamical Phenomena at Surfaces, Interfaces and Superlattices*, Springer Ser. Surf. Sci., Vol. 3 (Springer, Berlin, Heidelberg 1985)

2. Experimental Methods for Determining Surface Structures and Surface Corrugations

K. H. Rieder

With 30 Figures

A concise introduction into the newer developments of surface structural research using diffraction effects, microscopy techniques and ion scattering methods is given. Selected examples of results for clean unreconstructed and reconstructed surfaces as well as on adsorbate overlayers and thin films are presented in order to illustrate the current possibilities of the different methods discussed and the various research areas in which they are applied.

2.1 Basic Physical Principles

The aims of surface crystallographic investigations are manifold: the atomic species at and near the surface need to be identified, the positions of the atoms relative to each other and the bond lengths between them have to be determined, and – as the ultimate task – the character of bonding should be established [2.1,2]. In all real systems do the atoms at and near a surface exhibit a more or less modified arrangement, as compared to the crystal interior. The simplest rearrangement is surface relaxation, whereby the topmost layers retain the bulk symmetry, but the atomic distances perpendicular to the surface are different from the respective bulk value; for surfaces with low symmetry a given layer near the surface may also be shifted laterally against the layers in the bulk [2.2,3]. Surface reconstruction is a stronger disturbance giving rise to rearrangements of the topmost layers into different unit cell lengths and therefore into different symmetries, as compared to the respective bulk truncation. It may be worthwile to distinguish between displacive and reconstructive surface distortions; in the displacive case the surface atoms are shifted by not too large amounts from their ideal positions but no bonds are broken (although altered) or new ones formed [2.4]. Deposition of foreign species usually changes the geometrical and electronic arrangement of the substrate surface and can even lead to substrate reconstruction. Furthermore, the formation of (coverage dependent) regular adsorbate overlayers [2.5] in the region up to one monolayer due to direct or substrate mediated adatom-adatom interactions appears more to be the rule than the exception.

In many areas of physical and chemical technology, like in crystal growth, in catalytic and corrosion processes, in hardening and passivation treatments, in the formation of heterojunctions, thin film barriers and superlattices, it are the properties of solid and mostly crystalline surfaces which are of dominant influence. It is thus obvious that the necessity to gain experimental knowledge about surface structures and to understand the microcopic mechanisms which lead to their formation has given rise to the development of a wealth of different and complementary surface crystallographic methods.

This introduction attempts to outline the basic physical principles of the most important classes of surface structure methods, to provide some insight into the data evaluation procedures, and to illustrate their application by (a very personal but hopefully representative choice of) characteristic examples of recent results. Methods which basically aim for other information, but from which structural information can be obtained as a byproduct, are frequently mentioned in connection with the discussion of exemplary surface structural results. It is hoped that the reader can acquire some appreciation as to the respective possibilities, merits and limitations of the various methods and that he will get some insight into the different kinds of relaxations and reconstructions occuring on clean surfaces as well as of the ordering patterns of overlayers formed by foreign species. What the reader will in any case find is some guide through the jungle of acronyms used for the various techniques in this part of surface science.

2.2 Notation and Conventions

For the crystallographic characterization of a surface one simply uses its bulk Miller indices; starting point for a definition of the surface unit cell vectors is the idealized situation of a bulk truncated and unreconstructed surface (relaxation of the surface layers normal to the surface has no influence). The seventeen possible two-dimensional space lattices result in five categories of 2D Bravais lattices, namely the square, the primitive and centered rectangular, the hexagonal and the oblique lattice [2.6]. We denote the bulk-derived "ideal" surface unit mesh vectors by a_i^b and use the usual convention of x-ray crystallography $|a_1|\langle|a_2|$. In order to specify a certain reconstruction or adsorbate structure of the surface a commonly used notation was introduced by *Woods* [2.7]: the actual surface unit mesh vectors a_i are related to the a_i^b's by stating the ratios a_i/a_i^b and the rotation of a_1 against a_1^b. To refer even more closely to the ideal (bulk exposed) surface unit mesh, the $(\sqrt{2}\times\sqrt{2})R45$ frequently is called c(2×2) (c for centered); in the same spirit the (2×2) is called p(2×2) (p for primitive). In the matrix notation, which relates the a_i's and a_i^b's via a (2×2) matrix [2.1,2],

$$a = \underline{M}a^b, \qquad\qquad (2.1)$$

the c(2×2) may be described by the matrix

$$\begin{pmatrix} 1 & 1 \\ 1 & -1 \end{pmatrix}.$$

The reciprocal lattice vectors b_j (j=1,2) of the surface mesh are defined by

$$a_i b_j = 2\pi \delta_{ij} \qquad (2.2)$$

with δ_{ij} denoting the Kronecker symbol. Equation (2.2) implies that b_1 and b_2 are normal to a_2 and a_1, respectively. With β being the angle between a_1 and a_2 the lengths of b_1 and b_2 are determined by

$$|b_1| = \frac{2\pi}{a_1 \sin\beta}, \quad |b_2| = \frac{2\pi}{a_2 \sin\beta}. \qquad (2.3)$$

The relationship between direct and reciprocal lattices for a fcc bulk lattice and the corresponding (011) surface mesh is depicted in Fig. 2.1 [see Fig. 2.3 for a LEED pattern of a fcc(011)-surface].

A special notation is often used for high index surfaces in order to point out their build-up by low index terraces and regular steps [2.5]. For the fcc(755) for

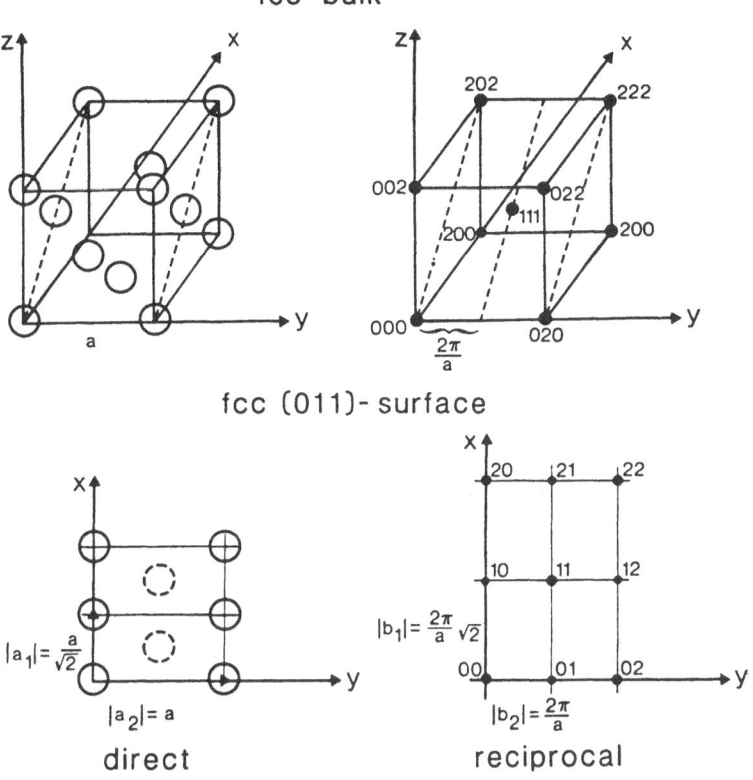

fcc- bulk

fcc (011)- surface

direct   reciprocal

Fig. 2.1. Graphical illustrations of the relation between direct bulk lattice and corresponding reciprocal lattice. *Upper part:* Bulk fcc-crystal; note that the reciprocal lattice points with even and odd indices are systematically missing. *Lower part:* direct mesh and reciprocal lattice rods of the fcc(001)-surface; note that also reciprocal rods with mixed indices are allowed and that only the rods indicated by larger circles would be allowed if the atoms in the second surface layer (indicated by dashed circles) would coincide in height with those of the first layer to form a centered rectangular lattice

example the notation fcc(S)-[6(111)×(100)] (S for stepped) shows immediately, that the surface consists of (111)-terraces six atoms wide and separated by steps with (100)-character of one atom height.

The convention on surface unit-vector notation is unproblematic as long as the crystal is monoatomic. More precise specifications are required for crystals with several different atoms in the unit cell. The unit cell vectors of the (100) surface of a Zincblende type crystal for example have the same length but they are inequivalent with respect to bonding towards the other kind of ions in the second layer. Assuming the origin of the bulk unit cell at a metal atom the convention seems to have been adopted (at least in the molecular beam epitaxy community) [2.8] to take a_1 along the $(\bar{1}10)$ direction, so that exactly below $a_1/2$ an anion is located in the second layer. We note also that for polar zincblende surfaces an further convention is used in the literature: the (111) surface has the metal atoms and the $(\bar{1}\bar{1}\bar{1})$ the anions in the topmost layer.

2.3 Surface Diffraction from Regular Gratings

2.3.1 Kinematical Interference Conditions for Surface Scattering

In all diffraction techniques a beam of well-defined wavelength λ ($\lesssim 1$ Å) impinges on the solid surface and the diffracted intensities are measured by appropriate movable detectors, whereby usually data for different angles of incidence, different surface azimuthal orientations and beam wavelengths are sampled. The necessary condition for observation of sharp diffraction beams is a sufficiently well ordered surface. The Laue condition for diffraction from a two-dimensional periodic array relates the incoming wavevector $k_i = (K, k_{iz})$ ($|k_i| = 2\pi/\lambda$) and the emerging wavevector $k_G = (K_G, k_{Gz})$ via

$$K - K_G = G \qquad (2.4)$$

with the reciprocal lattice vector $G = jb_1 + lb_2$ (the z-direction is chosen perpendicular to the surface). The wavelength remains unchanged during diffraction, so that $k_i^2 = k_G^2$ (elastic scattering). The Laue condition can be graphically represented by the Ewald construction which is illustrated schematically for a two-dimensional crystal in Fig. 2.2. Surface diffraction can occur into all those directions, where the Ewald sphere cuts a reciprocal lattice rod. It is important to notice the difference to the bulk case where diffraction does not occur in all these directions as the Laue condition is more stringent in this case: it requires the Ewald sphere to go through a reciprocal lattice point which in the example chosen is the case only for a single beam.

The measurement of the angular locations of the diffracted beams thus allows the determination of the dimensions of the surface unit cell as well as its orientation relative to the incoming beam. As surface adsorbate structures and reconstructions normally yield surface unit cells with larger dimensions than the corresponding truncated bulk structure, fractional Bragg beams (often called

(a)

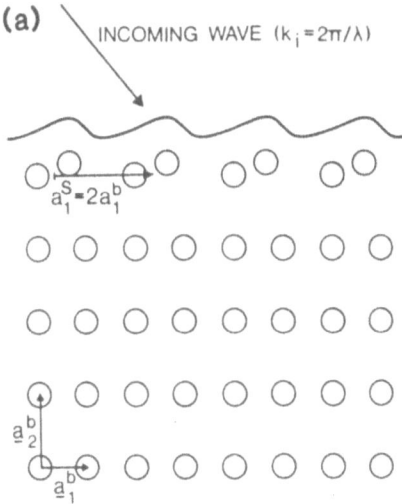

INCOMING WAVE $(k_i = 2\pi/\lambda)$

$a_1^S = 2a_1^b$

a_2^b

a_1^b

Fig. 2.2. (a) Sphere model of a (two-dimensional) crystal exhibiting a surface reconstruction with twice the corresponding bulk periodicity. A contour of constant surface electron density is schematically sketched. The bulk and surface unit cell vectors are indicated. **(b)** The Ewald construction for bulk and surface diffraction. For the case plotted bulk diffraction occurs only in one direction (*dashed arrow*) where the Ewald sphere goes through a reciprocal lattice point (the reciprocal lattice points are indicated as crosses). In contrast to this, surface diffraction can occur in all directions in which the Ewald sphere cuts the two-dimensional lattice rods. The Miller indices of the surface reciprocal lattice rods correspond to the surface unit cell

(b)

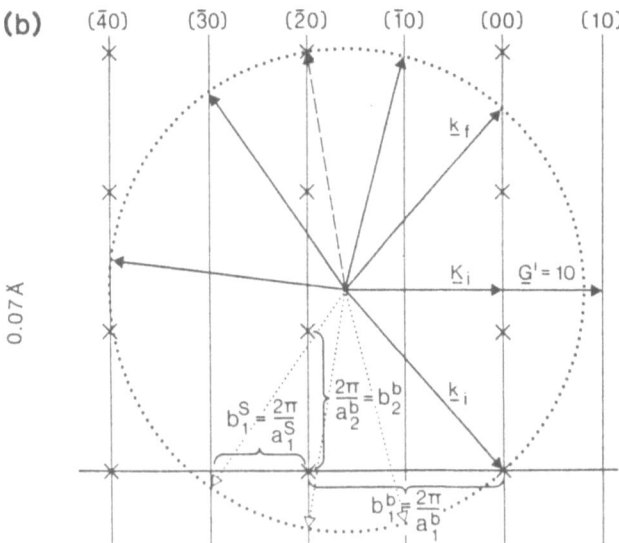

$(\bar{4}0)$ $(\bar{3}0)$ $(\bar{2}0)$ $(\bar{1}0)$ (00) (10)

0.07 Å

k_f

K_i $G' = 10$

K_i

$b_1^S = \dfrac{2\pi}{a_1^S}$ $\dfrac{2\pi}{a_2^b} = b_2^b$ k_i

$b_1^b = \dfrac{2\pi}{a_1^b}$

superspots) occur between the reflections characteristic for the ideal surface. In order to illustrate the relation of a diffraction pattern with the reciprocal lattice, Fig. 2.3a shows the LEED-diffraction pattern of the clean Cu(110) surface obtained with normal incidence, so that it resembles as closely as possible the surface reciprocal lattice of Fig. 2.1. Figure 2.3b shows the LEED-pattern of the (2×1)O-phase on Cu(110); notice that between every vertical line of diffraction spots characteristic of the clean surface another spot array is visible according to the doubling of the surface unit cell along the close packed metal rows. Determination of the distribution of the scattering centers within the unit cell requires measurement and analysis of the intensities I_G of the Bragg peaks.

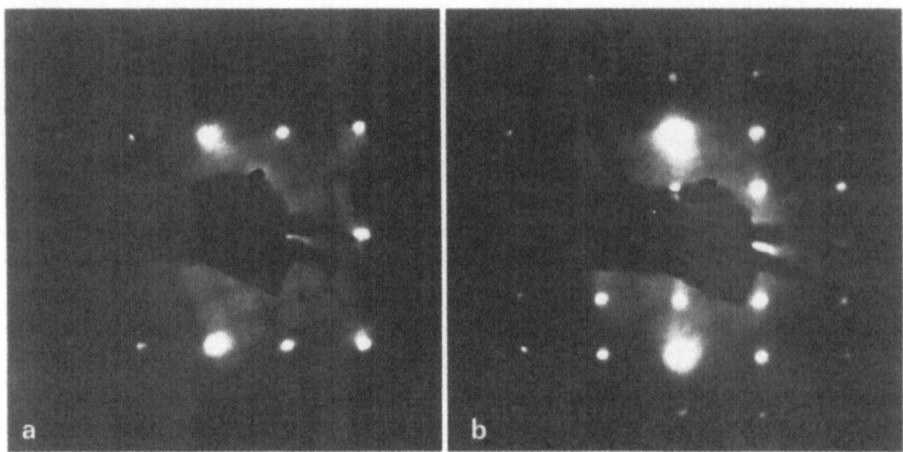

Fig. 2.3. (a) Low Energy Electron Diffraction pattern obtained for Cu(110) with a beam energy of 140 eV. (b) LEED pattern of the same surface with a (2×1) oxygen chemisorption phase. The beam is impinging at normal incidence in both cases. Courtesy of W. Stocker

Analysis is complicated by the fact as the scattering equations contain the (complex) scattering amplitudes F_G, whereas only intensities $I_G \propto |F_G|^2$ can be measured, so that the information on the beam phases is lost. This is the reason why for almost all methods discussed in the next Section for intensity analyses up to now trial and error procedures are used.

2.3.2 Experimental Methods Employing Diffraction from Gratings

a) Glancing Incidence X-Ray Diffraction

X-ray diffraction techniques are the most successful and important for investigations of bulk crystal structures. The scattering centers for x-ray photons are the regions of high electron density, i.e the atom cores. Since the interaction cross section of x-ray photons with matter is small, the well-known kinematical theory is usually sufficient for intensity data analyses [2.9]. X-ray scattering therefore represents the natural starting point for a description of surface diffraction methods. That x-ray scattering can be made surface sensitive has been demonstrated only a few years ago by *Marra* et al. [2.10] by going to glancing angles of incidence. The important advantage of x-ray surface scattering lies in the fact that the simple and well-known kinematical theory and the wealth of evaluation techniques for x-ray data inversion [2.11,12] can be applied. The method will be especially powerful in the future in connection with the high brightness (flux per unit area) synchrotron sources and has excellent prospects for structural studies not only of surface but also interface structures. It appears worthwile to point out in this connection that the synchrotron sources are superior to the most powerful commercial x-ray sources (60 kW rotating anode tubes) by a factor of thousand in flux and a factor of hundred in collimation.

X-ray beams, incident at grazing angles of a few tenths of a degree, are totally reflected out of the surface at small angles and diffracted parallel to the surface over large angles; due the small sampling depths of some ten Angstroms Bragg superreflections characteristic for surface structures can therefore be measured. Recall that the kinematical structure factor for 3D-diffraction can be written as [2.9]

$$F(g) = \sum_j f_j \exp[i(\mathbf{k_g} - \mathbf{k_j})\mathbf{r_j}] = \sum_j f_j \exp(i\mathbf{g}\cdot\mathbf{r_j}), \qquad (2.5)$$

where g denotes a particular 3D reciprocal lattice vector (connecting the origin with the corresponding reciprocal lattice point, Fig. 2.1b), f_j is the atomic scattering factor for the j'th species in the unit cell and the sum runs over all atoms. The Laue condition for 3D-diffraction $\mathbf{k_g} - \mathbf{k_i} = g$ is incorporated in (2.5). In the case of surface diffraction the Laue condition is relaxed according to (2.4), so that (2.5) has to be modified to

$$F(G, q_{Gz}) = \sum_j f_j \exp\{i[\mathbf{G}\cdot\mathbf{R_j} + (k_{Gz} - k_{iz})z_j]\} \qquad (2.6)$$

with $q_{Gz} = k_{Gz} - k_{iz}$ being the component of the scattering vector perpendicular to the surface (Fig. 2.1b). For conditions of grazing incidence and emergence k_{iz} and $k_{Gz} \simeq 0$, so that (2.6) reduces correspondingly.

As an example of a surface x-ray diffraction study we discuss the investigations by *Robinson* [2.13] of the (1×2) reconstruction of Au(110). With a sample deliberately cut under a small angle from the ideal surface orientation, Robinson observed half order superstructure peaks which were displaced from their ideal positions (Fig. 2.4), and concluded that the surface has a long range reconstruction incommensurate with the bulk in the direction perpendicular to the close packed metal rows; he explained the incommensurability by an overlayer that is locally a (1×2) reconstruction with interruptions by domain walls perpendicular to the close packed rows. The local structure was determined by using a crystal surface cut with high precision and measuring the intensities of several half order beams $(h/2, k, k)$. Calculating the Patterson function [2.11]

$$P(x, y, z) = \sum_{hkl} |F(hkl)^2| \cos[2\pi(hx + ky + lz)] \qquad (2.7)$$

for the observed in-plane reflections $F(h/2, k, k)$ the projection $P(x, y)$ onto the surface plane was obtained as shown in Fig. 2.5. As only half order reflections were included into the Patterson synthesis, its features correspond just to the reconstructed region of the crystal surface. The pronounced nonorigin peak in the Patterson projection of Fig. 2.5 at $(0.6, 0.25, 0.25)$ corresponds to a single interatomic spacing that could describe a structure with two or three atoms per surface unit cell; the second small peak at $(0.8, 0, 0)$ was argued to favour the three atom alternative. Indeed, least squares refinements with three atoms at $(0, 0, 0)$ and $(\pm x, 1/4, 1/4)$ gave a better R-factor [2.11,12] than the best two atom model; the optimum value for x was 0.53 (in units a_0). This result

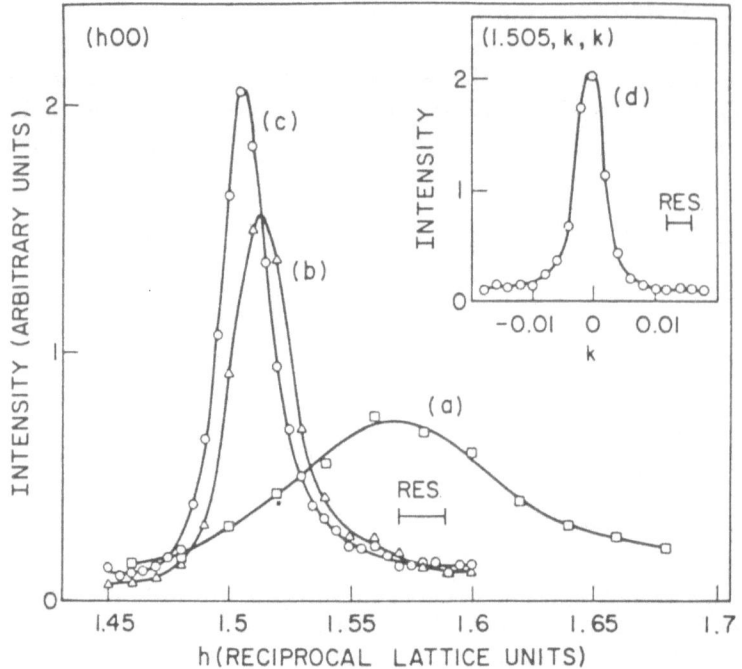

Fig. 2.4. X-ray diffractometer scans of the superlattice peak near (3/2,0,0) for the Au(110) (1×2) reconstructed surface. (*a–c*) Radial scans along the [100]-direction (perpendicular to the close-packed metal rows). (*a*) Sample cut at 1.5° from the ideal surface orientation. (*b,c*) Samples cut better than 0.1°. (*d*) Scan parallel to the close packed rows of the peak in (*c*). The instrumental resolution is indicated. After *Robinson* [2.13]. ©1983 American Physical Society

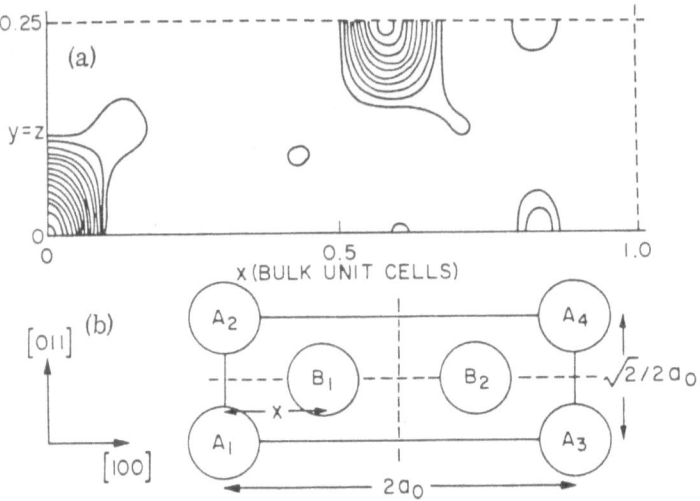

Fig. 2.5. (a) Contour map of the surface projection of the Patterson function obtained for Au(110)(1×2). The asymmetric unit corresponds to 1/4 of the (1×2) cell. (b) Top view of atomic positions in the first (*A*) and second layer (*B*) of the full (1×2) cell

together with a few auxiliary arguments confirmed the missing row model, in which every second close packed row in the topmost layer should be absent, and yielded also a lateral shift of the second layer atoms away from the topmost atoms. According to theoretical considerations [2.14] the intensity along reciprocal lattice rods should be sensitive to the distance perpendicular to the surface; measured intensity profiles led *Robinson* [2.13] to the conclusion that on Au(110) an expansion takes place between first and second layer. In contrast to the result on the lateral atomic displacements in the second layer, the normal expansion was not confirmed by the most recent LEED-investigations [2.15]; for further discussions of other results obtained with other methods on this prototype reconstruction, see Sects. 2.3.4 and 2.5.2. More details on the method of kinematical surface x-ray scattering as well as further possibilities to perform surface structure investigations with x-rays can be found in [Ref. 2.16, Chap. 6].

Since the refraction index of solid materials for neutrons is always smaller than unity, grazing incidence neutron scattering should have great promises especially in the investigation of surface magnetic structures [2.14,17]. Neutron sources with the required high brightness are, however, much more difficult to achieve than x-ray sources.

b) Atomic Beam Scattering

In the last few years, the development of high-pressure nozzle sources yielding highly monochromatic neutral particle beams of thermal energies (20–300 meV) and sufficient intensity [2.18] has allowed particle-surface scattering experiments to be performed under ultra-high vacuum conditions. Molecular beam scattering is presently being used in many branches of surface science like in studies of the particle-surface interaction potential [2.19], of surface phonon dispersions [2.20], of global energy exchange [2.21], of surface diffusion [2.22] and of substrate-mediated chemical reactions [2.23].

In the present context we restrict ourselves on the use of atom diffraction (mainly He and Ne) for surface structure studies [2.24]. The nature of the scattering centers in the case of atom scattering is revealed by a discussion of the atom/surface-interaction potential: At distances not too far from the surface the incoming atoms feel an attraction due to van der Waals forces, the depth of the potential ($D \leq 10\,\text{meV}$) being never large compared to the incoming particle energies. Closer to the surface, the particles are repelled due to the overlap of their electronic densities with that of the surface. This causes a steeply rising repulsive part of the potential. The classical turning points are usually farther away on top of the surface ions rather than between them, which gives rise to a periodic modulation of the repulsive part parallel to the surface. The modulation of the repulsive part of the atom-surface potential constitutes the corrugation function which follows closely that contour of constant total surface electron density for which the embedding energy corresponds to the normal component of the energy of the incoming particles [2.25]; although the mean distance of the measured surface density contour is with 2.5–4Å rather far away from the surface ion cores, the corrugation nevertheless often reflects directly the geometrical arrangement of the surface atoms. In analyzing intensities up to now one mostly resorted to the so-called hard corrugated wall model [2.26,27], in which the repulsive part is assumed to be infinitely steep and the lateral modulation is described by a corrugation function whose amplitude is energy dependent; this assumption has been shown to give reliable results for the corrugation shapes and amplitudes as long as small angles of incidence are involved [2.28]. With the hard corrugated wall model quick calculational procedures [2.26] have been developed; they are necessary for analyses of complicated diffraction patterns which may arise, if many particles contribute to forming the corrugation of a single surface unit cell. The main ingredients of an especially simple computational scheme, the so called eikonal approximation, can be directly derived from (2.5). Since every point on the corrugation is a scattering center and all points have identical scattering properties, the sum in (2.5) has to be simply replaced by an integral over the surface unit cell with area $S = a_1 a_2$, which yields

$$F(G) \propto \frac{1}{S} \int \exp\{i[G \cdot R + (k_{Gz} - k_{iz})z(R)]\} dR \tag{2.8}$$

As all atoms are scattered back from the surface, the sum of all diffracted beam intensities (including the specular) must be equal to the total incoming intensity. This can be fulfilled to a good degree of accuracy by taking into account kinematical prefactors which can be obtained from a more rigorous derivation of (2.8) [2.26]. The intensity analyses can be performed in a completely model-free manner by assuming for the corrugation the general form $z(R) = z(G)\exp(iG \cdot R)$ and searching for the Fourier-coefficients which give the best agreement with the measured intensity set.

As a beautiful and physically interesting example of surface structural investigations with He-diffraction we discuss the adsorbate system hydrogen on Ni(110), where six different ordered adsorption phases were identified with He scattering [2.29,30]. The arrangement of the adatoms in the different phases gives a fascinating picture of the coverage-dependent ordering of the hydrogen atoms on this particular surface. The phases observed are: three different c(2×6) phases corresponding to coverages θ of 1/3, 2/3 and 5/6 monolayers (ML), a c(2×4) with $\theta = 1/2$ML, a (2×1) with 1 ML and a (1×2) phase with the saturation coverage of 1.5 ML. Experimental diffraction scans for the lowest coverage c(2×6) are exhibited in Fig. 2.6 together with hard-wall best-fit curves. The corrugations of the clean surface, of two c(2×6) phases, the (2×1) and the (1×2) are shown together with sphere models of the surface structures in Figs. 2.7a–e. The corrugations in Figs. 2.7b–d provide a direct picture of the adatom configurations as every adsorbed hydrogen atom produces a distinct corrugation hill on the Ni(110) substrate. The hydrogens form zig-zag chains along the close-packed Ni rows even at very low coverages with the adatoms in three-fold coordinated sites. The exact location of the adatoms relative to the substrate is not directly visible from the corrugation functions, but has

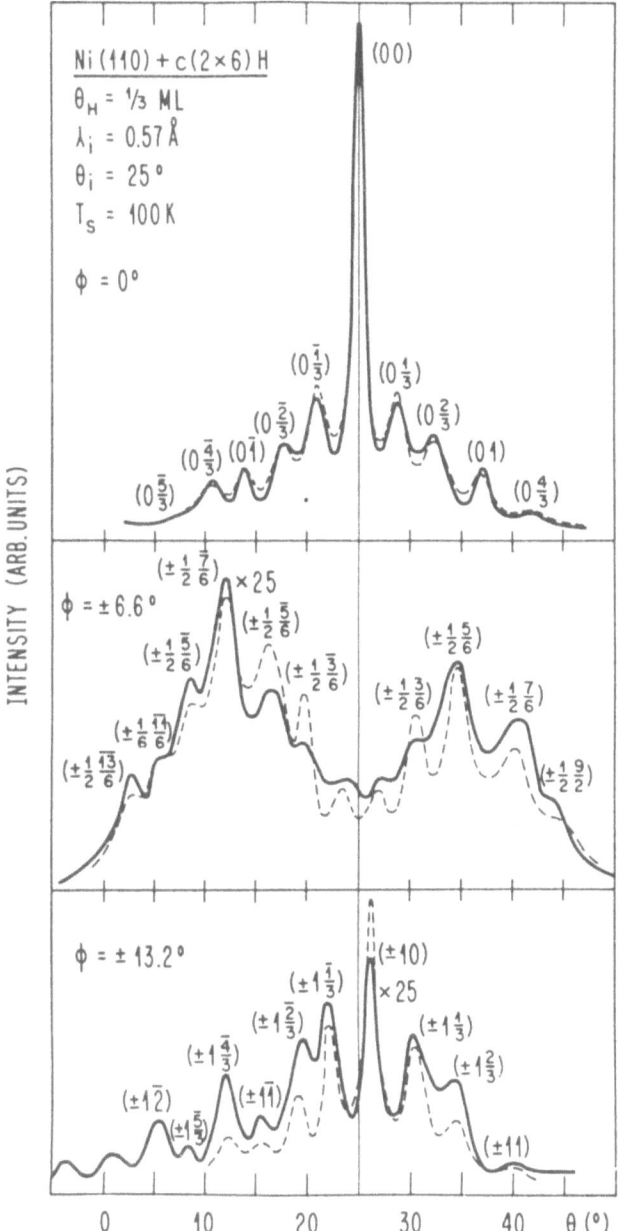

Fig. 2.6. In-plane and out-of-plane He-diffraction traces for the c(2×6) phase of hydrogen on Ni(110) corresponding to a coverage of 1/3 monolayer. The full line corresponds to the experimental result, and the broken line to the best-fit calculation. The corresponding corrugation function is shown in Fig. 2.7b. ©1983 The American Physical Society

been predicted by theoretical calculations [2.31] and was recently confirmed by dynamical LEED analyses [2.32]. The lateral interaction of the zig-zag H-chains is so long-ranged, that at 1/3 ML they form a sufficiently well-ordered c(2×6) phase with alternating zig-zag and zag-zig configurations at distances of 10.6 Å, whereby two close-packed Ni-rows between the hydrogen chains remain adsorbate free (Fig. 2.7b). In the next ordered c(2×4) phase the H-chains cover

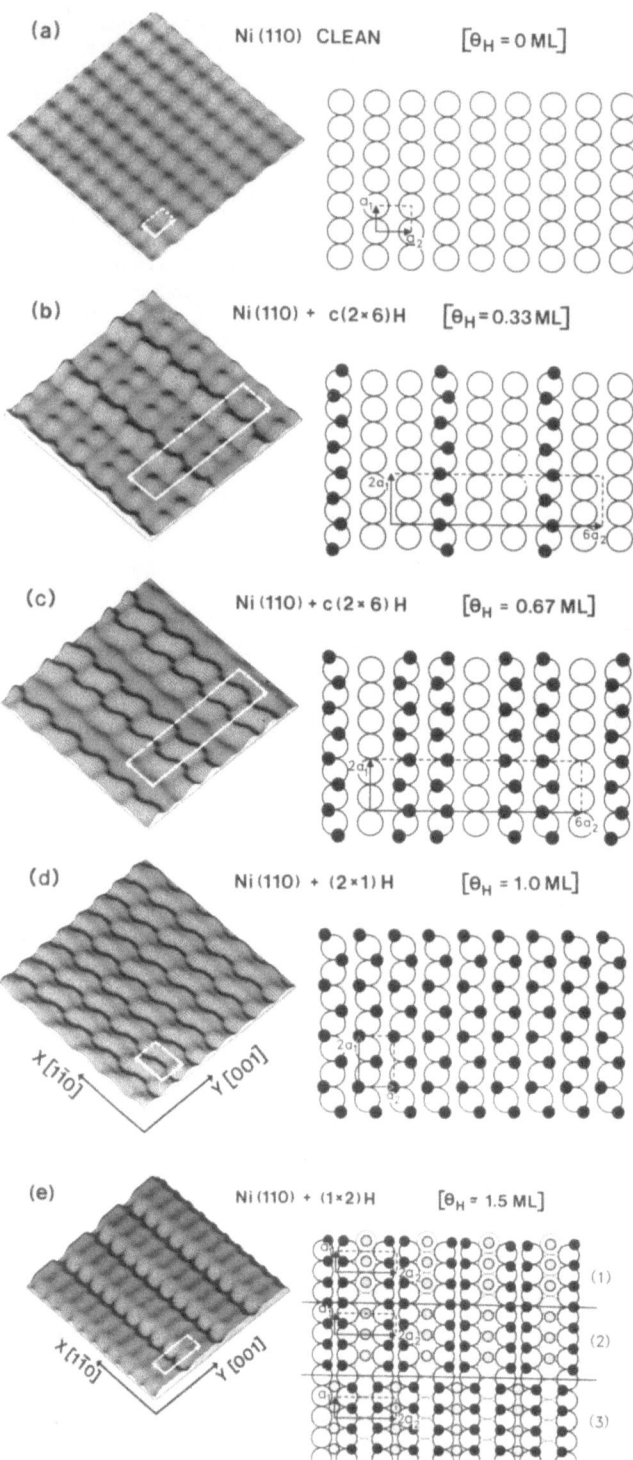

(a) Ni (110) CLEAN $[\theta_H = 0\,\text{ML}]$

(b) Ni (110) + c(2×6)H $[\theta_H = 0.33\,\text{ML}]$

(c) Ni (110) + c(2×6)H $[\theta_H = 0.67\,\text{ML}]$

(d) Ni (110) + (2×1)H $[\theta_H = 1.0\,\text{ML}]$

(e) Ni (110) + (1×2)H $[\theta_H \approx 1.5\,\text{ML}]$

Fig. 2.7
Caption see opposite page.

28

half the surface leaving each other Ni-row adsorbate free. The following ordered phase corresponding to 2/3 ML has c(2×6) periodicity; here, two zig-zag and two zag-zig chains alternate leaving one close-packed Ni row adsorbate-free (Fig. 2.7c). The fourth ordered structure has again c(2×6) periodicity and corresponds to 5/6 ML; the extra 1/6 ML of H goes into the energetically obviously less-favoured two-fold coordinated sites on the previously adsorbate-free Ni rows, forming in this way a distorted hexagonal overlayer of H-adatoms. This is the only phase up to 1 ML in which apparently two different adsorption sites are occupied. It may act as a precursor to build up the (2×1) with 1 ML, where the adatoms form a slightly denser distorted hexagonal pattern with all close-packed Ni rows covered with parallel hydrogen chains(Fig. 2.7d); the last c(2×6) may be necessary to facilitate the phase jump of the zag-zig chains along the close-packed Ni-rows in order to make all chains parallel. The transition to the (1×2) saturation phase (Fig. 2.7e) involves an adsorbate-induced substrate reconstruction [2.33] with partial hydrogen accommodation in the second layer. Despite numerous investigations of this phase with different methods there are still several possibilities for its structure as indicated in the models in Fig. 2.7e. A similar hydrogen-induced reconstruction leads to effective channels for hydrogen uptake into the bulk in the case of Pd(110) [2.34].

Atomic beam scattering has, due to the small energies of the incoming particles, the very important advantage of being far less destructive than any other surface method. As the surface electron density is drastically influenced by any adsorbate (including hydrogen), the method is very well suited and especially valuable in studying light adsorbates on heavy substrates, which is problematic with most of the other methods. In contrast to the methods using charged particles, atomic beams are – like x-rays – equally well suited for studies of metal, semiconductor [2.35] and insulator surfaces [2.36]. As the corrugation function is a replica of the electron density contour corresponding to the energy of the incoming He-beam, it was attempted to use simple surface charge density calculations by overlapping atomic charge densities to determine the location of adatoms. It was, however, found that due to the changes in the charge distributions upon surface bond formation appreciable errors may be made. Reliable results on bond lengths therefore require coupling to the best selfconsistent surface charge density calculations (which are very expensive) [2.37]. It was recently found that Ne "sees" larger corrugation amplitudes and is more sensitive to details of the corrugation shape than He [2.38]; that this effect seems to be restricted to metal surfaces still calls for theoretical explanation.

Fig. 2.7a–e. Corrugation functions (*left side*) and hard-sphere models (*right side*) of the clean Ni(110)-surface (**a**) as well as four of the six ordered phases of hydrogen on Ni(110) (**b**) to (**e**). Note that the corrugations (**b–d**) yield a direct picture of the adsorbate configurations: every pronounced hill corresponds to a H-atom. The formation of hydrogen zig-zag chains and their ordering with increasing coverage is clearly visible. The corrugations are expanded by a factor of five in the vertical direction. The maximum corrugation amplitude is $\simeq 0.25$ Å for all hydrogen phases. For the clean surface the corrugation is $\simeq 0.07$ Å perpendicular and $\simeq 0.03$ Å parallel to the close-packed Ni rows. ©1983 American Physical Society

For an example of a metal surface corrugation obtained with Ne-diffraction see Fig. 2.9c.

c) Low-Energy Electron Diffraction (LEED)

Besides being the oldest technique, LEED is without any doubt the most important and fruitful surface structural method up to now. Commercial equipment [2.1] is available from several manufacturers and hardly any ultrahigh-vacuum system dedicated for single crystal surface research is lacking a LEED retarding field analyzer [2.1,2]. Besides the routinely performed quick characterisations of surface order or the visual tracing of the development of regular overlayers upon adsorption (Fig. 2.3), careful LEED spot shape analyses can be used for characterisation of surface disorder [2.39] and in principle also for determinations of the site symmetry of adsorbed particles [2.40].

Furthermore also most of the crystallographically complete surface structure determinations by intensity analyses are based on LEED data [2.4,41]. As in the case of x-ray diffraction, the scattering centers of electrons are the regions of high electronic density, i.e. the atom cores, so that a complete structure analysis can yield the locations and mutual distances of all atom cores. Compared to x-rays, the scattering cross section for electrons is about three orders of magnitude larger. This fact has two important consequences. On the one side it causes the high surface sensitivity of low energy electrons (20–500 eV) due to their small penetration depths of 5–8 Å, but on the other side it gives rise to a severe complication for exact intensity calculations: whereas in the case of x-rays a kinematical theory [2.11,12], in which multiple scattering processes can be neglected, is applicable, LEED requires a full dynamical theory, in which the many different multiple scattering sequences must be added up. Layer by layer treatments are used in which the intermediate scattering processes are taken into account by allowing the atomic scattering cross sections to be position dependent; the crystal is divided into layers parallel to the surface and within one layer all atoms of a given species with the same local surrounding are equivalent. As there is no translational symmetry perpendicular to the surface the unit cell has to involve all crystal layers. However, as the electron wave is strongly damped with increasing distance to the surface, it usually suffices to perform the calculations for several layers only [2.42]. Full calculations are mostly restricted to small unit cells due to computing limitations. Good guides to the complete theory and (working!) computer programs have been published by *van Hove* and *Tong* [2.43], see also [2.3]. In most cases intensity versus voltage (energy) curves are fitted for different Bragg beams and various scattering geometries. The starting point is always a model assumption on the surface structure, whereby the potential around the individual atoms is assumed spherically symmetric and constant between the spheres ("muffin-tin"); a small number of fitting parameters (interlayer distances of the substrate, adsorbate bond lengths etc.) is left open and the degree of agreement between calculation and experiment is usually judged on the basis of differently defined reliability factors.

LEED intensity analyses performed independently by four different groups [2.44–47] using the full dynamical theory revealed the theoretically expected oscillatory behaviour [2.48] of the interlayer relaxation in the unreconstructed Ni(110)-surface: there is an appreciable decrease in the lattice spacing d_{12} between the first and second layer, a small increase between the second and third d_{23} and possibly also a small decrease between the third and fourth layer d_{34}. Table 2.1 compiles the data of the different groups and has also the result of a high energy ion scattering investigation [2.49] included. Whereas the LEED results show impressive agreement with one another, the ion scattering results yield appreciably smaller d_{12} values with, however, the same general trend in oscillatory behaviour. Similar results were obtained with LEED and ion scattering by *Adams* et al. [2.50] for Cu(110), Table 2.1. Comparison of the respective layer distances serves as an excellent example as to the degree of accuracy to which interlayer spacings at surfaces can be determined at present.

Table 2.1. Comparison of multilayer relaxation data obtained for Ni(110) and Cu(110) with LEED and High Energy Ion scattering (HEIS). d_{mn} gives the percentage change in layer distance between layer m and n, whereby the minus sign means contraction and the plus sign expansion with respect to the bulk value.

Material	d_{12}	d_{23}	d_{34}	Method	Ref.	Year
Ni(110)	−8.4	+3.1		LEED	[2.44]	1984
	−8.7	+3.0	−0.5	LEED	[2.45]	1985
	−9.8	+3.8		LEED	[2.46]	1985
	−9.6	+4.3		LEED	[2.47]	1985
	−4.8	+2.4		HEIS	[2.49]	1983
Cu(110)	−8.5	+2.0		LEED	[2.50]	1982
	−5.3	+3.3		HEIS	[2.50]	1982

More open surfaces as the saw-tooth-like Fe(310)were investigated with LEED by *Sokolov* et al. [2.51]; they exhibit more complex relaxations whereby atomic displacements perpendicular as well as parallel to the surface can occur; the results are in partial agreement with predictions of calculations minimizing the total energy of semiinfinite crystals with high-index surfaces [2.52]. The LEED result on Fe(310) as well as on other open Fe surfaces [2.51] shows that the lateral relaxation tends to symmetrize the location of the topmost layer relative to the second; the first layer contractions are much larger than in the case of the more close-packed surfaces (compare Table 2.1). According to LEED [2.53] the latter is also the case for Ni(311) whereas only negligible lateral relaxations of the topmost layer relative to the second has been observed (notice the structural similarities of the bcc(310) and the fcc(311)-surfaces, Figs. 2.8 and 2.9, respectively). However, as can be seen from Fig. 2.9c, the corrugation function obtained with Ne-diffraction suggests strongly that a charge density relaxation takes place on Ni(311), which tends to shift charge from the close-packed three-fold hollows to the more open four-fold ones, so that the corrugation of the latter is practically vanishingly small [2.54].

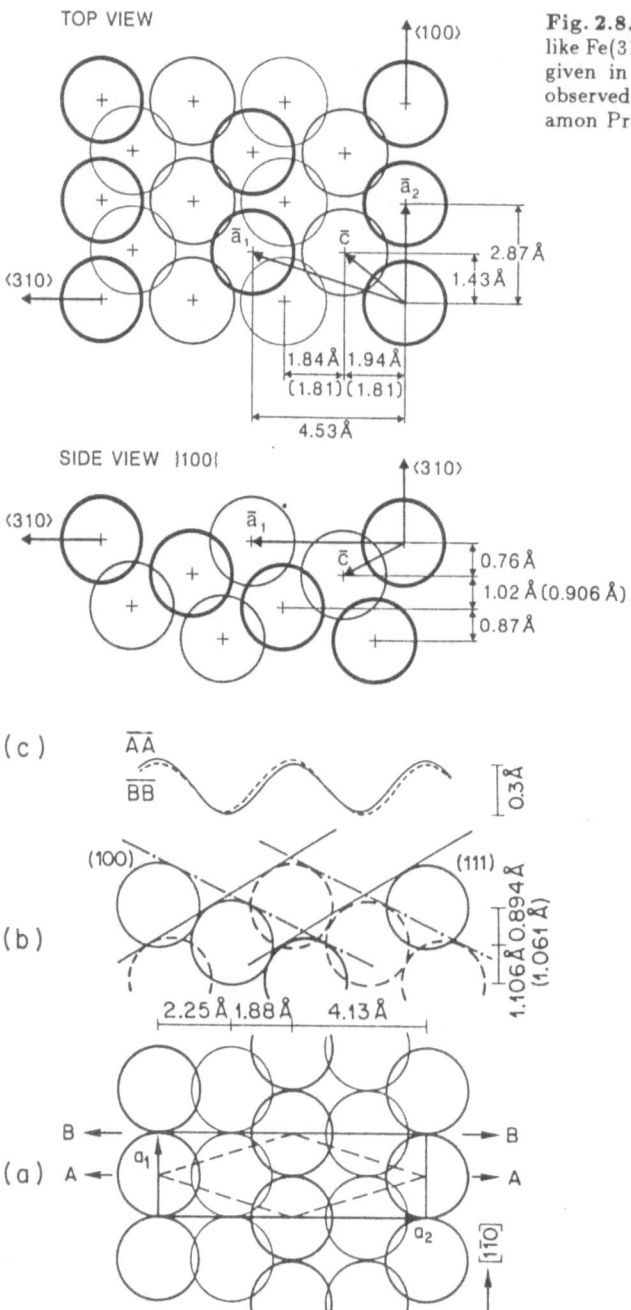

TOP VIEW

(100)

$\bar{a}_2$

(310)

$\bar{a}_1$ $\bar{c}$

2.87Å

1.43Å

1.84Å 1.94Å
(1.81) (1.81)

4.53Å

Fig. 2.8. Schematic drawing of the bulk-like Fe(310) surface. Ideal distances are given in brackets below the actually observed val ues [2.51]. ©1984 Pergamon Press

SIDE VIEW {100{

(310)

(310) $\bar{a}_1$ $\bar{c}$

0.76Å
1.02 Å (0.906 Å)
0.87Å

(c) $\overline{AA}$
      $\overline{BB}$

0.3Å

(100)   (111)

(b)

2.25Å 1.88Å   4.13Å

1.106Å 0.894 Å
(1.061 Å)

B ────── B

(a) A ────── A
     a_1

     a_2

[110]

[33$\bar{2}$]

Fig. 2.9. (a,b) Schematic drawing of top and side views of the bulk-like Ni(311) surface. Ideal distances are given in brackets below the values derived by *Adams* et al. [2.53] from LEED. **(c)** Contours of best-fit corrugations along cuts AA *(solid line)* and BB *(dashed line)* in Fig. 9a obtained with Ne-diffraction; this result points to a charge density relaxation causing appreciable charge flow from the (111)- to the (100)-microfacets. After [2.54]

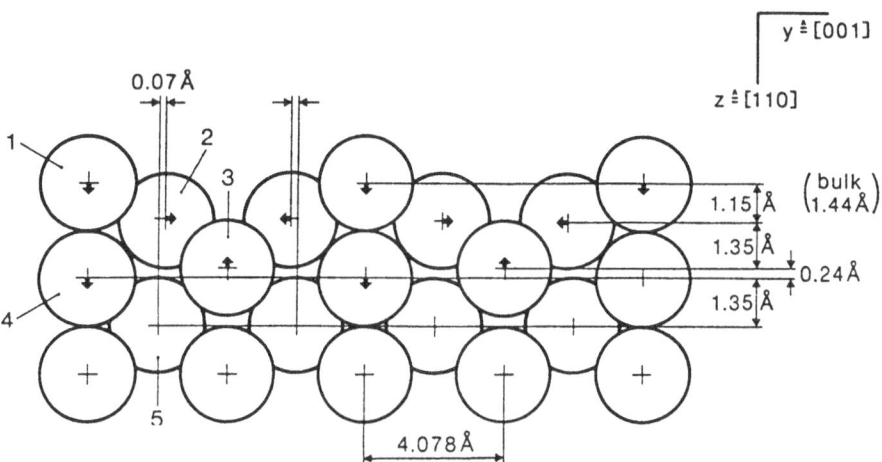

Fig. 2.10. Schematic drawing of the Au(110) surface with (1×2)-reconstruction [2.15]. The ideal normal distance between layers is given in brackets. ©Elsevier Science Publishers B.V

A side view of the most recent and most detailed result on the Au(110)(1×2) reconstruction obtained with LEED by *Moritz* and *Wolf* [2.15] is displayed in Fig. 2.10. Beside global changes of layer distances normal to the surface, these authors also took into account symmetry-compatible lateral and normal relaxations of atoms in deeper layers and found optimum agreement with experimental I/V-curves with the rather complex reconstruction pattern of Fig. 2.10.

Semiconductor surfaces in general exhibit complicated reconstructions, which may be explained by orbital rehybridization in order to saturate the "dangling bonds" left on a bulk truncated surface. It would be logic to deal here with the simplest – but nevertheless already very involved – semiconductor reconstruction, namely the (2×1) on the (111) cleavage plane of Si, but since LEED did not play as dominant a role in solving this structure as theory [2.55] and ion beam scattering [2.56], we leave the discussion of this structure to Sect. 2.5. Here we present the recent result on an even more complicated system, the (2×2) reconstruction of GaAs(111). Thorough LEED I/V-measurements and analyses applying dynamical theory were performed by *Tong* et al. [2.57] for a polished sample which was cleaned in UHV by ion bombardment and annealed at elevated temperatures. Figure 2.11 shows a top view and Fig. 2.12 a side-view along xx′ (viewing direction S) of the best-fit model. In this vacancy-buckling model one of the four Ga-atoms in the ideal surface unit cell is missing and the three remaining Ga-atoms undergo buckling distortions in which three As-atoms in the atomic plane below are involved. The Ga-atoms in the surface layer recede towards the bulk and their atomic orbitals rehybridize to form sp^2-bonds. The As-atoms are pushed sideways and outwards whereby they also produce modified bonds. Note that five atomic layers have the (2×2) periodicity. The most important feature of this model is buckling resulting from the creation of a cation vacancy. On the (111)-surface each Ga atom has one dan-

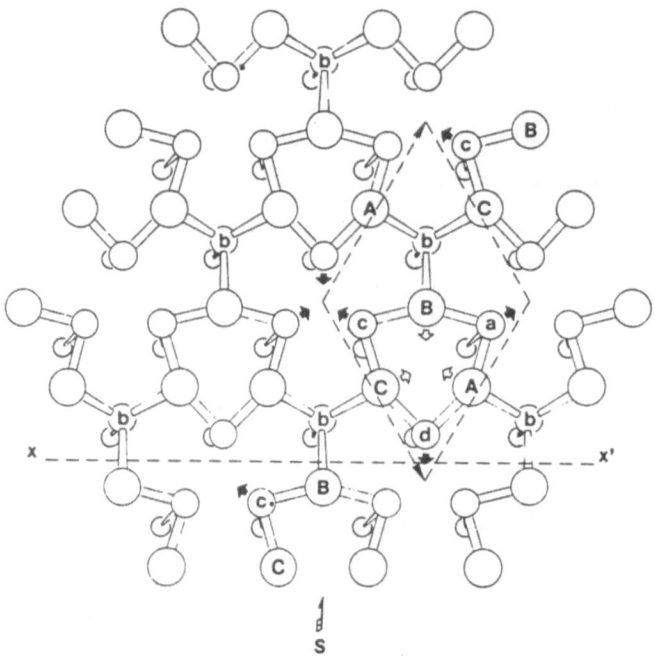

Fig. 2.11. Top view of the c(2×2) vacancy-buckling model of the GaAs(111) surface. The broken arrows indicate the surface unit cell. Three atomic layers are shown. (*A, B* and *C*) denote the Ga-atoms in the surface layer; (*a, b, c* and *d*) denote As atoms in the layer below and the Ga-atoms in the third layer are denoted as the smallest circles [2.57]. ©1984 The American Physical Society

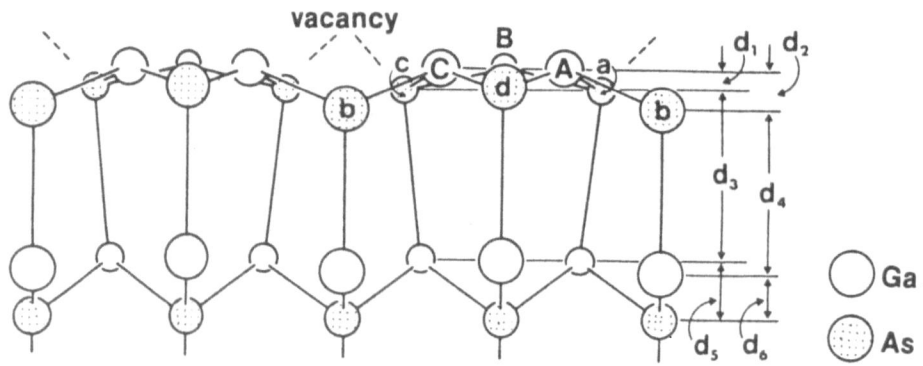

Fig. 2.12. Side view of the c(2×2) vacancy-buckling model of GaAs(111). The Ga- and As-atoms in the topmost two layers are denoted as in Fig. 2.11 [2.57]. ©1984 The American Physical Society

gling bond. By removing one Ga per surface unit cell, three As dangling bonds are created which can only be balanced by three surface Ga-atoms: this requires the (2×2) periodicity. It can be seen from Fig. 2.11 that in each reconstructed unit cell every Ga-atom with a dangling bond is bonded to an As-atom which has also a dangling bond. This configuration promotes orbital hybridization via

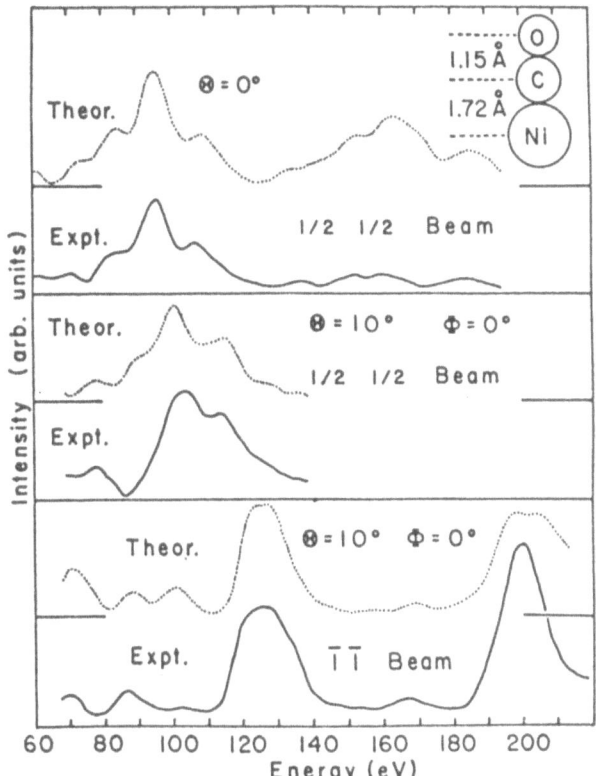

Fig. 2.13. Comparison of experimental and best-fit LEED I/V-curves for the c(2×2)CO overlayer on Ni(100). The structural model is indicated in the insert [2.60]. ©1979 American Physical Society

buckling and minimizes total energy [2.57]. A very similar structure was found very recently – at least for the top bilayer – for InSb(111)(2×2) by *Bohr* et al. [2.58] using glancing incidence x-ray diffraction and model-free Patterson synthesis (Sect. 3.2.1).

The probably most reliable LEED investigation of an adsorbate structure is that of the system Ni(100)+c(2×2)CO, for which two different groups independently measured and analyzed I/V data and came to very similar conclusions: CO stands perpendicular to the surface with the C-atoms on top of the Ni-atoms; the Ni-C bond length was determined to be 1.8±0.1 Å and the C-O distance 1.1±0.1 Å [2.59,60]. Figure 2.13 is reproduced to illustrate the degree of agreement between best-fit and experiment.

Very recently the stringent restrictions due to computer limitations have been relaxed by new methods which take advantage of the symmetry properties of the surface structure [2.61] or by use of the concept of "beam set neglect" in which third and higher order multiple scattering is neglected [2.62]. In this way also more complicated structures with larger unit cells like incommensurate overlayers [2.61] and adsorbate structures of organic molecules [2.62] seem to have been successfully tackled. Figure 2.14 shows as an example the case of a large surface unit cell formed by as large a molecule as benzene coadsorbed with CO on Rh(111).

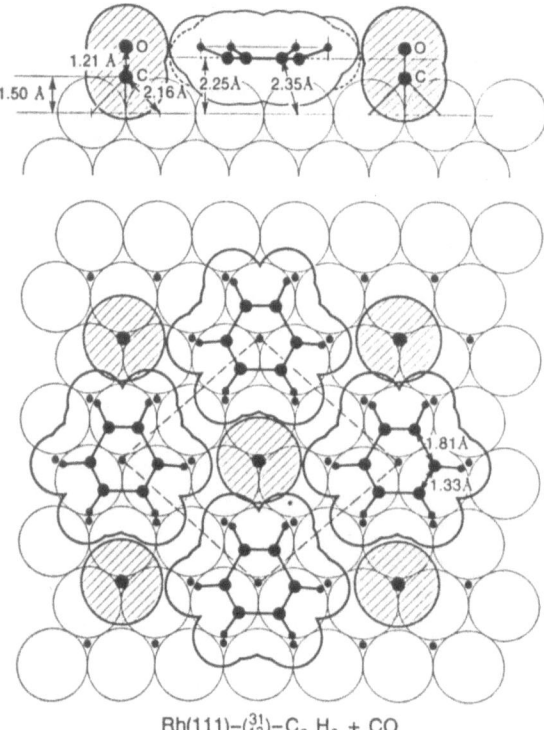

Rh(111)–$\binom{3\ 1}{1\ 3}$–C_6H_6 + CO

Fig. 2.14. Side and top views of the structure for C_6H_6 coadsorbed with CO on Rh(111) as proposed by *Van Hove* et al. [2.62] on the basis of LEED I/V analyses. The hydrogen positions are assumed. Van der Waals radii of 1.8 Å for C and 1.2 Å for H in the benzene and 1.6 Å for both C and O in the CO are indicated. The benzene shows an in-plane distortion as proposed by the authors (C-C bond lengths of 1.33 and 1.81 Å). The side view shows possible CH bending away from the surface. Dots between surface metal atoms denote second layer metal atoms. ©1986 American Chemical Society

TOP VIEW

C(2×2) C/Ni (001)

SIDE VIEW

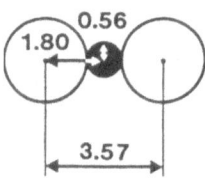

Fig. 2.15. Example of an adsorbate induced reconstruction found with LEED: Top and side views of the Ni(100)+(2×2)C structure [2.68]. The C-atoms reside ≃0.1 Å above the topmost Ni-plane. The Ni-atoms are laterally displaced by ≃0.35 Å and 0.2 Å outwards from the second layer. ©1983 North Holland Publishing Co.

Surface relaxations occuring at clean surfaces are usually changed and surface reconstructions can be removed or induced upon adsorption of (sometimes very small amounts of) foreign species [2.61–67,33,34]. As an example for an adsorbate induced surface reconstruction solved with LEED, Fig. 2.15 shows the structure of the (2×2)C phase (formed by cracking ethylene) on Ni(100) as determined by *Onuferko* et al.[2.68]. These researchers first realized the systematic absence of the $(\pm u0)$ and $(0\pm u)$ beams (u odd) at normal incidence, which implies the existence of glide lines along both close-packed directions of the clean surface and requires p4g space-group symmetry [2.6] for this structure. I/V-analyses established that the C-atoms occupy the fourfold hollows and that they distort the topmost layer of the Ni-substrate parallel to the surface such that nearest neighbour C-atoms are surrounded by differently oriented squares of Ni-atoms.

It must be emphasized that with x-rays and electrons light atoms in the presence of heavy ones are not easily "seen" as their scattering power is proportional to the square of the respective electron numbers [2.69]. Furthermore the relatively high energies may cause disorder in or damage of adsorbate phases; the latter problem can be circumvented by applying the Video-LEED technique developed recently by *Mueller* et al. [2.70].

d) Reflection High Energy Electron Diffraction (RHEED)

In RHEED electrons with energies between 1 and 10 keV are used. In this energy range the mean free path becomes as large as 30–100 Å so that the surface sensitivity is lost, unless grazing (1–3°) incidence and emergence is used. According to the large k-vectors and large angles of incidence used, the Ewald sphere is very large compared to the reciprocal lattice spacing. Due to the finite spatial divergence and energy spread of the electron beam this causes the (00) and (properly aligned) neighbouring lattice rods to be touched along a great length, so that the RHEED pattern of a flat surface shows long streaks normal to the surface [2.2,8]. Small hills on the flat surface through which the RHEED beam must pass give on the other hand rise to three-dimensional diffraction and spots are observed instead of streaks. It is clear that this method is particularly well suited to study growth processes on flat surfaces.

Thus RHEED was recently used by *Seguin* et al. [2.71] in a study of the growth mechanism of noble gases and molecular nitrogen on graphite. According to the relative strengths of the atom-atom interaction energies in the overlayer E_a and the adatom-substrate interaction energies E_s one expexts from theoretical considerations either layer-by-layer growth ($E_s \leq E_a$, "complete wetting") or cluster growth ($E_s \lesssim = E_a$, "incomplete wetting"). For layer-by-layer condensation the RHEED diffraction pattern is composed of parallel streaks, whereas spots appear when the adsorbate forms small bulk-like crystallites. Incomplete wetting was found to occur for He, Ne and N_2, whereas layer-by-layer growth was observed up to ten layers for Ar and Xe, in surprising contradiction to theory.

In earlier work, *Mitchell* et al. [2.72] applied RHEED in a study of the oxygen adsorption on and the oxidation of a Ni(110)-surface. As long as pure chemisorption phases occur, the RHEED pattern exhibits streaks. The inset of oxidation is signalled by the spontaneous formation of small oxide nuclei at a critical oxygen coverage which give rise to a pattern of broad spots. The widths of the spots [2.12] could be used to estimate the thickness and lateral extension of the oxide nuclei and their lateral growth upon further oxygen uptake could be followed.

The special scattering geometry allowed RHEED to become the most used tool for in-situ surface structure characterization in molecular beam epitaxy (MBE) research and applications [2.8,73]. Note, however, that full determination of the periodicity and symmetry of a surface requires observation of the streak patterns along different azimuthal orientations of the sample [2.8].

2.4 Other Methods Based on Diffraction Effects

2.4.1 Surface Extended X-Ray Absorption Fine Structure (SEXAFS)

The x-ray absorption coefficient of a free atom shows above a steeply rising core absorption edge a smooth decrease with increasing energy (or wavevector). If the same atom is built into a molecule or into condensed matter surrounding its absorption coefficient exhibits characteristic modulations at 50 to several 100 eV from the edge as can be seen in Fig. 2.16. The observed fine structure is due to interference effects between the outgoing wave of photoelectrons excited by the x-rays and the wave of electrons scattered back by the surrounding atoms. This interference modifies the ionization cross section of the photons and thus the x-ray absorption coefficient. Varying the x-ray energy changes the kinetic

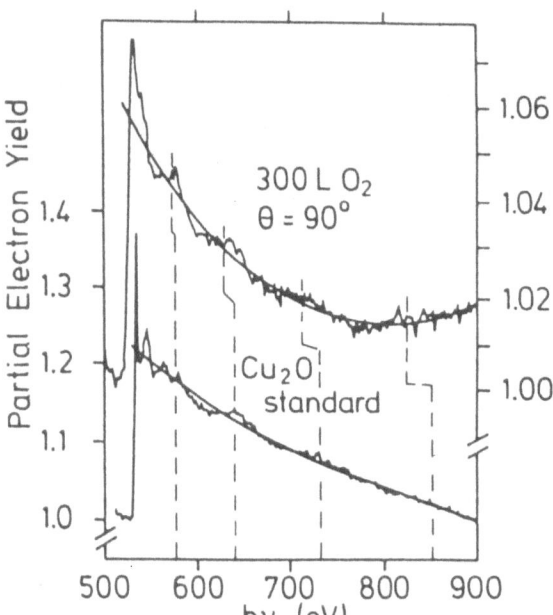

Fig. 2.16. SEXAFS electron yield for c(2×2)O on Cu(100) and for a Cu₂O standard sample. The dashed lines indicate the EXAFS oscillation periods [2.78]

energy and wavelength of the photoelectrons and therefore the interference conditions, so that the absorption coefficient is modulated and the period of oscillations establishes a measure of the distance between the reference atoms and the surrounding ones. Since the outgoing waves are strongly damped, only backscattering from the nearest shells of neighbors plays a role. The Fourier transform of the modulated part of the absorption coefficient shows peaks from which the interatomic distances of these shells from the absorbing atom can be determined. The amplitude of the interference oscillations increases the larger the number of atoms in a given shell [2.73-75]. The information obtained concerns the local arrangements and spacings around the absorbing atom, which makes the method equally applicable to crystalline and amorphous materials. The method is element specific since particular absorption edges can be chosen [2.74]. An accuracy of bond length values of 0.01–0.02 Å can be obtained.

Surface sensitivity in the sense that atoms in the topmost layers can be distinguished from deeper lying ones is not achieved with EXAFS, since the mean free paths of the photons are of the order of 1000 Å and the escape depth of the photoelectrons is 5–50 Å. Sensitivity for surface layers is, however, obtained by the chemical selectivity (K-edge energy) of adsorbates. Different detection modes are applied like electron yield (total, partial, Auger) and photon fluorescence yield (characteristic K-edge) measurements [2.73-75]. Since the energy must also be continously tunable, SEXAFS has been developed only since the high flux synchrotron sources are available [2.76,77]. It is worth emphasizing that in contrast to the surface diffraction methods discussed in the previous sections, SEXAFS does not depend on the formation of regular overlayers.

As a very recent example of the presently very rich SEXAFS literature we discuss the determination of the chemisorption site of oxygen on Cu(100) by *Doebler* et al. [2.78]. The upper part of Fig. 2.16 shows the SEXAFS raw data obtained at normal incidence ($\theta = 90°$) above the oxygen K edge for an oxygen coverage corresponding to the formation of a c(2×2) overlayer; the lower curve in Fig. 2.16 shows the spectrum of a Cu_2O sample which was used as a standard to allow empirical determination of the scattering phase shift between the O and Cu atom pairs. Since the SEXAFS oscillation periods appear at smaller distances in the chemisorption system, it is immediately evident that the O-Cu bond is larger than in the compound. After subtraction of the smoothly decaying curve in Fig. 2.16 the data are displayed again in Fig. 2.17b. The Fourier transform of this curve is shown in Fig. 2.17a; it exhibits a single prominent peak, which is characteristic for the nearest neighbor shell of Cu atoms. Use of the pase shift derived from the Cu_2O standard yields a O-Cu distance of 1.94 Å in good agreement with the result obtained for $\theta = 45°$, Fig. 2.17c. Fourier backtransformation with this value and the empirical phaseshift yields the smooth curve in Fig. 2.17b. The intrinsic anisotropy of the surface gives rise to a dependence of the oscillation amplitudes on the angle of incidence as well as – for low surface symmetries – on the azimuthal angle of the electric field vector in the surface plane. This "search-light" effect was used to derive the adsorption site of the O-adatoms on Cu(100) via the experimental amplitude

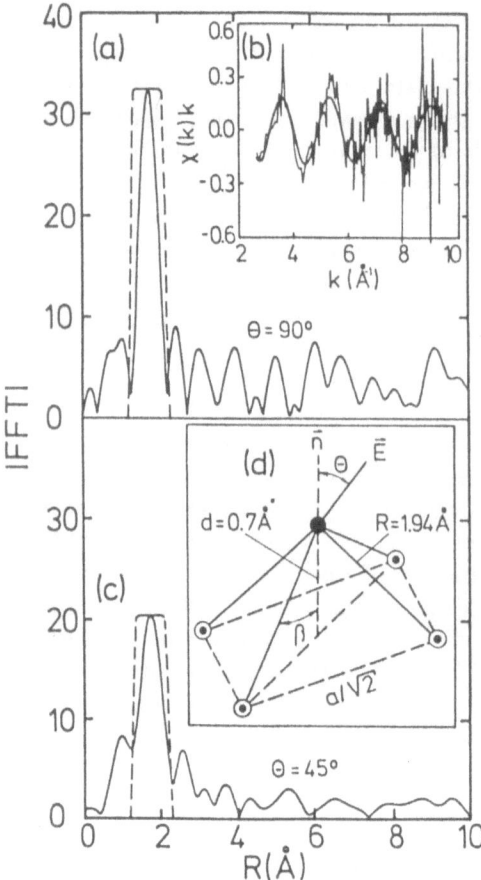

Fig. 2.17. Fourier transforms of the EXAFS data for c(2×2)O on Cu(100) for **(a)** normal ($\theta = 90°$) and **(c)** oblique ($\theta = 45°$) incidence. The dashed line is the filter function used for back transformation. **(b)** shows the measured and the filtered back transformed curves. **(d)** sketches the proposed oxygen chemisorption structure [2.78]. ©1985 The American Physical Society

ratio at the two angles of incidence of Fig. 2.17. The relevant analytical expression for the absorption cross-section $\chi(k)$ contains the effective coordination number of the absorbing atom as a function of the angle θ between the electric field vector of the x-rays and the surface normal z and the angle β between the adsorbate-substrate internuclear axis and z (Fig. 2.17d). For the fourfold hollow adsorption site the expected amplitude ratio is 1.5 in good agreement with the observed value 1.4 (Figs. 2.11a and c); for adsorption on a bridge site the ratio would be 0.6, which is clearly outside the experimental error.

2.4.2 Surface Extended Energy-Loss Fine Structure (SEELFS)

Closely related to SEXAFS is surface extended energy-loss fine structure [2.79] which uses electron beams in the range of several keV as excitation radiation. This method has been applied by *de Crescenti* et al. [2.80] to investigate the bond length of c(2×2)O on Ni(100). Figure 2.18a shows the SEELFS spectra above the oxygen K edge for the clean surface as well as for different oxygen exposures. The spectra were measured using 2 keV electrons with a cylindrical

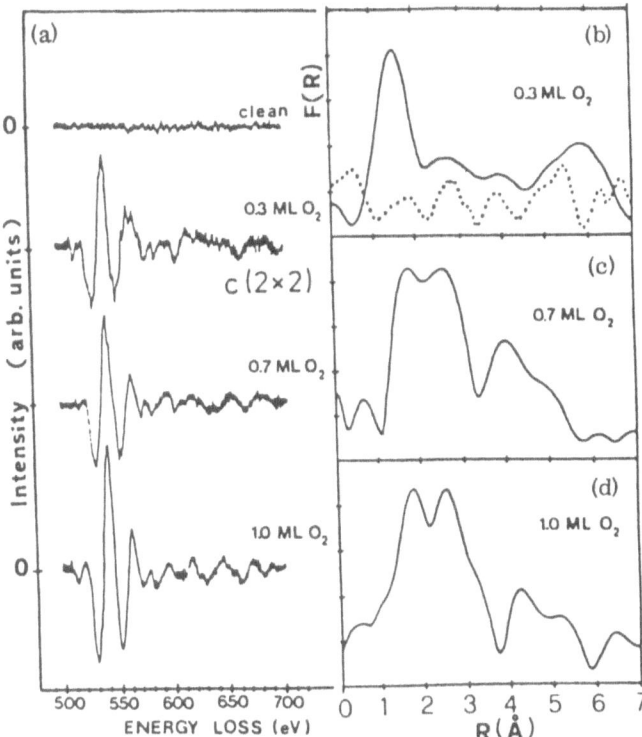

Fig. 2.18. (a) Electron reflection energy loss spectra of Ni(100) obtained with a 2 keV beam for the clean surface as well as for different oxygen coverages. (**b**–**d**) Fourier transforms of the data in (a). The dotted line in (**b**) corresponds to the Fourier transform of the spectrum obtained with the clean Ni-sample. The pronounced peak in (**b**) corresponds to the O-Ni bond length in the c(2×2) chemisorption phase, whereas the two adjacent dominant peaks in (**c**) and (**d**) are interpreted to correspond to O-Ni and O-O distances in the thin surface oxide film [2.80]. ©1983 The American Physical Society

mirror analyzer [2.1,2] in the second derivative mode. The lowest exposure corresponds to the formation of the c(2×2)O overlayer and the higher exposures correspond to the early stages of Ni-oxidation. Data evaluation is done as in SEXAFS by calculating the Fourier transforms shown in Figs. 2.18b–d, whereby the dotted line in Fig. 2.18b corresponds to the clean Ni-surface and provides a feeling as to the limits of reliability of the structures in the Fourier transforms. The Fourier transform corresponding to the c(2×2)O phase, Fig. 2.18b shows a pronounced peak at 1.6 Å, which yields (with the emprical phase shift obtained with a completely oxidized sample) a O-Ni bond length of 1.96±0.03 Å in excellent agreement with the results of LEED [2.81], SEXAFS [2.82], ion beam scattering [2.83] and photoelectron diffraction [2.84]. Although in the SEELFS work the adsorption site could not be determined, the fourfold hollow site was established by the other methods. The most prominent peak structures in the Fourier transforms of the higher exposure SEELFS curves, Figs. 2.12c and d, correspond to 2.1 and 3.0 Å and are interpreted as O-Ni and O-O distances in the thin NiO layer, since they are in good agreement with the corresponding bulk values. The future of this technique, which is at present by far not as refined as SEXAFS, can be regarded with optimism, since it is certainly a "poor man's SEXAFS alternative".

2.4.3 Photoelectron Diffraction

Photoelectron waves emitted at a particular atom in photoemission processes by ultraviolet and x-ray radiation can scatter off neighbouring surface atoms; the oscillations in the partial cross-section resulting from interference between different scattering paths contain information on the distances between the scattering centers [2.85]. For a regular adsorbate system, for example, by varying the photon energy and measuring photoelectron intensity emitted from a given core level with the detector fixed in a certain escape direction interference maxima and minima are observed which are related to the height of the adsorbate above the plane of substrate atoms (energy dependent photoelectron diffraction, EDPD) [2.86]. Another possibility consists in fixing the energy and changing the escape direction (angular dependent photoelectron diffraction, ADPD) [2.87]. Whereas the first is particularly sensitive to the distances between atomic planes, the second is more sensitive to site symmetry especially if the azimuthal angle is changed by rotating the surface about its normal [2.88].

As an example of EDPD we discuss again the system CO on Ni(100) (Sect. 2.3.4). Figure 2.19 shows in the inset the intensity of the C 1s core-level photoemission normal to the surface as a function of the kinetic energy of the photoelectrons [2.86]. The observed intensity oscillations have no analogue for the free CO molecule. They arise because of interference between the directly outgoing photoelectrons with back-reflected ones from the Ni substrate layers. The interference maxima occur, in a coarse approximation, when a multiple of the photoelectron wavelength matches the spacing between the C atom and the outermost plane of Ni-atoms. Due to phase shifts upon back reflection and especially because of multiple scattering similar complications as in LEED theory occur and a LEED type calculation was necessary to obtain the accurate distances. The best-fit C-Ni distance of 1.8 Å is in good agreement with the LEED results of [2.59 and 60] (Fig. 2.13). In order to determine the C-O distance backscattering of photoelectrons from the O1s level would have to be investigated.

The usual energy range of photoelectrons in EDPD is between 30 and 150 eV. It has been noted recently, that at higher energies (100–500 eV) single scattering theory is applicable [2.89], so that Fourier transform methods as used in SEXAFS can be applied [2.90]. This has led to the development of angle resolved photoemission extended fine structure (ARPEFS) [2.90,91], which has the advantage of much easier data analysis, although the measurements are much more difficult and require synchrotron radiation. The method was successfully tested for the system S/Ni(100) [2.89] and could not only establish the

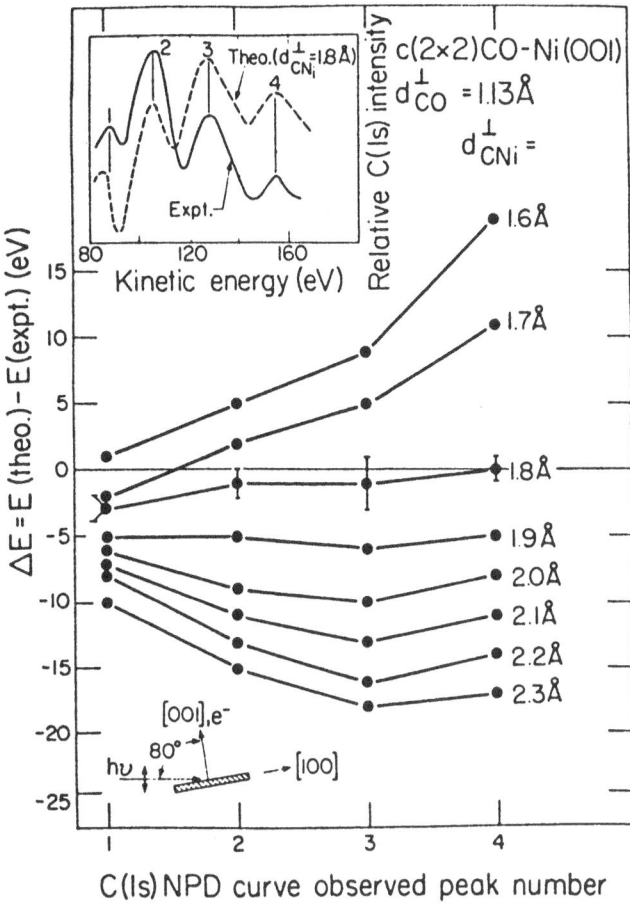

Fig. 2.19. Photoelectron diffraction from CO on Ni(100) is observed as modulation of the normal emission intensity against photon energy for C 1s emission. The interference maxima and minima yield a C-Ni distance of 1.8 Å in good agreement with LEED results (compare Fig. 2.13) [2.86]. ©1983 American Physical Society

bond length but also the coordination - by taking advantage of the search light concept as in SEXAFS, i.e. applying different light polarizations and measuring at different azimuthal orientations of the sample.

2.4.4 Near-Edge X-Ray Absorption Fine Structure (NEXAFS)

The energy region just above an x-ray absorption threshold up to $\simeq 50\,\mathrm{eV}$, referred to as near-edge x-ray absorption fine structure (NEXAFS, previously called x-ray absorption near edge structure XANES), is usually dominated by multiple scattering and is therefore discarded for the single-scattering analysis of the SEXAFS-regime [2.90]. Due to multiple scattering and because of the long mean-free paths of the low energy photoelectrons, which implies that not only nearest neighbours are involved in the scattering processes, quantitative evaluation of NEXAFS structures is very involved for chemisorbed atoms and requires complex calculations comparable to full dynamical LEED analyses. Such calculations were successfully performed by *Norman* et al. [2.92] for the case of O on Ni(100) and their results confirmed the conclusions obtained with other methods [2.80-84].

NEXAFS recently became, however, very important for investigations of molecular chemisorption structures, since it was found to be dominated for molecules and molecule fragments by intramolecular scattering resonances with surprisingly little influence from the substrate [2.93]. It should be noted that SEXAFS investigations of intramolecular distances in low-Z-molecules are difficult since the backscattering amplitudes decay quickly and since due to the short bond lengths only few EXAFS oscillations are observable. The observed NEXAFS resonances arise from transitions of a 1s core electron to empty or partially filled molecular orbitals (bound states) or to continuum states with an enhanced amplitude on the molecule (quasi-bound states or shape resonances) [2.75]. The transitions are governed by dipole selection rules and intensity measurements in dependence of the orientation of the electric field vector E allow accurate determinations of molecular orientations relative to the surface [2.94]. NEXAFS spectra of low-Z-molecules are dominated by σ- and π-resonances. σ-resonances are observed whenever E has a nonvanishing component along the internuclear axis between two atoms in the molecule in analogy to SEXAFS. It is a multiscattering resonance with the photoelectron trapped in the intramolecular field or – more pictorially – the photoelectron scatters back and forth between the excited atom and the neighbouring one. π-resonances are bound-state transitions and their excitation energy is always smaller than the 1s ionization potential. From a molecular orbital point-of-view the σ- and π-resonances correspond to transitions into σ^*- and π^*-antibonding orbitals, respectively. π-resonances are therefore only observed if E has a finite component along the π-orbital, i.e. perpendicular to the σ-bond axis. As a consequence σ-resonance intensities go with $\cos^2\alpha$ (α being the angle between E and the direction from the central atom to the scattering neighbour) and π-resonances with $\sin^2\alpha$. The value of NEXAFS for intramolecular bond length determination ("bond lengths with a ruler" [2.95]) rests on the applicability of the simple formula $(D - V)R^2 = C$ whereby the values of the "inner potential" V and the constant C could be empirically fixed (at least for molecules containing C,N and O). It should be noted that resonances associated with bonds to H-atoms are too weak to be observable.

The usefulness of these concepts is illustrated in Fig. 2.20, which shows NEXAFS spectra above the oxygen K edge for methoxy (CH_3O) and carbon monoxide chemisorbed on Cu(100) [2.93,75]. CH_3O is bonded by a single C-O bond, whereas CO is triple bonded with two orthogonal π-orbitals normal to the C-O axis. Consequently, the C-O bond lengths differ significantly in the two species: 1.43 Å for CH_3O and 1.13 Å for CO. The dominant feature of the spectrum for CH_3O/Cu(100) obtained at grazing incidence is a large σ-resonance

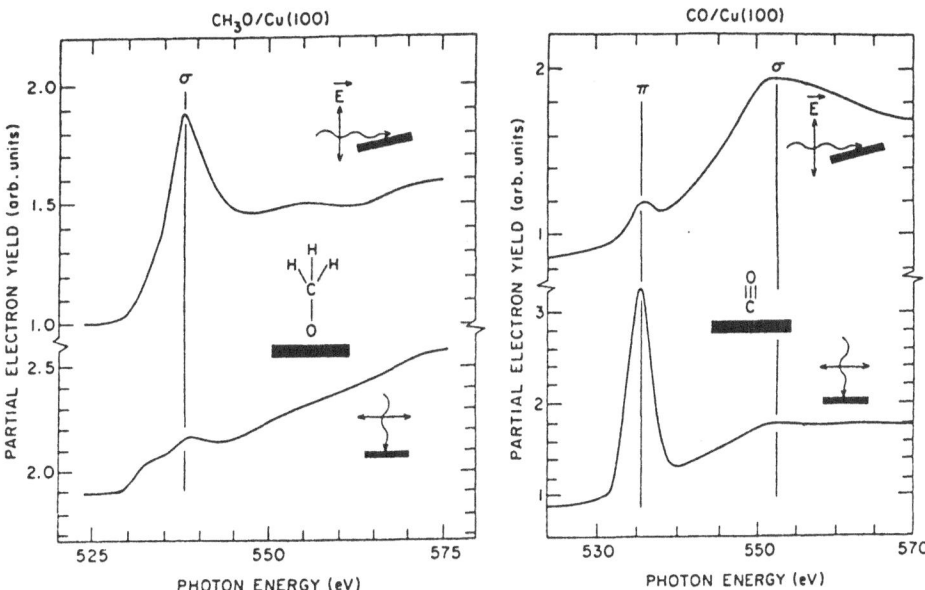

Fig. 2.20. NEXAFS spectra for CH_3O and CO on $Cu(100)$. Methoxy shows only a σ-resonance because of the C-O single bond, whereas because of the triple bonded bond with carbon monoxide π- and σ-resonances are observed. The polarization dependence establishes the molecular orientation and the energy position of the σ-bond the C-O bond length [2.75]

which vanishes at normal incidence; this establishes that CH_3O stands up on the surface. For $CO/Cu(100)$ both σ- and π-resonances are observed and the orientational intensity dependence shows that this molecule also stands up on the surface (the NEXAFS data allow, however, no determination whether the bond to the substrate is via the C- or the O-atom). The O-shape resonance appears at an appreciably smaller energy for methoxy than for CO because of the larger C-O bond length in this molecule. Using these data and assuming that the bond lengths in both species do not change upon chemisorption (which is supported ,as the vibrational frequencies in both states are found to be very similar with infrared absorption and high resolution energy loss spectroscopy) the parameters V and C in the above equation could be determined. Using these values and applying them to NEXAFS data on formate HCO_2 and molecularly adsorbed oxygen O_2, *Stöhr* et al. [2.93] found the formate C-O bond being very similar to the gas phase value and the O-O bond in $O_2/Cu(100)$ being stretched appreciably, again in accordance with vibrational spectroscopy data.

NEXAFS has in the mean time very successfully been used in studies of intermediate species and their bonding to the relevant active sites on catalytically active surfaces [2.96,97] as well as of the influence of promoters on stretching bond lengths and hence decreasing binding energies [2.98].

2.5 Microscopy Techniques with Atomic Resolution

2.5.1 Field Ion Microscopy

The field ion microscope (FIM), developed in 1951 by *Mueller* [2.99], was the first method which allowed direct observation of atomic positions. The somewhat philosophically sounding statement by Panitz in a recent review [2.100], that "for the first time, human beings could actually see individual atoms, a truly monumental accomplishment considering the simplicity of the technique", reveals the immediate fascination of real space images with atomic resolution in contrast to the indirect information obtained by diffraction techniques. In FIM, the sample is prepared in the form of a sharp metallic tip with a radius of several 100 Å; a positive potential is applied to the specimen so that a field of several Volts per Å at the tip surface is established. The particles of an imaging gas (mainly He or Ne at pressures of $10^{-4} - 10^{-6}$ Torr) are polarized in the electric field and therefore drawn towards the tip. In several collisions with the tip they loose their field-induced energy by inelastic processes and after having accommodated to the tip temperature they remain sufficiently long close to the surface, so that quantum mechanical tunneling, by which the particles loose an electron, can take place. Since electron tunneling can only occur into unoccupied metal states, there exists a critical distance for field ionization which corresponds to a position at which the energy of the valence electron in the atom is equal to the Fermi energy of the metal; the tunneling probability will be low at smaller distances to the surface because the atomic level is below the Fermi level and at larger distances because the barrier width is larger. The critical distance z_c is thus related to the field strength F, the work function ϕ of the tip and the ionization potential I of the imaging particle by $eFz_c = I - \phi$; z_c values are typically a few Angstroms with an ionization region around z_c of $\simeq 0.2$ Å.

The field-ionized positive ions are accelerated away from the tip in the electric field and give rise to a spot visible on a fluorescent screen. Modern systems use microchannel plate image intensifiers and image storage and reproduction techniques. The almost radial projection from the tip to the screen provides the necessary magnification of several million times, whereby the magnification is roughly given by the ratio of the tip diameter to the screen diameter. The imaging particles are ionized the better the higher the local electric field, i.e. in regions where the radius of curvature of the tip is highest. Thus, atoms with missing neighbours and especially isolated adatoms on flat terraces cause more ionization processes than atoms in a flat plane and are therefore most easily seen. A beautiful sphere model of a bcc(110) emitter tip with an impressive comparison of an optical visualization of the model tip contours and an actual FIM picture can be found in the review by *Panitz* [2.100]. Crucial for the optimum lateral resolution of the FIM are the size of the imaging particles as well as their velocity component parallel to the tip. The latter requires the tip temperature to be chosen as low as possible; a natural limitation in this

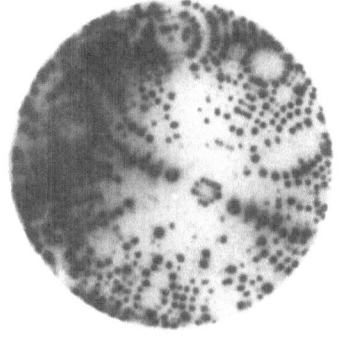

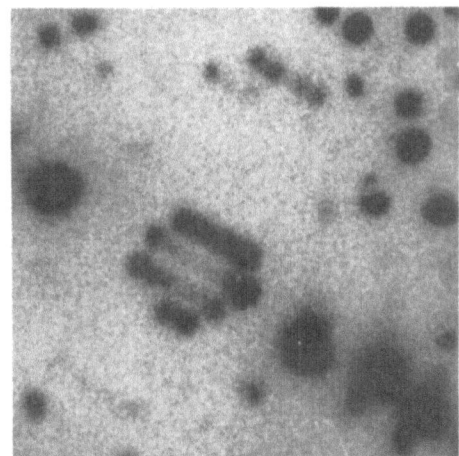

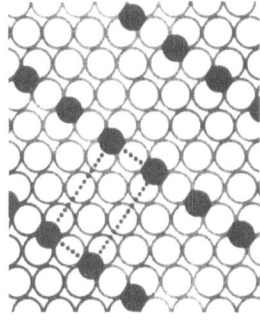

Fig. 2.21. *Upper left:* FIM-image of a (110)-oriented tungsten tip after deposition of Si at low temperature and anneal-ing at about 300 K. *Upper right:* Blow-up of the region of the (110)-plane showing the ordered Si-overlayer. *Lower left:* Sphere model of the Si-adatom configuration [2.103]. Cour-tesy of H.W. Fink

respect is the condensation temperature of the gas. Due to its small size and its low condensation temperature (<5 K), He is the optimum imaging gas. How-ever, its ionization requires large fields, so that appreciable mechanical stress is induced to the tip. Refractory metals are therefore particularly well suited for He imaging. Mixtures of gases often improve image contrast and resolution and allow also application of FIM to other metals and alloys and semiconductors [2.101,102].

Figure 2.21 shows as an example, the FIM image of a well ordered Si adlayer on W(110) obtained by *Fink* [2.103] together with a ball model of the specimen tip and the adsorbed Si-island. Note that the adatom-cluster is made up of well separated Si-chains aligned along the not close-packed [110]-direction of the W substrate.

The tip is usually chemically etched and then in the FIM-system prepared to its final form by raising the electric field to such a high value (usually appre-ciably above the "best image voltage") that surface atoms are removed by field evaporation. There is a self regulating effect in this procedure since protruding structures, sharp edges etc. are removed preferentially so that sufficiently large areas of low index surfaces are obtained. Applied in successive steps, field evaporation can be used for metallurgical studies of the distribution of vacancies, precipitates, grain boundaries as a function of depth [2.104,105]. Applied in a more subtle form, field evaporation leads to the atom-probe FIM [2.106], which allows also determination of the chemical nature of specific surface atoms: In the time-of-flight atom probe, a voltage pulse of a few nanoseconds duration is applied to the tip, so that field evaporation is restricted to a very short time. Using the leading edge of the voltage pulse as a time-marker, the flight-time of the evaporated species to the detector can be measured. From the kinetic energy, which is known from the total potential applied to the tip, and the flight distance the mass and thus the chemical identity can be determined; mass resolutions of $\Delta m/m \simeq 1/200$ can be achieved. By use of a small probe-hole as entrance aperture to the detector the field-of-view can be restricted to a preselected site on the tip from which evaporation is to be studied. For this reason the tip is made movable and rotatable. Low temperature field desorption is advantageous over conventional thermal desorption since surface migration is negligible and therefore does not lead to mistakes in data interpretation.

FIM was in the last years extensively used in quantitative investigations of the migration of single adatoms and small clusters on surfaces as well as of the interactions between adatoms [2.107,108]. The procedure is as follows: The tip morphology is established by field evaporation and a FIM image taken at low temperature. Adatoms are deposited by thermal evaporation from a hot filament and field desorption is used to restrict the number of adatoms to the amount desired. By heating the tip to an appropriate temperature for a certain time surface diffusion is allowed to take place, whereby the field is removed. After cooling the tip a FIM picture is taken and the new adatom position(s) determined. The mean square displacement is determined from a large number of such images. The new adatom configurations can be used to determine interaction energies. Measurements at several temperatures allow determination of the diffusion coefficient and attempt frequency.

Using atom-probe FIM, *Wrigley* and *Ehrlich* [2.109] have investigated the very interesting case of diffusion of W on Ir(110), which shows preferential movement of the adatoms across the close packed rows. This is not only in contrast to expectation but also to observations on other materials with groove structure, like Rh(110), where the preferred motion is along the grooves [2.110]. The Ir(110) clean surface exhibits (in contrast to Rh(110)) a (1×2)-reconstruction of the missing-row type in analogy to the respective surfaces of Au and Pt (Sect. 3.1 and Fig. 2.4). The (1×2) missing row structure was actually observed with FIM and constitutes one of the rare examples that metal reconstructions typical for extended surfaces are also present on the small area of the FIM tip. The c(2×2) reconstruction of W(100) was also seen [2.111] with FIM but the structure seems to be different from the one established by LEED and ion scattering [2.112]; this fact shows that the strong electric field may substantially influence the formation of surface structures. In the case of the cross channel diffusion on Ir(110) two different mechanisms were regarded possible: (a) Atomic jumps over the rows forming the ridges, which is much

more unlikely than jumps along the channels. (b) Exchange processes in which
the additional atom replaces an atom in the close packed row by pushing it into
the adjacent channel. It is clear that atom probe FIM provides the optimum
method to decide: A single W adatom is deposited to the cooled (110) plane
and the tip is then heated to allow for a cross channel jump. After cooling,
the atom appearing in the neighbouring channel is pulse desorbed and its mass
determined. The results showed that indeed Ir atoms are found in the adjacent
channels rather than W atoms. Further field evaporation proved that the W
atoms were indeed incorporated into the substrate lattice.

2.5.2 Scanning Tunneling Microscopy

Although the earliest theoretical work on quantum tunneling concerned electron tunneling through a vacuum barrier [2.113], mankind had to wait half a
century until in 1982 when *Binnig* et al. [2.114] provided a clear experimental
verification of this conceptually simplest tunneling phenomenon. Before this,
exploitation of quantum tunneling for spectroscopic and technological investigations was restricted to solid tunnel barriers [2.115]. Metal-vacuum-metal
tunneling requires a space gap held constant at a few Angstroems. At such distances it is extraordinarily difficult to control the gap width and to make sure
that surface irregularities do not give rise to undesired direct contact across
the gap. The most direct evidence of genuine metal-vacuum-metal tunneling is
the observation of the theoretically predicted exponential increase in resistance
with increasing gap width, whereby the appropriate work function governs the
slope. For an applied voltage U, the tunnel current J_T is given, in a first approximation, by [2.116]

$$J_T \propto (U/d) \cdot \exp(-A\phi^{1/2}d), \tag{2.9}$$

whereby $eU \ll \phi$ and E_F, the latter being the Fermi energy, is assumed; A is
$\simeq 1.025\,(eV)^{-1/2}\text{Å}^{-1}$ for a vacuum gap, ϕ is the average of the two electrode
work functions and d the minimum distance between the electrodes. Such a
dependence was, for the first time experimentally established by *Binnig* et al.
[2.114] by means of a novel instrument that allows to control of the distance
between the tunnel electrodes with a precision of 0.05 Å. Had earlier attempts
to demonstrate and use vacuum tunneling been plagued by insufficient sup-
pression of vibrations in the experimental apparatus, *Binnig* et al. succeeded
in achieving the necessary protection against external and internal vibrations
by placing their tunneling device into a two-stage system of quartz glass frames
suspended on very soft springs. Furthermore, residual vibrations were almost
ideally damped via eddy currents induced in copper counter-plates by powerful
magnets. A further important step, namely the ability to control precisely the
lateral position of the electrodes by means of a three-legged piezoelectric support has enabled the group to scan with one electrode, in the form of a sharp
tip, over the surface of the other. The resulting scanning tunneling microscope
[2.117–122] reaches a lateral resolution of $\simeq 2$ Å and a resolution normal to the
surface of several tenths of an Angstrom and can thus resolve single atoms. The

instrument quickly became the most celebrated and awarded newer method for studies of surface topographies. Whereas the first studies with STM were performed under high and ultra-high vacuum conditions, it was recently shown to work also in air.

Depending on the tip shape, the lateral resolution can be varied in a wide range, so that a given surface can be looked at with different magnifications. For scans with optimum lateral resolution it was claimed that the top of the tip consists of a single atom. For tip preparation the following recipe was found applicable: Wires of proper electrode material (Ir or W, for example) of $\approx 1\,\mathrm{mm}$ diameter are ground at one end at roughly $60°$. Scanning electron micrographs show overall tip radii of $<1\,\mu\mathrm{m}$, but due to the rough grinding process a few rather sharp minitips on each tip are formed. The extreme sensitivity of the tunnel current on the gap width selects the minitip closest to the sample for the tunneling. In this way lateral resolutions of $\approx 20\,\text{Å}$ can be obtained. An important advantage of the overall quite blunt tips is the mechanical stability and the much higher resonance frequencies than those of slim field emission tips. Sharpening of the minitip in action by touching the surface can reduce the resolution to $\simeq 10\,\text{Å}$, still with quite stable tips. The resolution can be increased considerably by exposing the tip for several tens of minutes to electric fields of $10^8\,\mathrm{V/cm}$; it is presently not known whether this procedure extracts or migrates atoms at the tip or adsorbs atoms on it. These very sharp tips are

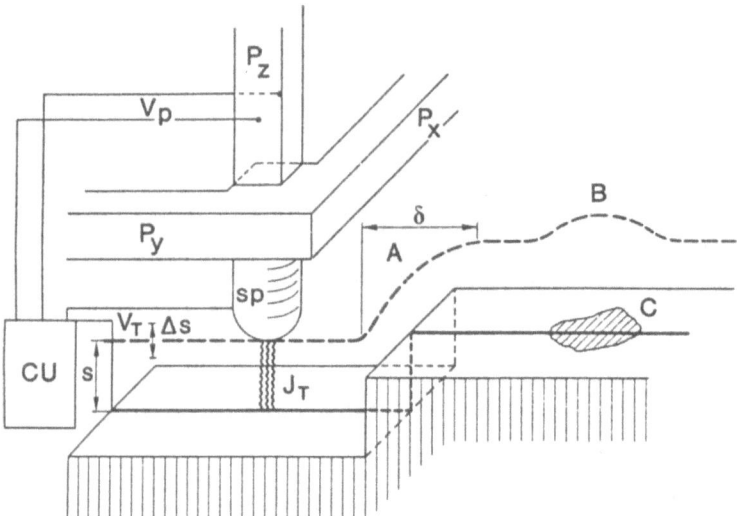

Fig. 2.22. Schematic drawing showing the principle of operation of the STM. The piezodrives P_x and P_y move the sample tip over the surface. The control unit CU delivers the appropriate voltage U_p to the piezodrive P_z for constant tunnel current J_T at constant tunnel voltage U_T. For surface regions with constant work function, the voltages applied to the piezodrives P_x, P_y and P_z yield the surface topography directly, whereas modulation of the tunnel distance d by Δd allows local measurement of the workfunction. The broken curve indicates normal displacement of the tip in a lateral scan *(A)* over a surface step and *(B)* over a contamination area *(C)* with lower workfunction [2.117]. ©1982 The American Physical Society

usually not very stable, so that the optimum resolutions cannot be maintained over long times and the field sharpening procedure has to be repeated.

A schematic picture of the STM is shown in Fig. 2.22. The tunnel tip is mounted to a rectangular piezodrive P_x, P_y, P_z, made of commercial piezoceramic material. P_z and P_y scan the tip over the surface. A voltage U_z (usually several tenths of a Volt) is applied to the piezo P_z by the control unit CU such that the tunnel current J_T remains constant. Thus, at constant workfunction, U_z in dependence of U_x and U_y gives the topography of the surface in the form of the corrugation function $z(x, y)$ as illustrated at the surface step at A, i.e. the tunnel tip is moved at constant distance d over the surface. The value of d is $\simeq 6$–$10\,\text{Å}$ and J_T changes by about an order of magnitude for every Å change of d. By keeping the tunnel current constant, changes in the work function are compensated by corresponding changes in d. Thus a lower work function at C induces the apparent surface structure B. Such artificial structures in the surface morphology, however, can be accounted for by measuring ϕ separately. This reqires measurement of the slope of the tunnel charcteristic and can be obtained during scanning by modulating the tunnel distance d. Thus, although separation of surface topography and local average work function changes is involved, it is possible and has actually been applied by *Binnig* et al. [2.123] to distinguish Au-islands on a Si-substrate from regions without adsorbate (Fig. 2.23).

In the mounting and cleaning stages the sample has to be kept sufficiently far away from the tip (typically a few millimeters). To approach the tip within its working range of the piezodrives (several $1000\,\text{Å}$) without mechanical contact between the outside and the tunnel unit, a rough drive, the so called "louse", was developed by the STM-inventors. It consists of a piezoplate, which carries the sample holder on top and rests on three metal feet, separated by high-dielectric constant insulators from the metal groundplates. The feet are

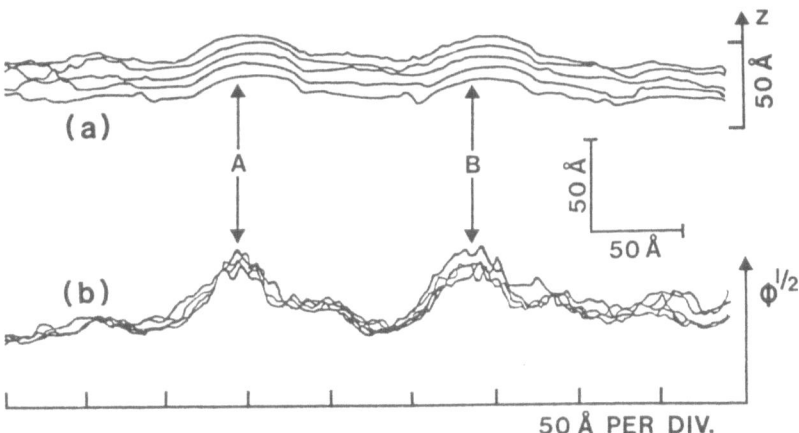

Fig. 2.23a,b. Au islands on Si. STM topography (a) shows a rough surface with two smooth hills which are identified by the corresponding average work function scans (b) to be Au islands [2.123]. ©1983 North-Holland, Publishing Co.

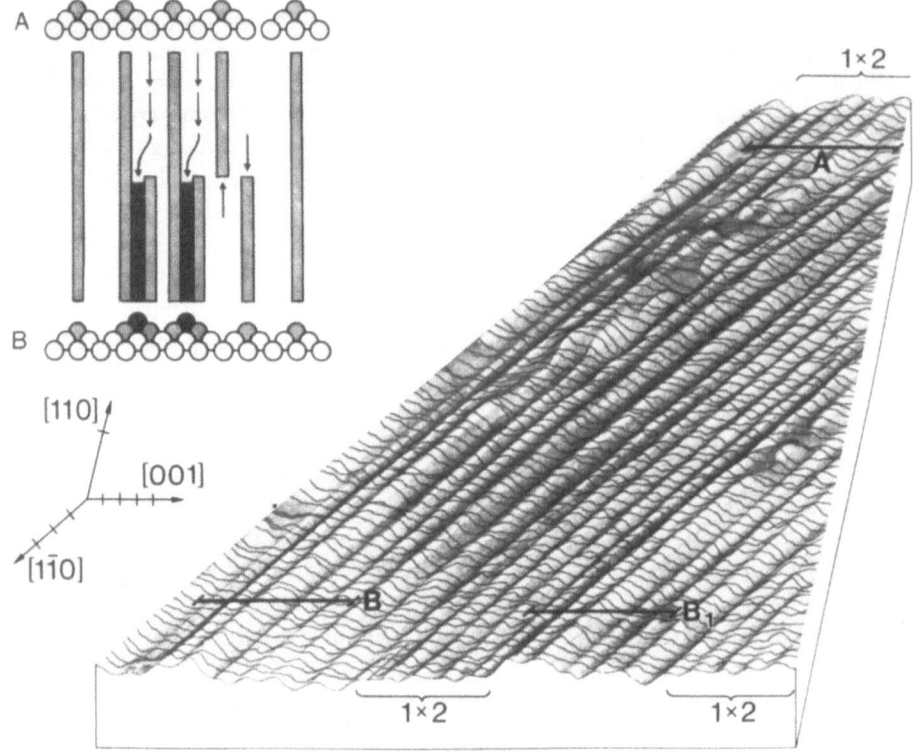

Fig. 2.24. STM recording lines obtained for Au(110). (1×2) reconstructed ribbons are interrupted by monolayer steps and (1×3) channels. The skewed directions of the [1$\bar{1}$0] and [110] axes are determined by thermal drift and crystal misalignment, respectively. Divisions on all axes correspond to 5 Å. The disorder structure at *(B)* repeats at *(B1)*. The insert suggests a possible transition from the structure at *(A)* to that of *(B)* by shifting rows along the close packed [1$\bar{1}$0]direction. ©North Holland Publishing Co.

clamped electrostatically to the groundplate by applying a voltage. Elongating and contracting the piezobody of the louse with an appropriate clamping sequence of the feet moves the device in any direction in steps between 100 Å and 1 μm with up to 50 steps per second.

As examples of surface structure investigations with atomic resolution we discuss the results of *Binnig* et al. on the reconstructions of the clean Au(110) [2.119] and Au(100) surfaces [2.120] as well as that on the famous (7×7) reconstruction of Si(111) [2.118].

Figure 2.24 shows STM scans of the Au(110) surface which confirm the conclusions of other methods, that missing rows and build-up of larger close-packed (111)-facets are the basic ingredients of its reconstruction. The STM picture was not obtained with an ideally ordered (1×2) reconstruction but also shows patches with (1×3) and (1×4) channels. It is well known that the latter reconstruction unit cells can also be stabilized on this surface with a good order; the (1×2) structure can certainly be obtained with a much better order, since disorder of the extension shown in the STM image would give rise to fuzzy diffraction patterns which would be refused for intensity analyses. Deviations from perfect order in a real space image of a surface corrugation on the other hand makes the picture interesting and, more important from a scientific point of view, some significant details may be derived from them. As a possible example of the latter, we refer to the adsorption of foreign species which seem to go preferentially to defect or edge sites at low coverages [2.124].

The reconstruction of the Au(100) is of a much more complicated nature. Whereas LEED work has stated a unit cell as large as (26×68) [2.125] it was found with He-diffraction that the dominant features correspond to a (1×5) periodicity [2.126], as actually observed with LEED for the Ir(100) surface. It is worth mentioning that the Pt(100) surface with its (5×20) reconstruction shows an intermediate complexity between Ir and Au [2.125]. The STM corrugation [2.120] showed that the topmost Au layer forms a hexagonal close-packed pattern (and thus confirmed a suggestion based on LEED observations [2.125]) with slightly contracted surface atoms (3.8%); the top layer is rotated

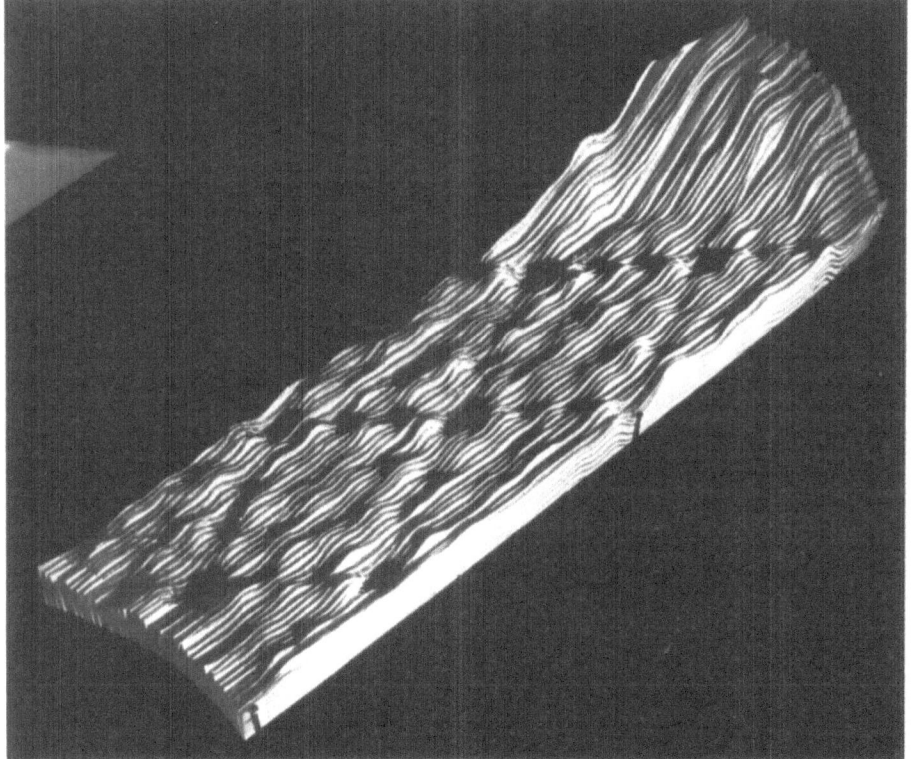

Fig. 2.25. STM relief of the (7×7) reconstructed Si(111) surface. The two complete unit cells are easily discernible by the deep corner minima. The $[\bar{2}11]$ direction points from right to left, along the diagonal. Each unit cell exhibits nine minima and twelve maxima. Surface irregularities as the hill at the right do not seem to influence the topography of the (7×7) areas. ©1983 The American Physical Society

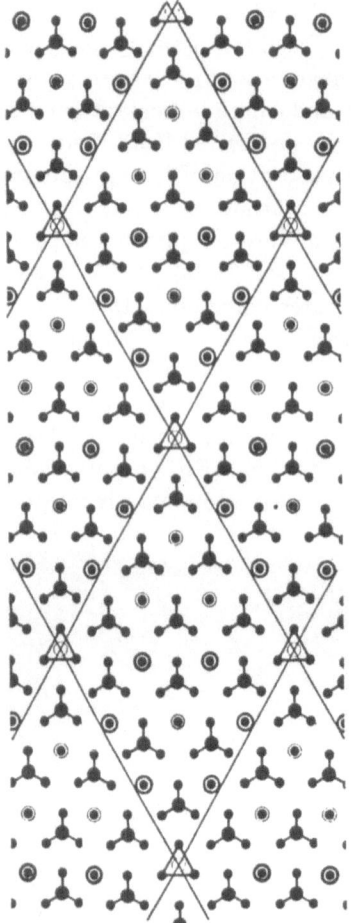

Fig. 2.26. The modified adatom model proposed on the basis of the STM result. The atoms in the layer below the adatoms are shown by dots and the atoms with unsatisfied dangling bonds are marked by circles, whose thickness indicates the minima depths measured with the STM. The adatoms are plotted as large dots with the corresponding bonding arms. The possible adatom position found to be empty with the STM is shown in the triangle of adjacent rest atoms (deep minima in Fig. 2.25). The grid indicates the (7×7) unit cell. ©1983 The American Physical Society

by 0.1° against the underlying cubic layer. The structure of the second layer determines: (a) the vertical position of the top-layer atoms whereby the resulting buckling pattern shows dominant (1×5)-regions which are very similar to the two basic corrugations observed with He-scattering [2.126]; and (b) the large repeat distances of the surface structure. As an interesting byproduct the real space STM pictures suggest that the surface reconstruction nucleates at steps.

The (7×7) reconstruction of the Si(111) surface with its large unit cell constitutes one of the most intriguing and most investigated problems in surface science. Numerous and mostly contradictory models had been proposed on the basis of different observations and theoretical considerations. Although the STM-results could not establish precise structural details, at least many of the existing propositions could be discarded. The original STM scans are reproduced in Fig. 2.25 and the model favoured by *Binnig* et al. [2.118] is shown in Fig. 2.26. It is a modification of the adatom model of *Harrison* [2.127] designed to reduce the unsatisfied bonds in that it contains only 12 instead of 13 adatoms; in this way the deep minima in the unit cell corners are accounted for. More recent STM-results [2.128] show that the two halves of the unit cell are not equivalent since the heights of the maxima are slightly different and the symmetry is therefore threefold in agreement with LEED-results [2.129]. Systematically missing maxima in a p-type Si-sample were attributed to surface boron dopants, whereby no noticeable changes of the morphology in the rest of the unit cell were found. The simple adatom structure of Fig. 2.26 was stated not to be consistent with ion scattering [2.130] and electron microscopy results [2.131]; new models have been proposed on the basis of energetic considerations [2.132].

It has to be emphasized that in contrast to the corrugation function obtained with atomic beam diffraction, which is a replica of charge density contours corresponding to the total density, in STM it are only states near the Fermi level which contribute [2.133–135]. This may lead to important and enlightening differences in the respective corrugations in cases where the surface electronic structure of the system investigated is very involved near the Fermi edge.

2.5.3 High Resolution Transmission Electron Microscopy (TEM)

Modern high-resolution electron microscopes were recently shown by *Marks* and collaborators to be capable of imaging atomic structures directly [2.136,137]. First experimental results confirmed the missing row nature of the (1×2)-reconstruction of Au(110) [2.136]. The specimens used were small (200–400 Å) Au particles epitaxially grown by evaporation in ultra-high vacuum onto cleaved alkali-halide substrates and coated with amorphous carbon films. The TEM images were taken along the close-packed Au-rows of (110)-surface facets so that every atomic row produced a distinct contrast. The (1×2) missing row pattern was clearly visible in regions where the carbon coating had been etched away, whereas (1×1) unreconstructed areas were found under carbon covered areas. Simulations of the spot widths and brightnesses as well as distances between the top and deeper layers led to the conclusion that the distance between top and second layers in the (1×2) increases by $\simeq20\%$, but it has been pointed out that this conclusion is not unique [2.138] and it is indeed in clear contradiction to the most recent LEED-results [2.15] (Sect. 2.3.4, Fig. 2.10). Since the size of the particles investigated with TEM is of the order of typical particle sizes in supported catalysts, the method will be very valuable for determining the distributions of active areas like steps, facets etc. in "real world" catalysis. Another kind of electron microscopy technique is currently being developed, in which a very narrow (5 Å diameter) electron beam is scanned over the surface and the yield of Auger electrons etc. is used to produce a surface image characteristic for different chemical species; this scanning transmission electron microscope (STEM) will be very valuable for studies of compositional variations at surfaces and chemically active surface sites. A detailed account of the possibilities and problems with TEM can be found in Chap. 3 by L.D. Marks; basic principles of electron microscopy have been discussed in recent books by *Reimer* [2.139].

2.6 Ion Scattering Spectroscopy (ISS)

The surface structure methods based on ion beam scattering rely heavily on the effects of shadowing, channeling and blocking [2.140,141]. The basic concepts may be recognized in discussing the example of the (2×1)-reconstruction of the Si(111), which forms if the crystal is cleaved in UHV [the (2×1) transforms irreversibly into the (7×7) upon heating the crystal to about 800°C]. The reorganisation of the broken bonds in the cleaved surface has given rise to a number of theoretical considerations attempting to explain the (2×1) surface periodicity. In the oldest, the buckling model by *Haneman*, alternate Si-rows were proposed to be displaced up and down [2.142]. The π-bonded chain model was proposed by *Pandey* [2.143,55] on the basis of total energy calculations and chemical considerations, and was especially successful in reproducing electronic surface state energy bands measured with angular resolved ultra-violet photoemission spectroscopy (ARUPS) [2.144] and core level shifts of surface atoms relative to atoms in the bulk [2.145]; in the Pandey model the surface geometry is changed to allow formation of π-bonded chains with large displacements of the surface atoms parallel to surface in the $[\bar{1}\bar{1}2]$ direction. *Chadi* [2.146] put forward the "molecule"-model in which the π-bonding should occur between pairs of atoms with atomic displacements out of the bulk $(\bar{1}10)$-planes.

The upper part of Fig. 2.27 shows schematically the scattering configuration used by *Tromp* et al. [2.147,148] in their investigations of the Si(111)(2×1): a cut through the ideal (111)-Si-surface is plotted along a $(\bar{1}10)$-plane (the $[\bar{1}12]$-direction is perpendicular to the paper plane). The parallel medium-energy (99.2 keV) H^+-ion beam impinges normal to the surface in a so-called channeling configuration: the first atom of each atomic row casts a shadow cone because of Coulomb repulsion by the nucleus of the atom, so that deeper atoms are hit by the ions with less probability. Only that part of the spatial probability distribution (caused by its vibrational motion) of an inner atom outside the shadow cone can contribute to the scattering probability and the sum of hitting probabilities of all atoms in a row is called "number of atoms per row" visible to the beam. If the first atom in a row is statically displaced so that it is not exactly in line with the deeper atoms, shadowing is less effective and the number of atoms in a row is larger. Thus, for the buckling model, for which only perpendicular displacements should occur, only the bulklike rows would be seen by the beam, whereas in the π-bonded chain model at least one extra double layer should be exposed. Absolute determination of the number of atoms per row is achieved by relating the backscattered intensity to the intensity obtained with a random crystal orientation (or a polycrystalline or amorphous standard).

Further important information on surface geometry is obtained by measuring the intensity of the backscattered ions as a function of scattering angle with the incoming beam fixed along a given direction. As indicated in the upper part of Fig. 2.27 ions scattered in deeper layers can reach the detector only if they are not blocked in their way out by some atom closer to the surface. In directions in which blocking occurs a so-called surface blocking minimum is

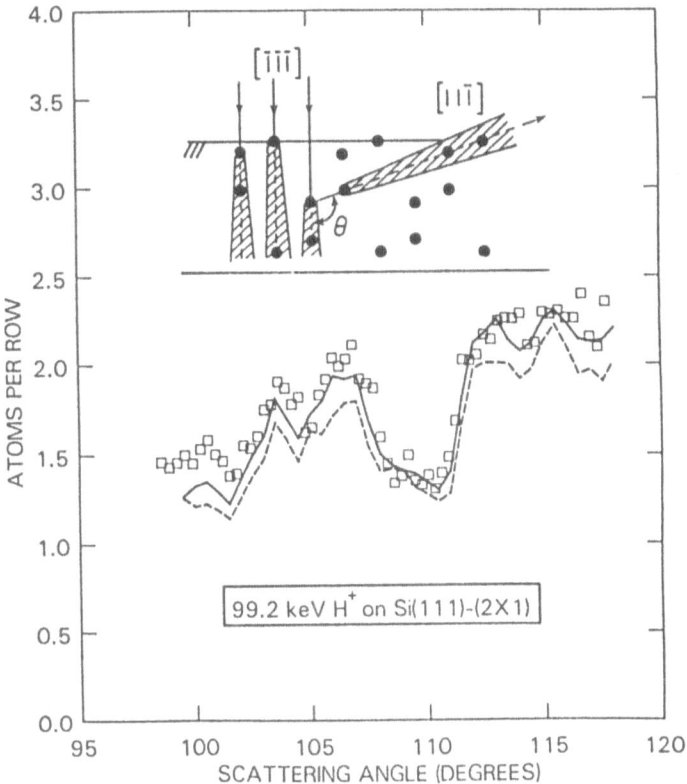

Fig. 2.27. Ion beam scattering from single domain Si(111)(2×1). *Upper part:* schematic sketch of scattering geometry illustrating the concepts of shadowing, channeling and blocking. *Lower part:* Number of atoms per [$\bar{1}1\bar{1}$] row visible to the beam. The solid line corresponds to the blocking dependence calculated with the model of Fig. 2.28, the dashed line corresponds to the same structure but to smaller thermal vibration amplitudes [2.147]. ©1984 The American Physical Society

measured; in Fig. 2.27 a blocking direction along (11$\bar{1}$] is indicated. Note that surface relaxation changes the blocking angle; for a reconstructed surface additional blocking minima are observed which can be used to infer the location of the surface atoms. Note further that backscattered ions are observed at lower energies because of kinematical energy loss, which allows identification of the chemical nature of the collision partner.

The lower part of Fig. 2.27 shows the surface blocking pattern of a single domain Si(111)(2×1) surface; beside the deep minimum at 109,48°at smaller angles there are several partly overlapping minima, whereas at larger angles the number per row is fairly constant. With the sample misaligned by 2°from the geometry shown in Fig. 2.27 an almost flat blocking pattern was observed which allowed the conclusion that 2.4–2.7 ML Si are displaced parallel to the surface; thus, the buckling model is to be discarded. Since for the ideal ($\bar{1}10$) geometry the yield is about 0.9 ML smaller than in the rotated geometry along

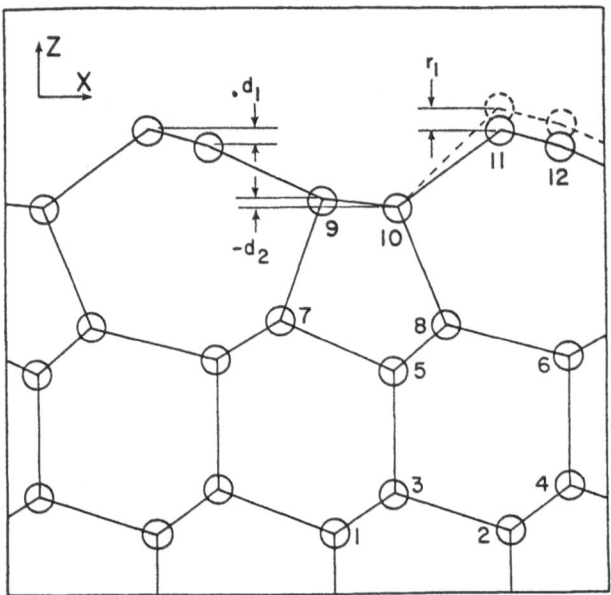

Fig. 2.28. Side view of the optimized π-bonded chain model of the (2×1) reconstruction of Si(111) in $[\bar{1}10]$-projection [2.147]. ©The American Physical Society

the $(\bar{1}10)$ plane, blocking is effective, so that the displaced surface atoms are still located in the $(\bar{1}10)$-plane; therefore the molecule model is not consistent with these findings either. These conclusions were corroborated and refined by Monte Carlo trajectory calculations of blocking curves [2.141] which confirmed the π-bonded chain model with, however, a modification by a tilt in the first and second layer chains [2.147] as shown in Fig. 2.28. It is interesting to note that the optimum agreement between experimental and calculated curves required appreciable thermal displacement enhancements for the atoms in the first few surface layers [2.147]. The most recent LEED analyses seem to confirm these results [2.148].

The next example demonstrates the value of high-resolution energy analyses of scattered ions. It concerns the old and fascinating problem of melting. According to recent theoretical propositions the surfaces of crystals should act as nucleation centers of bulk melting and should therefore melt at lower temperatures than the bulk [2.149–151]. This hypothesis was quite recently tested by *Frenken* and *van der Veen* [2.152] by proton scattering studies on the (110)-surface of lead. The idea behind the experiment is sketched in Fig. 2.29: For the Pb(110)-crystal a beam aligned along the $(\bar{1}01)$-axis will shadow atoms in the second and deeper layers effectively and only due to thermal motion atoms in near surface layers are hitten with a small probability. Blocking of the backscattered ions in the (011)-direction further reduces the backscattering yield in this direction. The corresponding energy spectrum therefore consists of a rather sharp peak containing the signal from the surface layer and a small

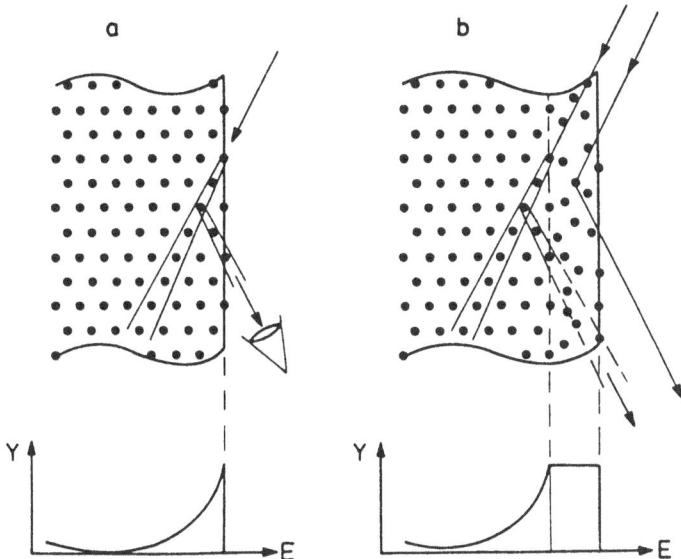

Fig. 2.29. Schematic pictures of energy spectra obtained with a shadowing and blocking geometry for **(a)** a well ordered crystalline surface and **(b)** the same solid covered with a liquid film [2.152]. ©The American Physical Society

contribution at smaller energies from nonshadowed and nonblocked deeper layers as indicated in Fig. 2.29a. With the crystal covered by a liquid film, the coherent shadowing and blocking effect is only effective underneath the liquid-crystal interface, whereas all atoms in the liquid film contribute to the surface signal thus increasing the width and height of the surface peak (Fig. 2.29b). *Frenken* and *van der Veen* achieved an energy resolution of 390 eV for their 97.5 keV H^+-beam [2.152]. By absolute calibration (as outlined above) the surface peak can be converted into the number of monolayers visible to the beam. The latter is shown as a function of temperature in Fig. 2.30 (above the bulk melting temperature $T_m = 600.7$ K no shadowing and blocking is indeed observed as expected for a liquid). Curve I in Fig. 2.30 corresponds to the result of Monte Carlo trajectory calculations expected for solid lead; the increase of thermal amplitudes causes the slight increase of the curve with rising temperature. Whereas Curve I is obtained by using bulk vibration amplitudes and therefore systematically underestimates the number of visible layers, Curve II accounts for the increased vibrational amplitudes of the surface atoms and fits the experimental result quite well up to 500 K. However, at a value of 600.5 K just below bulk melting the number of visible layers has risen to about twenty; this cannot be accounted for by reasonable solid surface vibrational amplitudes and would not reproduce the width of the measured energy spectrum either. Fitting both the intensity and the shape (width) of the measured curves allowed *Frenken* and *van der Veen* to convert the number of visible layers into the number of molten layers and led them to conclude that the surface melting

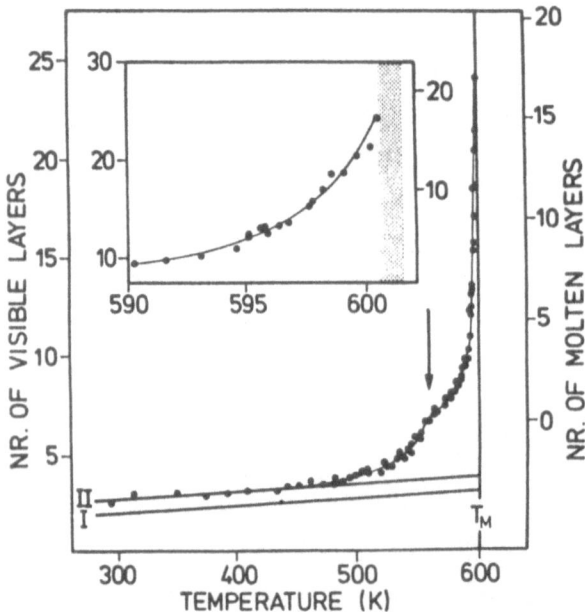

Fig. 2.30. Number of visible layers on Pb(110) as a function of temperature. The vertical line gives the bulk melting temperature. The arrow identifies the surface melting temperature at 560 K (*see text*). The insert shows an expanded view of the region near the bulk melting temperature [2.152]. ©1985 The American Physical Society

temperature is at $\simeq 560$ K. The region between 500 and 560 K is interpreted as a very disordered thin solid surface layer [2.152].

As a further remarkable recent result we mention the work of *van Loenen* et al. [2.153] concerning the detailed interface structure of a 25Å thick epitaxial $NiSi_2$-film on Si(111). TEM-results [2.154] had left open the question whether Ni at the interface occupies fivefold or sevenfold coordinated sites. By applying the same concepts of shadowing and blocking as outlined above, *van Loenen* et al. [2.153] were able to unambiguosly decide for the sevenfold structure and also to determine lattice contractions across the interface, thus establishing the usefulness of ion scattering for structure investigations of rather deeply buried interfaces in the technologically so important thin-film overlayers.

Another variant of ion scattering, transmission channeling, was recently used by *Steensgard* and *Jakobsen* [2.155] for the determination of the adsorption site location of deuterium on Ni(100). The ions of a beam impinging along a low index axial direction undergo correlated collisions with the atomic rows, so that they get steered into the open channels: the flux in a plane transverse to the beam is very low close to the rows and large in the center of the channels. Measuring the yield of a characteristic nuclear reaction induced by the incoming ions as a function of angle of incidence yields minima if the beam is exactly aligned with the steering rows and increases upon slight misorientations. If the substrate crystal is sufficiently thin and covered at the exit side with an adsorbate located in the channeling center the characteristic adsorbate reaction yield would exhibit a maximum where the substrate shows a minimum. Angular scans along different azimutal orientations allow qualitative conclusions on the adsorbate site symmetry, whereas determination of the adsorbate height requires comparison of experimental yield curves with Monte Carlo trajectory calculations. For the case of D on Ni(100) *Steensgard* and *Jakobsen* [2.155] used a 800 keV ^{3}He$^+$-beam to achieve maximum cross section for the reaction $D(^3He, p)^4He$. They determined the adsorbate site to be the fourfold coordinated hollow in agreement with earlier conclusions based on vibrational spectroscopy [2.156]; the height of the adatoms above the top Ni-layer was found to be 0.5 Å [2.154]. This result was very recently confirmed by theoretical work [2.157,158]. Notice that this method does not require ordered adsorbate layers.

The steering effect of atomic rows running along a surface also can be exploited for surface structure research if grazing angles of incidence of the ion beams is used. This method has been used by *Sailer* and *Varelas* [2.159] for studies of deuterium chemisorption on Ni(111) with similar beam conditions as in [2.155]; the adatoms were found to occupy the threefold hollows with a normal distance to the top Ni layer of 1.15Å. This result compares very well with the first result obtained for a H-chemisorption phase with LEED: *Behm* et al. [2.160] found for Ni(111)+ (2×2) H with a coverage of 1/2 ML the same normal distance; the hydrogens form a graphite-like layer whereby both fcc and hcp threefold sites are occupied.

All examples discussed above referred to medium (20–200 keV) and high energy (200 keV – 2 MeV) ion scattering (MEIS and HEIS, the latter often also being referred to as Rutherford backscattering, RBS [2.141]). For low energy ion scattering (LEIS, 1–20 keV) recently a variant called impact collision scattering spectroscopy (ICISS) was put forward by *Aono* et al. [2.161,162]. Here the scattering angle is fixed, as close as possible, to 180°so that only ions undergoing head-on collisions are observed and multiple scattering can be expected to be minimized. At grazing angles of incidence the backscattered intensity from a surface is zero since each atom is in the shadow cone of its preceding neighbour. The intensity increases strongly when going to less grazing angles at a critical angle which corresponds to the stepping out of the neighbour atoms from the shadow cone. Thus the relative positions of the atoms in the surface row can be straightforwardly determined provided the shape of the shadow cone is known. Whereas *Aono* et al. used noble gas ions, *Niehus* and *Comsa* [2.162–164] in their studies of clean and adsorption induced reconstructed metal surfaces used alkali ions, which, because of their smaller neutralization probabilities, allow for smaller beam intensities and therefore guarantee less surface damage [2.165]. More details on ion beam scattering techniques [2.166] in general as well as on the surface channeling methods may be found in Chap. 4 by C. Varelas.

2.7 Other Methods

2.7.1 Stimulated Desorption

Structural – and bonding – information can also be obtained by exciting adsorbed molecules with electrons or photons and observing the desorption of atomic or molecular ions and/or neutrals (electron and photon induced desorption, ESD/PSD). Measurements of the angular distribution of desorbing ions (electron stimulated desorption ion angular distribution, ESDIAD) show that desorption takes place into emission cones whereby the particle trajectories are determined by the bonds broken [2.167–170]. Experimental evidence and theoretical considerations show that azimuthal angles are preserved whereas polar angles are generally – due to image force effects – increased. Scanning the ion desorption patterns yields direct information on the local geometric structure of surface molecules and the method finds lively application in chemical reaction studies, especially for investigations of reaction intermediates and molecule fragments and their orientation relative to the substrate. As an example we mention the recent work of *Benndorf* et al. [2.171] who investigated the adsorption of water on Ni(110). They found that the presence of oxygen promotes adsorption and decomposition of H_2O whereby adsorbed OH forms on the substrate, probably by the hydrogen abstraction reaction: $H_2O + O(ad) \rightarrow 2OH(ad)$. ESDIAD shows that the adsorbed OH is bound to the surface via the oxygen atoms and that it is inclined against the surface normal and oriented along the $[00\pm1]$-direction. Angle resolved ultraviolet photoemission spectroscopy (ARUPS) data confirmed these conclusions.

2.7.2 Vibrational Spectroscopies

Many important contributions to surface structure questions were also delivered by vibrational spectroscopies, particularly high resolution electron energy loss spectroscopy (HREELS) [2.172]. Modern HREELS intruments can achieve energy resolutions between 3 and 10 meV for incoming energies between several eV and a few 100 eV. The main scattering mechanism exploited up to now is "dipole scattering" which, due to the long-range Coulomb interaction between the incoming electron and the vibrating surface, is mainly observed near the specular direction. Measurement of characteristic frequency losses allows identification of surface species (and thus decision between associative or dissociative chemisorption). Application of surface dipole selection rules (the vibration must have associated a dynamical dipole moment perpendicular to the surface) may be used to find out the orientation and location of molecules and fragments relative to the substrate. Variations of characteristic vibration frequencies also allow distinction of different adsorption sites [2.174] and frequency changes allow determination of bond length changes due to the presence of other surface species like catalyst promoters [2.175]. Very similar principles apply to infrared absorption, which for surface applications has become feasible in the reflection-absorption infrared specrocopy (RAIRS) geometry to gain the necessary sensitivity [2.176]. Due its high resolution ($\simeq 0.15$ meV) RAIRS is suited for studying small frequency shifts resulting from isotopic substitutions or from coverage variations and thus to discriminate between dipole and chemical interactions [2.177]. It is worthwile mentioning that with both HREELS and RAIRS sensitivities have been reached which allow coverages of less than 0.01 ML to be observed.

In "impact scattering" [2.178] the electrons are scattered by the local atomic potential as in LEED and the thermal motion of the ion cores gives rise to inelastic scattering in directions away from the Bragg peaks. Although impact scattering intensities are much lower than those of dipole scattering, measurements of acoustical and optical surface phonon dispersion relations could be performed with HREELS for both clean and adsorbate covered metal surfaces [2.179,180]. The only other technique capable of measuring surface phonon dispersions at present is inelastic atomic beam scattering [2.20,181]. With atomic beams a better energy resolution (<0.5 meV) can be obtained than with HREELS, but it is difficult to observe high frequency optical modes [2.182]. Measured dispersion curves not only give insight into the force constants acting at and near the surface but can also be used to derive structural information as discussed in Chap. 5 by J.E. Black.

The relation of "resonance scattering", in which the electron is captured to form a compound state with the adsorbed molecule [2.183] to the shape resonances used in NEXAFS (Sect. 2.4.4) is obvious and with the observation in electron resonance scattering from chemisorbed molecules [2.184,185] exciting new applications in surface science can be expected in the near future. Most of the topics of this section will be treated in Chap. 7 by M. Rocca et al.

2.7.3 Surface Electronic Properties and Optical Methods

Structural information on surface reconstructions and chemisorbed overlayers can in an indirect way also be inferred from the electronic surface structure. X-ray and ultraviolet photoemission (XPS, UPS) may help to identify chemisorbed molecules and fragmented species [2.186] and also to determine their orientation relative to the substrate [2.187]. Core level shifts measured with XPS for bulk and surface atoms give information on the coordination number of surface atoms [2.188,145]. On the latter grounds the proposition of a hexagonal surface layer in the reconstructions of the (100)-surfaces of Ir and Au, which was put forward on the basis of LEED observations [2.125], found a valuable support: smaller core level shifts (indicating larger coordination) were measured for the reconstructed than for the unreconstructed (metastable) surfaces [2.189,190]. Electronic surface state dispersions measured by angle-resolved UPS (ARUPS) can also be used to decide between different reconstructions as pointed out in Sect.3.6 in connection with the question of the $Si(111)(2 \times 1)$-reconstruction [2.144].

With the rapid development of laser techniques, optical methods by which electronic surface transitions are measured and selection rules based on surface symmetry properties are exploited, also become of increasing importance. Several variants of measuring surface absorption coefficients [2.191,192] as well as second harmonic generation [2.193,194] have been applied in studies of the Si(111)(2×1) and the results are all in support of the *Pandey*-model [2.55,143].

The refined nuclear magnetic resonance (NMR) techniques in recent years have been applied by *Slichter* [2.195] and collaborators to study the electronic properties of small metal particles, the bonding and structure of adsorbates to them as well as their diffusive motion on them, and also the rates of prototypic reactions. These studies, which can be applied to "real" supported catalysts are especially interesting since they will provide the necessary experience, to decide which of the knowledge obtained in the UHV surface research may be transferred to the areas of (high-pressure) chemotechnical applications [2.196].

2.8 Outlook

The field of surface crystallography saw rapid development since about fifteen years and at the moment there are about twenty different methods actively pursued for surface structure studies and several more are likely to come [2.197]. It is not astonishing therefore that a remarkable wealth of information has been obtained. Some prototype systems have been studied (as shown by a few examples in this chapter) with several methods and the results seem to converge to agreement; this is an important development in view of the sometimes confusing and contradictory earlier claims. Reliable results are at the moment available for metal surface relaxations perpendicular and parallel to the surface, for metal and semiconductor reconstructions as well as for atomic and molecular chemisorption systems (contradictory results at present seem to concern mainly chemisorption induced surface reconstructions). The role of theory in obtaining our present secure knowledge has been twofold: first, it helped to understand the different experimental methods as well as to develop the necessary evaluation techniques and second, it provided (sometimes together with some good intuition) the basic insight into the dominant physical processes responsible for the various surface phenomena. In the future even more complicated systems will likely be tackled than have been solved so far larger surface unit cells, more complicated molecular adsorbates as well as coadsorption systems. The diffraction techniques with their sensitivity to long range order (and its decay) will find wide applications in the fields of surface phase transition phenomena [2.198–200], x-ray and ion beam scattering will reveal many details of interface formation, and the short-range probes and imaging techniques will be very valuable for studies of local surface morphology, the ocurrence of surface defects and their role in nucleation phenomena, chemical reactions and build-up of thin films. With all these prospects, there is no better way than to conclude with the optimistic words of *Tong* in his recent short survey [2.201], which was frequently and gratefully consulted by the author as an excellent guide when putting together the present chapter: "If the last years provide any indication, then the next decade promises to be a most exciting period, and will likely mark the golden age of surface science".

Acknowledgements. The author wishes to express his thanks to K. Baberschke, A. Baratoff, M. Baumberger, G. Binnig, E. Courtens, H.W. Fink, E. Latta, H.P. Meier, B. Reihl, H. Rohrer, E. Stoll, W. Stocker, E. Tosatti for many clarifying discussions, for pointing out references, providing original material, giving hints and thoughts and last but not least for critical comments on the manuscript. Thanks are also due to Mrs. D. Bruellmann, Mrs. J. Gygax, Mr. U. Bitterli, and Mr. M. Wagner, of the IBM Zurich Research Laboratory, for careful preparation of the paper.

References

2.1   G. Ertl, J. Kueppers: *Low Energy Electrons and Surface Chemistry* (Chemie, Weinheim 1985)

2.2   M. Prutton: *Surface Physics*, 2nd ed. (Clarendon, Oxford 1983)

2.3   R.N. Barnett, U. Landman, C.L. Cleveland: Phys. Rev. Lett. **51**, 1359 (1983)

2.4   P.J. Estrup: In *Chemistry and Physics of Solid Surfaces V*, eds. R. Vanselow, R. Howe, Springer Ser. Chem. Phys., Vol. 35 (Springer, Berlin, Heidelberg 1984) p. 205

2.5   G.A. Somorjai, M.A. Van Hove: *Adsorbed Monolayers on Solid Surfaces*, Structure and Bonding, Vol. 38 (Springer, Berlin, Heidelberg 1979)

2.6   *International Tables for X-Ray Crystallography*, Vol. 1–3 (Kynoch Press, Birmingham 1952–1962);
       B. K. Vainshtein: *Modern Crystallography I*, Springer Ser. Solid-State Sci., Vol. 15 (Springer, Berlin, Heidelberg 1981)

2.7   E.A. Wood: J. Appl. Phys. **35**, 1306 (1974)

2.8   P. K. Larsen, B.A. Joyce, P.J. Dobson: In *Dynamical Phenomena at Surfaces, Interfaces and Superlattices*, eds. F.Nizzoli, K.H. Rieder, R.F. Willis, Springer Ser. Surface Sci., Vol. 3 (Springer, Berlin, Heidelberg 1985) p. 196

2.9   C. Kittel: *Introduction to Solid State Physics*, 5th ed. (Wiley, New York 1976)

2.10   W.C. Marra, P. Eisenberger, A. Cho: J. Appl. Phys. **50**, 6927 (1979);
       P. Eisenberger, W.C. Marra: Phys. Rev. Lett. **46**, 1081 (1981)

2.11   E.R. Woelfel: *Theorie und Praxis der Roentgen-Strukturanalyse*, 2nd ed. (Vieweg, Braunschweig 1981)

2.12   M.J. Buerger: *Contemporary Crystallography* (McGraw-Hill, 1970)

2.13   I. K. Robinson: Phys. Rev. Lett. **50**, 1145 (1983)

2.14   G.H. Vineyard: Phys. Rev. B**26**, 4146 (1982)

2.15   W. Moritz, D. Wolf: Surf. Sci. **163**, L641 (1985)

2.16   J. Als-Nielsen: In *Structure and Dynamics of Surfaces*, ed. by W. Schommers, P. von Blanckenhagen, Topics Current Phys. (Springer, Berlin, Heidelberg 1986) to be published

2.17   S. Dietrich, H. Wagner: Phys. Rev. Lett. **51**, 1469 (1983) and Z. Phys. B **56**, 207 (1984)

2.18   J.P. Toennies, K. Winkelmann: J. Chem. Phys. **66**, 3965 (1977)

2.19   H. Hoinkes: Rev. Mod. Phys. **52**, 933 (1980);
       G. Boato, P. Cantini: In *Advances in Electronics and Electron Physics* (Pergamon, Oxford 1982)

2.20   P. Toennies: J. Vac. Sci. Technol. A**2**, 1055 (1984)

2.21   A.W. Kleyn, A.C. Luntz, D.J. Auerbach: Phys. Rev. Lett. **47**, 1169 (1981);
       J.A. Barker, D.J. Auerbach: Surf. Sci. Repts. **4**, 1 (1984)

2.22   B. Poelsema, L. K. Verheij, G. Comsa: Phys. Rev. Lett. **49**, 1731 (1982)

2.23   T. Engel, G. Ertl: J. Chem. Phys. **69**, 1267 (1978);
       M.P. D'Evelyn, R.J. Madix: Surf. Sci. Repts. **3**, 413 (1984)

2.24   K.H. Rieder: In *Dynamics of Gas-Surface Interaction* , eds. G. Benedek, U. Valbusa, Springer Ser. in Chem. Phys., Vol. 21 (Springer, Berlin, Heidelberg 1982), p. 61;
       T. Engel, K.H. Rieder: In *Springer Tracts Mod. Phys.*, **91** (Springer, Berlin, Heidelberg 1982) p. 55

2.25   N. Esbjerg, J. Norskov: Phys. Rev. Lett. **45**, 807 (1980);
       J. Harris, A. Liebsch: J. Phys. C**15**, 2275 (1982); Phys. Rev. Lett. **49**, 341 (1982)

2.26   U. Garibaldi, A.C. Levi, R. Spadacini, G.E. Tommei: Surf. Sci. **48**, 649 (1975)

2.27   N. Garcia: J. Chem. Phys. **67**, 897 (1977)

2.28   K.H. Rieder: Surf. Sci. **117**, 13 (1982);
        K.H. Rieder, N. Garcia: Phys. Rev. Lett. **49**, 43 (1982)
2.29   K.H. Rieder, T. Engel: Phys. Rev. Lett. **43**, 373 (1979); ibid. **45**, 824 (1980); Surf. Sci. **109**, 140 ('1981)
2.30   K.H. Rieder: Phys. Rev. B**27**, (RC) 7799 (1983)
2.31   J.P. Muscat: Surf. Sci. **110**, 389 (1981)
2.32   W. Reimer: Strukturbestimmung in geordneten Adsorbatschichten mit Hilfe von LEED, Diploma Thesis, University of Munich (1985)
2.33   J. Demuth: J. Colloid Interface Sci. **58**, 184 (1977)
2.34   K.H. Rieder, M. Baumberger, W. Stocker: Phys. Rev. Lett. **51**, 1799 (1983)
2.35   M.J. Cardillo: In *Dynamics of Gas-Surface Interaction*, eds. G. Benedek, U. Valbusa, Springer Ser. Chem. Phys., Vol. 21 (Springer, Berlin, Heidelberg 1982) p. 40
2.36   G. Boato, P. Cantini, L. Mattera: Surf. Sci. **55**, 141 (1976);
        K.H. Rieder: Surf. Sci. **118**, 57 (1982)
2.37   A.J. Freeman: In *Dynamical Phenomena at Surfaces, Interfaces and Superlattices*, eds. F.Nizzoli, K.H. Rieder, R.F. Willis, Springer Ser. in Surf. Sci., Vol. 3 (Springer, Berlin, Heidelberg 1985) p. 162
2.38   K.H. Rieder, W. Stocker: Phys. Rev. Lett. **52**, 352 (1984)
2.39   M. Henzler: In *Dynamical Phenomena at Surfaces, Interfaces and Superlattices*, eds. F.Nizzoli, K.H. Rieder, R.F. Willis, Springer Ser. Surf. Sci., Vol. 3 (Springer, Berlin, Heidelberg 1985) p. 14
2.40   H. Richter, U. Gerhardt: Phys. Rev. Lett. **51**, 1570 (1983)
2.41   F. Jona, J.A. Strozier Jr., W.S. Yang: Rep. Progr. Phys. **45**, 527 (1982)
2.42   J.B. Pendry: *Low Energy Electron Diffraction* (Academic, London 1980)
2.43   M.A. Van Hove, S.Y. Tong: *Surface Crystallography by LEED*, Springer Ser. Chem. Phys., Vol. 2 (Springer, Berlin, Heidelberg 1979)
2.44   Y. Gauthier, R. Baudoing, Y. Joly, C. Gaubert, J. Rundgren: J. Phys. C**17**, 4547 (1984)
2.45   D.L. Adams, L.E. Petersen, C.S. Sorensen: J. Phys. C**18**, 1753 (1985)
2.46   M.L. Xu, S.Y. Tong: Phys. Rev. B**31**, 6332 (1985)
2.47   W. Reimer, R.J. Behm, G. Ertl, V. Penka: Verhadl. DPG(IV) O **130** (1985)
2.48   U. Landmann, R.N. Hill, M.Mostoller: Phys. Rev. B**21**, 448 (1980)
2.49   R. Feidenhans'l, J.E. Sorensen, I. Steensgard: Surf. Sci. **134**, 329 (1983)
2.50   D.L. Adams, H.B. Nielsen, J.N. Andersen, I. Steensgard, R. Feidenhans'l, J.E. Sorensen: Phys. Rev. Lett. **49**, 669 (1982)
2.51   J. Sokolov, F. Jona, P.M. Marcus: Solid St. Commun. **49**, 307 (1984)
2.52   R.N. Barnett, U. Landman, C.L. Cleveland: Phys. Rev. Lett. **51**, 1359 (1983)
2.53   D.L. Adams, W.T. Moore, K.A.R. Mitchell: Surf. Sci. **149**, 407 (1985)
2.54   K.H. Rieder, M. Baumberger, W. Stocker: Phys. Rev. Lett. **55**, 390 (1985)
2.55   K.C. Pandey: Phys. Rev. Lett. **49**, 223 (1982)
2.56   R.M. Tromp, L. Smit, J.F. van der Veen: Phys. Rev. Lett. **51**, 1672 (1983)
2.57   S.Y. Tong, G. Xu, W.N. Mei: Phys. Rev. Lett. **52**, 1693 (1984)
2.58   J. Bohr, R. Feidenhans'l, M. Nielsen, M. Toney, R.L. Johnson, I. K. Robinson: Phys. Rev. Lett. **54**, 1275 (1985)
2.59   S. Andersson, J.B. Pendry: Phys. Rev. Lett. **43**, 363 (1979)
2.60   M. Passler, A. Ignatiev, F. Jona, D.W. Jepsen, P.M. Marcus: Phys. Rev. Lett. **43**, 360 (1979)
2.61   W. Moritz: Analyse von Oberflächenstrukturen mit großen Einheitszellen und modulierten Überstrukturen durch Beugung niederenergetischer Elektronen, Habilitationsschrift, University of Munich (1983)
2.62   M.A. Van Hove, Rongfu Lin, G.A. Somorjai: J. Am. Chem. Soc. (submitted)
2.63   H.D. Shih, F. Jona, D.W. Jepsen, P.M. Marcus: Phys. Rev. Lett. **37**, 1622 (1976)
2.64   J.F. van der Veen, R.M. Tromp, R.G. Smeenk, F.W. Saris: Surf. Sci. **82**, 468 (1979)
2.65   T. Narusawa, W.M. Gibson, E. Toernquist: Phys. Rev. Lett. **47**, 417 (1981)
2.66   W. Ho, R.F. Willis, E.W. Plummer: Phys. Rev. Lett. **40**, 1463 (1978)
2.67   K. Griffiths, P.R. Norton, J.A. Davies, W.N. Unertl, T.E. Jackman: Surf. Sci., in press
2.68   J.H. Onuferko, D.P. Woodruff, B.W. Holland: Surf. Sci. **87**, 357 (1979)
2.69   R.J. Behm, K. Christmann, G. Ertl: Surf. Sci. **99**, 320 (1980)

2.70 K. Heinz, K. Mueller: In *Springer Tracts Mod. Phys. 91*, (Springer, Berlin, Heidelberg 1982) p. 1

2.71 J.L. Seguin, J. Suzanne, M. Bienfait, J.G. Dash, J.A. Venables: Phys. Rev. Lett. **51**, 122 (1983)

2.72 D.F. Mitchell, P.B. Sewell, M. Cohen: Surf. Sci. **69**, 310 (1977)

2.73 K. Ploog, K. Graf: *Molecular Beam Epitaxy of III–V Compounds* (Springer, Berlin, Heidelberg 1984);
      H.-J. Gossmann, L.C. Feldman: Appl. Phys. A**38**, 171 (1985)

2.74 J. Stöhr, R. Jaeger, S. Brennan: Surf. Sci. **117**, 503 (1982)

2.75 J. Stöhr: In *Chemistry and Physics of Solid Surfaces V*, eds. R. Vanselow, R. Howe, Springer Ser. in Chem. Phys., Vol. 35 (Springer, Berlin, Heidelberg 1984), p. 231; J. Stöhr: In *Principles, Techniques an Applications of EXAFS, SEXAFS and XANES*, ed. by D. Koningsberger, R. Prins (Wiley, New York 1985)

2.76 P.H. Citrin, P. Eisenberger, R.C. Hewitt: Phys. Rev. Lett. **41**, 309 (1978)
      P.H. Citrin, P. Eisenberger, J.E. Rowe: Phys. Rev. Lett. **48**, 802 (1982)

2.77 J. Stöhr, D. Denley, P. Perfetti: Phys. Rev, B**18**, 4132 (1978);
      R. Jaeger, J. Feldhaus, J. Haase, J. Stöhr, Z. Hussein, D. Menzel, D. Norman: Phys. Rev. Lett. **45**, 1870 (1980);
      U. Doebler, K. Baberschke, J. Haase, A. Puschmann: Phys. Rev. Lett. **52**, 1437 (1984)

2.78 U. Doebler, K. Baberschke, J. Stöhr, D.A. Outka: Phys. Rev. B**31**, 2532 (1985)

2.79 L. Papagno, M. DeCrescenci, G. Chiarello, E. Colavita, R. Scarmozzino, L.S. Caputi, R. Rosei: Surf. Sci. **117**, 525 (1982)

2.80 M. DeCrescenci, F. Antonangeli, C. Bellini, R. Rosei: Phys. Rev. Lett. **50**, 1949 (1983)

2.81 P.M. Marcus, J.E. Demuth, D.W. Jepsen: Surf. Sci. **53**, 501 (1975)

2.82 J. Stöhr, R. Jaeger, T. Kendelewitz: Phys. Rev. Lett. **49**, 142 (1982)

2.83 R.G. Smeenk, R.M. Tromp, J.W.M. Frenken, F.W. Saris: Surf. Sci. **112**, 261 (1981)

2.84 D.H. Rosenblatt, J.G. Tobin, M.C. Mason, R.F. Davis, S.D. Kevan, D.A. Shirley, C.H. Li, S.Y. Tong: Phys. Rev. B**23**, 3828 (1981)

2.85 F.J. Himpsel: Adv. Phys. **32**, 1 (1983)

2.86 S.D. Kevan, R.F. Davis, D.M. Rosenblatt, T.G. Tobin, M.G. Mason, D.A. Shirley, C.H. Li, S.Y. Tong: Phys. Rev. Lett. **46**, 1629 (1981)

2.87 H.H. Farrell, M.M. Traum, N.V. Smith, W.A. Roger, D.P. Woodruff, P.D. Johnson: Surf. Sci. **102**, 527 (1981);
      M.W. Kang, C.H. Li, S.Y. Tong: Solid State Commun. **36**, 149 (1980)

2.88 S. Kono, S.M. Goldberg, N.F.T. Hall, C.S. Fadley: Phys. Rev. B**22**, 6085 (1980);
      J.G. Tobin, L.E. Klebanoff, D.H. Rosenblatt, R.F. Davis, E. Umbach, A.G. Baca, D.A. Shirley, Y. Huang, W.M. Kang, S.Y. Tong: Phys. Rev. B**26**, 7076 (1982)

2.89 J.J. Barton, C.C. Bahr, Z. Hussain, S.W. Robey, J.G. Tobin, L.E. Klebanoff, D.A. Shirley: Phys. Rev. Lett. **51**, 272 (1983)

2.90 J. Haase: Appl. Phys. A**38**, 181 (1985)

2.91 J.J. Barton, C.C. Bahr, Z. Hussain, S.W. Robey, L.E. Klebanoff, D.A. Shirley: J. Vac. Sci. Technol. A**2**, 847 (1983);
      J.J. Barton, S.W. Robey, C.C. Bahr, D.A. Shirley: In *The Structure of Solid Surfaces*, eds. M.A. Van Hove, S.Y. Tong, Springer Ser. Surf. Sci., Vol. 2 (Springer, Berlin, Heidelberg 1985) p. 191

2.92 D. Norman, J. Stöhr, R. Jaeger, P. Durham, J.B. Pendry: Phys. Rev. Lett. **51**, 2052 (1983)

2.93 J. Stöhr, J.L. Gland, W. Eberhardt, D. Outka, R.J. Madix, F. Sette, R.J. Koestner, U. Döbler: Phys. Rev. Lett. **51**, 2414 (1983)

2.94 J. Stöhr, K. Baberschke, R. Jaeger, R. Treichler, S. Brennan: Phys. Rev. Lett. **47**, 381 (1981)

2.95 J. Stöhr, F. Sette, A.L. Johnson: Phys. Rev. Lett. **53**, 1684 (1984)

2.96 F. Sette, J. Stöhr, E.B. Kollin, D.J. Dwyer, J.L. Gland, J.L. Robbins, A.L. Johnson: Phys. Rev. Lett. **54**, 935 (1985)

2.97 A. Puschmann, J. Haase, M.D. Crapper, C.E. Riley, D.P. Woodruff: Phys. Rev. Lett. **54**, 2250 (1985)

2.98 J. Stöhr, D. Outka, R.J. Madix, U. Döbler: Phys. Rev. Lett. **54**, 1256 (1985)

2.99   E.W. Müller: Z. Physik **131**, 13 (1951)
2.100 J.A. Panitz: J. Phys. E (Sci. Instr.) **15**, 1281 (1982)
2.101 A.J. Melmed, R.J. Stein: Surf. Sci. **49**, 645 (1975)
2.102 R.J. Walko, E.W. Müller: Phys. Stat. Sol. (a) **9**, K9 (1972)
2.103 H.W. Fink, G. Ehrlich: Surf. Sci. **110**, L611 (1981)
2.104 D.N. Seidman: Surf. Sci. **70**, 532 (1978)
2.105 R. Wagner: Physik. Bl. **36**, 65 (1980) and *Field-Ion Microscopy in Materials Science Crystals*, Vol. 6 (Springer, Berlin, Heidelberg 1982)
2.106 E.W. Müller, J.A. Panitz, S.B. McLane: Rev. Sci. Instr. **39**, 83 (1968)
2.107 For recent reviews see: D.W. Bassett: In *Surface Mobilities on Solid Materials*, ed. V.T. Binh (Plenum 1983) p. 63,83;
       G. Ehrlich: In Proc. 9th Int. Vac. Congr. and 5th Int. Conf. on Solid Surf., Invited Speakers Volume, ed. J.L. de Segovia (A.S.E.V.A., Madrid 1983) p. 3
2.108 H.W. Fink, G. Ehrlich: Phys. Rev. Lett. **52**, 1532 (1984); Surf. Sci. **143**, 125 (1984)
2.109 J.D. Wrigley, G. Ehrlich: Phys. Rev. Lett. **44**, 661 (1980)
2.110 D.W. Bassett: Surf. Sci. **53**, 74 (1975);
       G. Ehrlich: Surf. Sci. **63**, 422 (1977)
2.111 A.J. Melmed, R.T. Tung, W.R. Graham, G.D.W. Smith: Phys. Rev. Lett. **43**, 1521 (1979)
2.112 M. K. Debe, D.A. King: Phys. Rev. Lett. **39**, 708 (1977);
       J.A. Walker, M. K. Debe, D.A. King: Surf. Sci. **104**, 504 (1981)
2.113 A. Sommerfeld, H. Bethe: *Handbuch der Physik*, 2nd ed., ed. by H. Geiger, K. Scheel (Springer, Berlin 1933) Vol. 24, Pt. 2, p. 333
2.114 G. Binnig, H. Rohrer, Ch. Gerber, E. Weibel: Appl. Phys. Lett. **40**, 178 (1982)
2.115 E.L. Wolf: In *Solid State Physics*, Vol. 30, (Academic, New York 1975) p. 2
2.116 R.H. Fowler, L. Nordheim: Proc. Roy. Soc. (London) Ser. A**119**, 173 (1928);
       J. Frenkel: Phys. Rev. **36**, 1604 (1930)
2.117 G. Binnig, H. Rohrer, Ch. Gerber, E. Weibel: Phys. Rev. Lett. **49**, 57 (1982)
2.118 G. Binnig, H. Rohrer, Ch. Gerber, E. Weibel: Phys. Rev. Lett. **50**, 120 (1983)
2.119 G. Binnig, H. Rohrer, Ch. Gerber, E. Weibel: Surf. Sci. **131**, L379 (1983)
2.120 G. Binnig, H. Rohrer, Ch. Gerber, E. Stoll: Surf. Sci. **144**, 321 (1984)
2.121 G. Binnig, H. Rohrer: Surf. Sci. **152/153**, 17 (1985)
2.122 A.M. Baro, G. Binnig, H. Rohrer, Ch. Gerber, E. Stoll, A. Baratoff, F. Salvan: Phys. Rev. Lett. **52**, 1304 (1984)
2.123 G. Binnig, H. Rohrer: Surf. Sci. **126**, 236 (1983)
2.124 G.A. Somorjai: *Chemistry in Two Dimensions: Surfaces* (Cornell U. Press, Ithaca 1981)
2.125 M.A. Van Hove, R.J. Koestner, P.C. Stair, J.P. Biberian, L.L. Kesmodel, I. Bartos, G.A. Somorjai: Surf. Sci. **103**, 189,218 (1981)
2.126 K.H. Rieder, T. Engel, R.H. Swendsen, M. Manninen: Surf. Sci. **127**, 223 (1983)
2.127 W.A. Harrison: Surf. Sci. **55**, 1 (1979)
2.128 G. Binnig, H. Rohrer, F. Salvan, Ch. Gerber, A. Baro: Surf. Sci. **157**, L373 (1985)
2.129 W.S. Yang, F. Jona: Solid State Commun. **48**, 377 (1983)
2.130 M. Aono, R. Souda, C. Oshima, Y. Ishizawa: Phys. Rev. Lett. **51**, 801 (1983);
       R.M Tromp, E.J. van Loenen, M. Iwami, F.W. Saris: Solid State Commun. **44**, 971 (1982);
       R.J. Culbertson, L.C. Feldman, P.J. Silverman: Phys. Rev. Lett. **45**, 2043 (1980)
2.131 K. Takayanagi, Y. Tanishiro, M. Takahashi, H. Motoyoshi, K. Yagi: In Proc. 10th Int. Congr. on Electron Microscopy, Hamburg, FRG (Deutsche Gesellschaft f. Elektronenmikroskopie, Franfurt 1982) p. 285
2.132 D.J. Chadi: Phys. Rev. B**30**, 4470 (1984);
       T. Yamaguchi: Phys. Rev. B**30**, 1992 (1984)
2.133 A. Baratoff: Physica **127**B, 143 (1984)
2.134 E. Stoll: In *Physique des Surfaces*, Proc. 26e Course de perfectionnement de l'Associacion Vaudoise des Chercheurs en Physique (1984) p. 88
2.135 J. Tersoff, D.R. Hamann: Phys. Rev. Lett. **50**, 1998 (1983)
2.136 L.D. Marks: Phys. Rev. Lett. **51**, 1000 (1983)
2.137 L.D. Marks, V. Heine, D.J. Smith: Phys. Rev. Lett. **52**, 656 (1984)
2.138 J.M. Gibson: Phys. Rev. Lett. **53**, 1859 (1984)

2.139   L. Reimer: *Transmission Electron Microscopy*, Springer Ser. Opt. Sci., Vol. 36 (Springer, Berlin, Heidelberg 1984);
        L. Reimer: *Scanning Electron Microscopy*, Springer Ser. Opt. Sci., Vol. 45 (Springer, Berlin, Heidelberg 1985)
2.140   R.M. Tromp: J. Vac. Sci. Technol. A1, 1047 (1983)
2.141   W.M. Gibson: In *Chemistry and Physics of Solid Surfaces V*, ed. by R. Vanselow, R. Howe, Springer Ser. Chem. Phys., Vol. 35 (Springer, Berlin, Heidelberg 1984), p. 427
2.142   D. Haneman: Phys. Rev. **121**, 1093 (1961)
2.143   K.C. Pandey: Phys. Rev. Lett. **47**, 913 (1981); **49**, 223 (1082)
2.144   F.J. Himpsel, P. Heiman, D. Eastman: Phys. Rev. B**24**, 2003 (1981)
2.145   F.J. Himpsel, P. Heiman, T.C. Chiang, D. Eastman: Phys. Rev. Lett. **45**, 1112 (1980); S. Brennan, J. Stöhr, R. Jaeger, J.E. Rowe: Phys. Rev. Lett. **45**, 1414 (1980)
2.146   D.J. Chadi: Phys. Rev. B**26**, 4762 (1982)
2.147   R.M. Tromp, L. Smit, J.F. van der Veen: Phys. Rev. B**30**, (RC) 6235 (1984)
2.148   I.P. Batra, F.J. Himpsel, P.M. Marcus, R.M. Tromp, M.R.Cook, F. Jona, H. Liu: In *The Structure of Solid Surfaces*, ed. by M.A. Van Hove, S.Y. Tong, Springer Ser. Surf. Sci., Vol. 2 (Springer, Berlin, Heidelberg 1985) p. 285
2.149   L. Pietronero, E. Tosatti: Solid State Commun. **32**, 255 (1979)
2.150   J. K. Kristensen, R.M.J. Cotterill: Philos. Mag. **36**, 437 (1977)
2.151   J.Q. Broughton, G.H. Gilmer: J. Chem. Phys. **79**, 5105, 5119 (1983)
2.152   J.W.M. Frenken, J.F. van der Veen: Phys. Rev. Lett. **54**, 134 (1985)
2.153   E.J.van Loenen, J.W.M. Frenken, J.F. van der Veen, S. Valeri: Phys. Rev. Lett. **54**, 827 (1985)
2.154   R.T. Tung, J.M. Gibson, J.M. Poate: Phys. Rev. Lett. **50**, 429 (1983)
2.155   I. Steensgard, F. Jakobsen: Phys. Rev. Lett. **54**, 711 (1985)
2.156   S. Andersson, Chem. Phys. Lett. 55, 185 (1978)
2.157   M. Weinert, J.W. Davenport: Phys. Rev. Lett. **54**, 1547 (1985)
2.158   C. Umrigar, J.W. Wilkins: Phys. Rev. Lett. **54**, 1551 (1985)
2.159   E. Sailer, C. Varelas: Nucl. Instr. Meth. Phys. Res. B**2**, 326 (1984)
2.160   M.A. van Hove, G. Ertl, K. Christmann, R.J. Behm, W.H. Weinberg: Solid State Commun. **28**, 373 (1979)
2.161   M. Aono, Y. Hou, C. Oshima, Y. Ishizawa: Appl. Phys. Lett. **40**, 178 (1982) M Aono, Y. Hou, R. Souda, C. Oshima, S. Otani, Y. Ishizawa: Phys. Rev. Lett. **50**, 1293 (1983)
2.162   H. Niehus, G. Comsa: Surf. Sci. **140**, 18 (1984)
2.163   H. Niehus: Surf. Sci. **145**, 407 (1984)
2.164   H. Niehus, G. Comsa: Surf. Sci. **151**, L171 (1985)
2.165   G. Comsa, B. Poelsema: Appl. Phys. A**38**, 153 (1985)
2.166   J.P. Toennies: Appl. Phys. **3**, 91 (1974)
2.167   T.E. Madey: In *The Structure of Solid Surfaces*, ed. by M.A. Van Hove, S.Y. Tong, Springer Ser. Surf. Sci., Vol. 2 (Springer, Berlin, Heidelberg 1985) p. 264
2.168   D. Menzel: J. Vac. Sci. Technol. **20**, 538 (1982)
2.169   *Desorption Induced by Electronic Transitions DIET-I*, ed. by N.H. Tolk, M.M. Traum, J.C. Tully, T.E. Madey, Springer Ser. Chem. Phys., Vol. 24 (Springer, Berlin, Heidelberg 1983)
2.170   *Desorption Induced by Electronic Transitions DIET-II*, eds. W. Brenig, D. Menzel, Springer Ser. Surf. Sci., Vol. 4 (Springer, Berlin, Heidelberg 1985)
2.171   C. Benndorf, C. Noebl, T.E. Madey: Surf. Sci. **138**, 292 (1984)
2.172   H. Ibach, D.L. Mills: *Electron Energy Loss Spectroscopy and Surface Vibrations* (Academic, New York 1982)
2.173   Ph. Avouris, J. Demuth: IBM Res. Rept.RC 10321 (1984)
2.174   H. Ibach: In *Vibrations in Adsorbed Layers*, Proc. Int. Conf. KFA-Jülich, ed. by H. Ibach, S. Lehwald, Report Jül-Conf-26 (1978) p. 64
2.175   F.M. Hoffmann, R.A. de Paola: Phys. Rev. Lett. **52**, 1697 (1984)
2.176   F.M. Hoffmann: Surf. Sci. Repts. **3**, 107 (1983)
2.177   Y.J. Chabal: In *The Structure of Solid Surfaces*, ed. by M.A. Van Hove, S.Y. Tong, Springer Ser. Surf. Sci., Vol. 2 (Springer, Berlin, Heidelberg 1985) p. 70
2.178   C.H. Li, S.Y. Tong, D.L. Mills: Phys. Rev. B**21**, 3057 (1980)
2.179   S. Lehwald, J.M. Szeftel, H.Ibach, T.S. Rahman D.L. A Mills: Phys. Rev. Lett. **50**, 518 (1983)

2.180   J.M. Szeftel, S. Lehwald, H. Ibach, T.S. Rahman, J.E. Black, D.L. Mills: Phys. Rev. Lett. **51**, 268 (1983)
2.181   R.B. Doak, U. Harten, J.P. Toennies: Phys. Rev. Lett. **51**, 578 (1983)
2.182   G. Brusdeylins, R. Rechsteiner, J.G. Skofronick, J.P. Toennies, G. Benedek, L. Miglio: Phys. Rev. Lett. **54**, 466 (1985)
2.183   G.J. Schulz: Rev. Mod. Phys. **45**, 378,423 (1973)
2.184   J.E. Demuth, D. Schmeisser, Ph. Avouris: Phys. Rev. Lett. **47**, 1166 (1981)
2.185   Ph. Avouris, D. Schmeisser, J.E. Demuth: Phys. Rev. Lett. **48**, 199 (1982)
2.186   D.E. Eastman, F.J. Himpsel: Phys. Today **34**, 64 (May 1981)
2.187   J.W. Davenport: Phys. Rev. Lett. **36**, 945 (1976)
2.188   D.E. Eastman, F.J. Himpsel, J.F. van der Veen: J. Vac. Sci. Techn. **20**, 609 (1982)
2.189   J.F. van der Veen, F.J. Himpsel, D.E. Eastman: Phys. Rev. Lett. **44**, 189 (1980)
2.190   F.J. Himpsel: Appl. Phys. A**38**, 205 (1985)
2.191   A. Cricenti, F. Ciccacci, S. Selci, P. Chiaradia, G. Chiarotti: In *Dynamical Phenomena at Surfaces, Interfaces and Superlattices*, ed. by F. Nizzoli, K.H. Rieder, R.F. Willis, Springer Ser. Surf. Sci., Vol. 3 (Springer, Berlin, Heidelberg 1985) p. 60
2.192   M.A. Olmstead, N.M. Amer: Phys. Rev. Lett. **52**, 1148 (1984)
2.193   T.F. Heinz, M.M.T. Loy, W.A. Thompson: Phys. Rev. Lett. **54**, 63 (1985)
2.194   Y.R. Shen: In *The Structure of Solid Surfaces*, ed. by M.A. Van Hove, S.Y. Tong, Springer Ser. Surf. Sci., Vol. 2 (Springer, Berlin, Heidelberg 1985) p. 77
2.195   C.P. Slichter: In *The Structure of Solid Surfaces*, ed. by M.A. Van Hove, S.Y. Tong, Springer Ser. Surf. Sci., Vol. 2 (Springer, Berlin, Heidelberg 1985) p. 84
2.196   C.D. Makowka, C.P. Slichter, J.H. Sinfelt: Phys. Rev. Lett. **49**, 379 (1982)
2.197   K.F. Canter, K.G. Lynn: J. Vac. Sci. Technol. A**2**, 916 (1984)
2.198   R.F. Willis: In *Dynamical Phenomena at Surfaces, Interfaces and Superlattices*, ed. by F. Nizzoli, K.H. Rieder, R.F. Willis, Springer Ser. Surf. Sci., Vol. 3 (Springer, Berlin, Heidelberg 1985) p. 196
2.199   R.J. Birgeneau, P. Horn, D.E. Moncton: In *The Structure of Solid Surfaces*, ed. by M.A. Van Hove, S.Y. Tong, Springer Ser. Surf. Sci., Vol. 2 (Springer, Berlin, Heidelberg 1985) p. 404
2.200   S.C. Fain, Jr., Hoydoo You: In *The Structure of Solid Surfaces*, ed. by M.A. Van Hove, S.Y. Tong, Springer Ser. Surf. Sci., Vol. 2 (Springer, Berlin, Heidelberg 1985) p. 404
2.201   S.Y. Tong: Phys. Today **37**, 50 (August 1984)

3. High-Resolution Electron Microscopy of Surfaces

L.D. Marks

With 29 Figures

This chapter reviews the current status of high-resolution electron microscopy of surfaces, primarily the profile imaging technique. First the background to the technique is described, in particular the electron scattering process, the effects of the microscope imaging system, the importance of one-to-one mapping between the object and the image and the role of image simulations in obtaining quantitative data. A number of different applications are then discussed, illustrated by a detailed description of results on the gold (110) surface. Finally, two other techniques, namely plan-view imaging and reflection imaging are briefly described, and some of the possible future directions of electron microscopy of surfaces are indicated.

3.1 Background

Over the years a wide range of different probes have been used to investigate surfaces. As a rule the procedure has been to bounce a beam of photons, electrons or ions off the surface and then monitor the results. These are implicitly diffraction experiments. The results of diffraction experiments from a perfectly ordered specimen can be inverted to yield atomic-structure information provided that there is no phase problem, and the multiple scattering is well understood. However, if the specimen is not well ordered there are problems. Disorder, as a rule, produces diffuse scattering away from the main diffraction spots which is well-nigh impossible to interpret because of the phase problem; the location of the disorder is contained in the phase of the wave, and the final diffuse elastic scattering is a complicated sum of the diffuse scattering from different places on the surface. To further confuse the issue, there is also, as a rule, diffuse scattering arising from inelastic processes such as phonon and plasmon excitations.

In the study of bulk materials, this breakdown of diffraction techniques for disordered regions stimulated the development of imaging techniques. In an image the phase information is retained, and by judicious choice of the experimental approach one can frequently determine uniquely the absolute amplitude and phase of the scattered wave at each point. Disordered regions clearly show as such, distinct from the regions of well ordered structure.

Because of the relative simplicity of constructing magnetic lenses to focus swift (10–1000 keV) electrons, electron microscopy has emerged as the prime imaging technique for microstructural investigations of inhomogeneous materials. Furthermore, due to the presence of chemically sensitive inelastic processes such as core excitations, analytical information on the nanometer scale is available. A full description of electron microscopical techniques which range from determining the spatial location of impurity levels in semiconductors to measuring the symmetry and structure of a material to a thousandth of an Angstrom lies outside the scope of this review. Some details can be found in [3.1–8].

As a technique for investigating surface, until very recently conventional electron microscopy {transmission electron microscopy (TEM) not scanning electron microscopy (SEM) or scanning Auger microscopy(SAM) [3.4]} has not been particularly useful. The electron microscope had a poor vacuum (10^{-5} torr), fairly rich in hydrocarbons from diffusion pump oil, which automatically made any surface results dubious. In addition, the standard electron microscopical specimen is a thin (1–100 nm) sheet from which any surface diffraction is small compared to the bulk diffraction.

Over the last ten years or so the situation has changed dramatically following improvements in the vacuum of the operating microscope and the development of techniques with improved surface sensitivity. Ultra-high vacuum scanning electron and scanning Auger microscopes, for example, are now widely available. For structural analysis the pioneering work of the group at the Tokyo Institute of Technology [3.9–12] has led to the development of reflection techniques, essentially reflection high energy electron diffraction (RHEED) inside an imaging ultra high vacuum (UHV) electron microscope, and a growing number of transmission techniques. The field is growing rapidly at present, and for some recent reviews see [3.12–15]. (Progress is sufficiently rapid at present that these reviews may be dated by the time this book appears.)

However, techniques such as scanning electron, Auger of reflection electron microscopy have at best a resolution of about 1 nm. This is too coarse to see atomic details. Electron microscopical techniques to image the atomic structure of materials have also been developed over the last ten years, a branch called high-resolution electron microscopy (HREM). For some recent reviews see [3.16–19]. Under the appropriate conditions images can be obtained which directly show the atomic structure. Unfortunately, under the wrong conditions, the images still resemble the atomic structure of the material under investigation, even though they, in fact, do not portray it correctly. The possibility of such artifacts which can mislead the unwary is one of the complications of the technique, soluble by means of numerical image simulations to check the veracity (or not) of the results.

Over the period 1983–1985 we have been able to utilise this high-resolution technique to directly image the atomic structure of surfaces in profile [3.20–29]; for instance, Fig. 3.1 part of a collaborative project with Dr. D.J. Smith using the Cambridge High-Resolution Electron Microscope [3.31]. The essence of the technique is to employ a beam of swift (500 keV) electrons grazing the surface of interest (as shown in Fig. 3.2). We then obtained high-resolution images showing both the bulk structure and that of the surface in profile. By care-

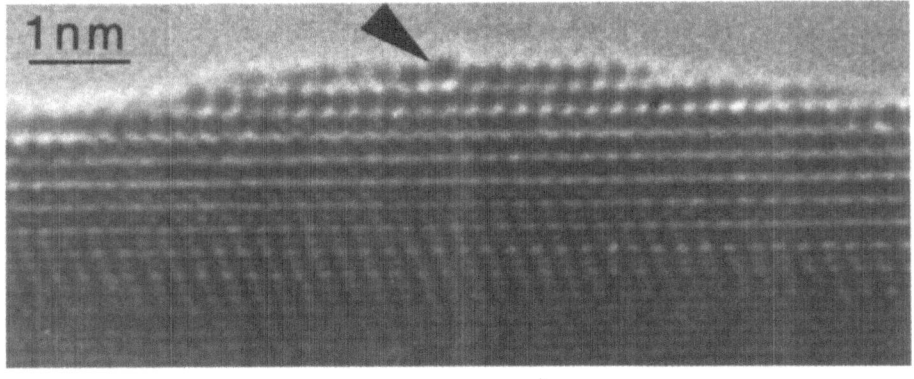

Electron Beam

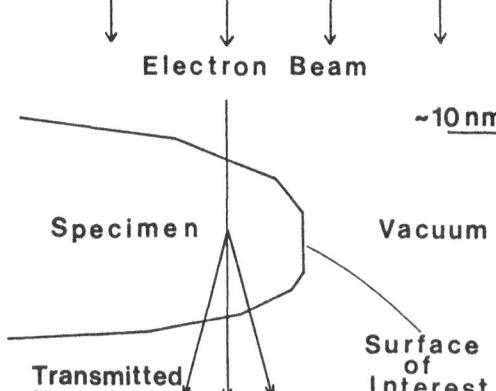

Fig. 3.1. High-resolution image of a region of gold(111) surface with a surface partial dislocation (*arrowed*) [3.23,25]

~10 nm

Fig. 3.2. Diagram to illustrate the geometry of the profile imaging technique

ful synergistic comparisons between experimental and numerically calculated images it proved possible to determine conditions such that there was a true one-to-one mapping between the object and the image. This meant that the experimental images faithfully showed the structure of the outermost atomic columns, making the technique one of true monolayer sensitivity. More recently, the profile-imaging technique has been applied to a number of oxide materials [3.32–38].

In this chapter we describe some aspects of high-resolution electron microscopy as a technique to image surfaces. Unfortunately, high-resolution electron microscopes are all different, and true experimental reproducibility between different machines is very difficult to achieve. However, the underlying physics does not change, and the variance between the results can be calculated and accounted for by well understood and proven theoretical methods. We will therefore concentrate upon the underlying science, using experimental images where appropriate to illustrate the nature of the results. The emphasis will be on the profile method which, at least at present, has been investigated more than alternative techniques and has interfaced to conventional surface science.

Other approaches, currently in the development stage, will also be briefly described. In addition, the technique is very new, and as yet the only detailed study has been for gold specimens. The complications that can arise with other specimens and microscopes are as yet unknown. Thus in some areas such as specimen preparation we cannot answer many of the potentially important questions.

3.2 Image Formation

3.2.1 Basics

Today's high-resolution electron microscope is a very sophisticated instrument, machined in places to tolerances of less than one micrometer, inside of which a variety of different experiments can be performed to study the same region of a specimen. Operating such a machine is somewhat akin to playing a violin – gently coaxing it to obtain the best possible performance. Often, however, the performance seems to depend more upon intangibles than any actions of the operator. A number of texts are now available covering the practical details of operating a high-resolution electron microscope (e.g., [3.8]), although many of the useful tricks remain part of the folklore rather than being available in the literature. We will not here become involved in any of these practical details,

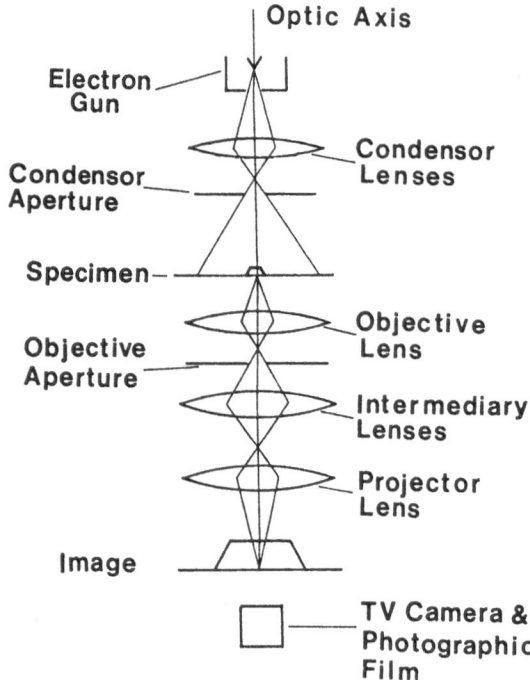

Fig. 3.3. Sketch of the column of a high-resolution electron microscope

but instead briefly sketch the basics of an electron microscope and then deal with the processes which lead to the final image.

A conventional high-resolution electron microscope is really the same as an optical microscope, using magnetic rather than glass lenses and electrons rather than photons. A sketch of the column of a high-resolution instrument is shown in Fig. 3.3. Like an optical microscope, we have a series of condensor lenses, an immersion objective lense and a number of projector lenses. Starting with an electron source, typically a heated LaB_6 rod, the electrons are concentrated onto the region of interest. For high-resolution imaging it is necessary to balance the coherence of the incident beam against its brightness. Reducing the coherence by using larger condensor apertures and thus increasing the energy range of the incident electrons and the beam convergence, reduces the image contrast (Sect. 3.2.3). Increasing the coherence and reducing the convergence, however, reduces the number of electrons and thus the signal strength. The balance between these two is met in most machines by an energy spread of about 1 eV and a convergence semi-angle of about 0.5 mrad, although these figures will vary from machine to machine.

After being scattered by the specimen (Sect. 3.2.2), the objective and projector lenses operate in tandem to magnify the image, typically 500,000 times or more. Compared to optical lenses today's magnetic lenses [3.39] are highly aberrated with a spherical aberration (C_s) and chromatic aberration (C_c) of about 1 mm. On a Rayleigh-principle basis, the attainable resolution in an electron microscope would be better than 0.01 nm, the wavelength of a 100 keV electron being 0.0037 nm. However, with these imperfect lenses the best that can currently be achieved is about 0.18 nm. At the bottom of the microscope the electrons are detected by a photographic film or a TV camera with an image intensifier [3.40]. Sophisticated image pickup systems have recently become fairly inexpensive and are now standard equipment. They allow good images to be displayed on a TV monitor, better than can be seen by focusing a pair of binoculars onto a phosphor viewing screen. In practice, one can easily see 0.2 nm detail on the monitor (or the screen). However, simple photographic film has an excellent quantum efficiency and a very small grain size coupled with a comparatively large area [3.41]. It is still the staple means of collecting the final image and is likely to remain so for some years.

One characteristic of high-resolution electron microscopy is the use of very thin specimens, typically less than 10 nm. Swift electrons will transverse substantially thicker specimens, but inelastic scattering reduces the resolution of the images (chromatic aberrations become large) and complicates the interpretation; at present, no fast numerical techniques exist for calculating the combined elastic and inelastic scattering from thicker specimens. Currently this is an area which most high-resolution electron microscopists tend to avoid.

Specimen preparation is frequently a long and tedious process, the precise methods being highly dependent upon the nature of the specimen. It is generally simplest for brittle materials such as ceramics which can be crushed, creating thin wedge shaped regions. Other standard techniques are chemical or

electrochemical thinning processes, ion-beam milling, and evaporation to produce small particles or a thin film. Descriptions of typical procedures can be
found in [3.1,8,42], although as a rule it is best to consult the literature for the
material of interest.

The preparation of specimens for surface profile imaging is at present a
grey area; as yet too little work has been done for any generalisations. What is
required is a thin specimen which can either be cleaned externally and then introduced via a UHV transfer, or cleaned in-situ inside the microscope. Cleaning
may turn out to be the trickiest part; any violent procedure such as argon-beam
bombardment followed by thermal annealing will probably destroy the thin regions. At present the only technique that has been successful relies upon the
electron beam either to induce a cleaning chemical reaction or to sublime away
part of the specimen to produce a fresh, clean surface. For instance, in the presence of an electron beam water vapor will etch carbonaceous contaminants on
the surface of gold [3.25], whilst *Briscoe* et al. [3.32,33], have used an intense
electron beam to remove surface layers on a zinc-chromium oxide. Preliminary
results [3.35,36] have indicated that the reactions taking place in the electron beam are similar to those that occur in desorption induced by electronic
transitions or DIET [3.43]. An electronic excitation, either the excitation of a
plasmon, an interband transition, or the ejection of a core electrons, puts the
solid in an excited state. At a surface this can activate a chemical reaction such
as the reaction of carbon with water [3.25], or lead to the ejection of anodic
elements such as oxygen [3.38]. In a ceramic the exposed metal may be able to
sublime off, leaving a fresh and, hopefully, clean surface. A series of different
gas treatments may work for many materials, for instance, treating the specimens with hydrogen to remove oxide layers. Hopefully, answers to the problem
of specimen preparation will appear over the next few years.

We now turn to some of the more specific details of the image formation
process. It is convenient to divide this into two sections, dealing separately with
the scattering (diffraction) of the electrons as they pass through the specimen,
then with the pseudo-optical image formation processes due to the magnetic
lenses.

3.2.2 Electron Scattering

The elastic scattering of swift electrons is fortunately a fairly well understood process [3.1–4]. Unfortunately, essentially independent of energy, electron
diffraction is a multiple scattering process and the simple kinematical analysis
used for x-ray scattering is not applicable. A standard benchmark is that at
100 keV, one atom of gold is a dynamical scatterer. We do not need to become
involved in any complicated and exotic analysis of relativistic effects. Although
the electrons are relativistic (a 100 keV electron travels at about half the speed
of light and 511 keV is the threshold for pair production), it is well established
that relativistic effects can be accounted for by correcting the electron mass
and wavelength [3.1,44,45].

To obtain some physical insight, we start by looking at the diffraction from a perfect solid and extend the analysis at a later stage to consider surfaces. We write Schrödinger's equation in the form

$$\{\nabla^2 + (8\pi^2 me/h^2)[E + V(r)]\}\psi(r) = 0, \tag{3.1}$$

where E is the electron energy, $V(r)$ the crystal potential, m the relativistic electron mass, and we are looking for solutions for $\psi(r)$ the electron wave. As in solid-state theory, we know that any solution must have the same periodicity as the lattice, i.e. it must be in the form of a Bloch wave. Therefore we expand $\psi(r)$ as a linear sum of Bloch waves

$$\psi(r) = \sum_j A_j B_j(k_j, r), \tag{3.2}$$

where the Bloch waves B_j have the form

$$B_j(k_j, r) = \sum_g C_g^j(k_j) \exp[-2\pi i(k_j - g) \cdot r] \tag{3.3}$$

and the g vectors are reciprocal lattice vectors. Since the potential also has the lattice periodicity, we may similarly expand

$$V(r) = \sum_g V_g \exp(2\pi i g \cdot r). \tag{3.4}$$

Combining these various equations and then collecting the terms with common exponential multipliers we obtain

$$[1 - (k_j + g)^2 h^2/(2meE)]C_g^j(k_j) + 1/E \sum_h V_h C_{g-h}(k_j) = 0. \tag{3.5}$$

This is the standard matrix equation for high-energy diffraction. So far we have not done anything to distinguish the analysis from that of low-energy electrons. We are simply dealing with a band-structure problem, concentrating upon very high energy levels rather than those for the conduction or valence electrons. As in solid-state theory, the k_j vectors cannot take on arbitrary values, only a selected range. In this sense they are pseudo-eigenvalues with the corresponding $C_g^j(k_j)$ being pseudo-eigensolutions. The permitted values of k_j fall on a dispersion surface, as in solid-state theory and illustrated in Fig. 3.4. (Unfortunately, it is conventional in electron microscopy to draw the 'y' axis inverted with respect to the solid-state usage). The momentum along the beam direction, $k_j \cdot \hat{z}$, is equivalent to a transverse energy, the beam propagating with z like a wave propagating with time in a two-dimensional solid [3.2,46,47]. Each Bloch wave, however, corresponds to the *same* total energy, unlike the atomic eigenvalue problem.

A number of important physical results come directly from the dispersion surface. We need to consider two properties; how the propagation of electrons is related to the form of the dispersion surface, and the form of the surface; in particular, whether it represents the solid-state case of nearly free or tightly bound electrons. The electrons (more properly the Bloch waves) propagate

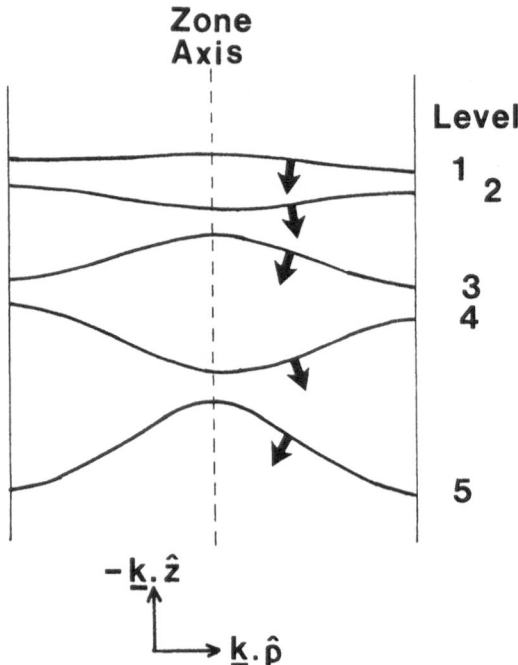

Zone Axis

Level
1
2

3
4

5

$-\underline{k}.\hat{z}$

$\underline{k}.\hat{p}$

Fig. 3.4. First few levels of the dispersion surface for swift electrons, illustrating the variation of $k \cdot \hat{z}$ with the transverse value of k. The term $k \cdot \hat{z}$ can be considered as the transverse energy of the electron wave, i.e. a transversed band structure diagram (with the 'y' axis inverted relative to the standard solid state usage). The Bloch waves propagate along the vector normals to the dispersion surface, as illustrated in the figure

normal to the dispersion surface, as illustrated in Fig. 3.4. Because of symmetry, on a zone axis the electrons will channelled parallel to the atomic columns. The spreading normal to this direction is determined by the curvature of the surface, i.e., in solid-state terms by the magnitude of the effective mass tensor. For the case of high-energy electrons the dispersion surface is very flat, with a far smaller curvature inside the solid than for the corresponding case of an electron in vacuo. Compared to the electron wavevector, the splitting between the various levels of the dispersion surface is small. However, what we are interested in is the curvature of the levels perpendicular to the electron-beam direction, since it is in this plane that most of the diffraction takes place. (Diffraction along the beam direction is very weak and can, in general, be ignored except for rather thick crystals.) Normal to the beam, the electron levels are analogous to tightly bound states, for heavy atoms in fact more tightly bound than the conduction/valence electrons in most large-band-gap insulators. This comes about as a consequence of the relativistic effects. The relativistic electron wavelength is [3.48]

$$\lambda = \lambda_c (1 - \beta^2)^{1/2}/\beta, \tag{3.6}$$

with λ_c the Compton wavelength and $\beta = v/c$. The wavelength is shorter than one would expect and tends to zero at infinite energy. However, the interaction with the specimen potential tends to a finite value at infinite energy. Thus relativistic effects mean that the diffraction does not become more kinematical as the energy increases (large curvature of the Ewald sphere relative to the potential), but instead more dynamical and tightly bound. Thus the spreading transverse to the atomic columns is very small on a zone axis for a high-atomic-weight material; and for a thin (<10 nm) specimen, in fact, negligibly small.

With such small transverse spreading we can consider the scattering by a Muffin-tin approach [3.49]. Outside the Muffin-tin for each column of atoms the electron wave passes essentially unchanged with all the diffraction taking place inside. Analytically, in this approximation the exit wave can simply be written in the Eikonal [3.50] or phase grating [3.2] form

$$\psi(\mathbf{r}) = \psi(z = 0, \varrho) \exp\left[i\sigma \int V(\mathbf{r}) dz\right],$$ (3.7)

where $\sigma = 2\pi \lambda_c m_0 e / \beta h^2$, m_0 being the electron rest mass, and $\psi(z = 0, \varrho)$ the wave on the specimen entrance surface. Including the transverse spreading as a perturbation with a time-like analysis [3.51], we get the slightly more accurate form for a thicker specimen of

$$\psi(\mathbf{r}) = \left\{1 - \alpha \int [|\nabla_\varrho V(\mathbf{r})|^2 + \nabla_\varrho^2 V(\mathbf{r})] dz\right\}$$
$$\times \exp\left[i\sigma \int V(\mathbf{r}) dz\right] \psi(z = 0, \varrho),$$ (3.8)

where

$$\alpha = \sigma^2 \lambda / 2\pi.$$ (3.9)

Because all the scattering takes place within each column of atoms independent of its neighbours, to a good approximation we can use precisely the same analysis at a surface. Therefore, as far as the diffraction is concerned we have minimum cross-talk between the columns and thus a true monolayer sensitivity.

Before leaving the subject of specimen scattering, it is useful to consider an approximate form for the scattering from a surface in profile. As a rule the potential seen by the swift electron can be represented as a sum over lattice sites l of the single-atom potential, i.e.

$$V(\mathbf{r}) = \sum_l v_a(\mathbf{r} - l), \quad l \cdot \hat{x} > 0$$ (3.10)

for a surface in a profile. In the kinematical limit with σt small, the diffraction pattern would be

$$|\bar{\psi}(\mathbf{u})|^2 = \delta^2(0) + \sigma^2 t^2 \sum_g |\tilde{v}_a(\mathbf{u})/(g - u_x)|^2,$$ (3.11)

where $\bar{\psi}(\mathbf{u})$, $\tilde{v}_a(\mathbf{u})$ are the Fourier transforms of $\psi(\mathbf{r})$, $v_a(\mathbf{r})$.

We see that the diffuse scattering due to the surface is largest around the g reciprocal lattice beams. This is an important point which substantially simplifies the physical interpretation of the lens imaging process, as we will see later.

It is also worth considering the effect of a surface relaxation. In reciprocal space the wave for an unrelaxed surface is

$$\overline{\psi}(u) = \delta(0) - i\sigma t \sum_{g} v_a(u)/(g - u_x)$$ (3.12)

if σt is small. With a surface expansion of d in the x direction, the modified wave $\overline{\psi}'(u)$ will be

$$\overline{\psi}'(u) = \overline{\psi}(u) + i\sigma t[1 - \exp(2\pi i u \cdot d)] v_a(u).$$ (3.13)

Even with a small relaxation, say 0.01 nm, there is still an appreciable phase change acting on low-order beams. Thus, we do not need 0.01 nm resolution to see the effect of an 0.01 nm surface relaxation[1].

3.2.3 Lens Aberrations

Once the electron wave leaves the specimen, it is magnified by the magnetic lenses to produce an image. As mentioned earlier, magnetic lenses are relatively imperfect compared to optical lenses possessing large spherical and chromatic aberrations [3.39]. In addition, in practice we do not use a perfectly coherent plane wave to illuminate the specimen but in order to achieve a reasonable signal must work with a range of electron energies and a slightly convergent beam. Certainly with the higher-voltage (e.g., 500 keV) instruments it is the energy spread of the electrons both from the thermal spread of the LaB_6 source and from instabilities in the high voltage which (at present) limit the microscope performance. (There are some indications that vibrations are limiting some of the newer instruments). For the lower-voltage machines (100 keV) the incident beam convergence is the limiting term.

The theoretical treatment of the lens imaging process is handled along the lines of an optical analysis [3.2,4,8]. For an objective-lens defocus of Δz, the wave undergoes a phase shift $\chi(u)$ as a function of the wavevector u transverse to the optic axis (in a well aligned microscope the incident beam direction), i.e.,

$$\chi(u) = \frac{\pi}{\lambda}\left(\Delta z u^2 \lambda^2 + \frac{1}{2} C_s u^4 \lambda^4\right),$$ (3.14)

so that the reciprocal-space wave $\overline{\psi}(u)$ is modified:

$$\overline{\psi}(u) \rightarrow \overline{\psi}(u) \exp[-i\chi(u)].$$ (3.15)

In the absence of any convergence or energy spread, we would obtain an image

$$I(r) = |\mathcal{F}^{-1}\{\overline{\psi}(u) \exp[-i\chi(u)]\}|^2,$$ (3.16)

where $\mathcal{F}$ stands for a Fourier transform.

The convergence and energy spread are inherently incoherent effects; we need to sum the final image intensity over a range of convergence and energy-

---

[1] Resolution, as used by electron microscopists, is a conventional term which does not have the strict dictionary definition. As with any scientific instrument, one can determine some parameter to a very high accuracy provided that it has a substantial effect on the signal, and the signal-to-noise ratio is sufficiently large. For instance, one can determine the absolute location of some spectral feature to an accuracy substantially better than the Gaussian response factor or the "resolution" of the spectrometer.

spread values. The latter is quite straightforward; the change of energy ΔE is related to a defocus shift ε via

$$\varepsilon = C_c \frac{\Delta E}{E}. \tag{3.17}$$

The variation in incident angle is a little more complicated. Dynamical electron diffraction is a strong function of the incident wave direction, so in principle we should allow the exit wave $\psi(\mathbf{r})$ to vary. Fortunately the change in $\psi(\mathbf{r})$ for a thin specimen (<20 nm) is very small, and as a rule can safely be neglected [3.52]. (There are certain classes of structures, primarily those containing screw axes, where this is not true). The main effects arise from the tilt off the optic axis, i.e. the change in $\chi(\mathbf{u})$. In this approximation it is quite straightforward [3.8,53–55] to show that integrating the intensity (3.16) over a range of incident directions and lens defoci is approximately equal to the reciprocal-space integration:

$$I(\mathbf{r}) = \mathcal{F}^{-1} \int \overline{\psi}^*(\mathbf{u}-\mathbf{v})\overline{\psi}(\mathbf{v}) \exp\left(i[\chi(\mathbf{u}-\mathbf{v}) - \chi(\mathbf{v})]\right) T(\mathbf{u}-\mathbf{v},\mathbf{v}) d^2\mathbf{v}, \tag{3.18}$$

where

$$T(\mathbf{u}-\mathbf{v},\mathbf{v}) = \exp\left(-\pi^2 S_0^2 \lambda^2 [(\mathbf{u}-\mathbf{v})^2 - \mathbf{v}^2]^2/2 - \pi^2 d_0^2 |\Delta z \lambda \mathbf{u} \ldots \right.$$
$$\left. \ldots + C_s \lambda^3 [(\mathbf{u}-\mathbf{v})^3 - \mathbf{v}^3]|^2\right). \tag{3.19}$$

Both the energy spread (in terms of the defocus change) and the convergence have been included as Gaussians of halfwidth at halfheight S_0[nm] and d_0[rad], respectively. The term $T(\mathbf{u}-\mathbf{v},\mathbf{v})$ is generally referred to as a nonlinear envelope term, and is the main term limiting the microscope performance.

Before moving on to consider the effects of the imaging for surfaces, it is worthwhile to briefly consider some of the general features of the imaging equations. Many of the basic characteristics can more clearly be seen for the case of a very weakly scattering specimen. Here $\overline{\psi}(0)$, the transmitted beam, will be substantially stronger than any of the scattered $\overline{\psi}(\mathbf{u})$ beams, and we obtain the approximation

$$I(\mathbf{r}) = 2 \operatorname{Re}\left\{ \mathcal{F}^{-1} \overline{\psi}^*(0)\overline{\psi}(\mathbf{u}) \exp\left[i\chi(\mathbf{u})\right] T(0,\mathbf{u}) \right\} \quad . \tag{3.20}$$

With very weak scattering we may also assume kinematical diffraction in which case

$$\overline{\psi}(\mathbf{u}) = iA(\mathbf{u}) + \overline{\psi}(0)\delta(\mathbf{u}), \tag{3.21}$$

where $A(\mathbf{u})$ is conjugate symmetric, and $\overline{\psi}(0)$ is a real number. We can then simplify (3.20) to the form

$$I(\mathbf{r}) = |\overline{\psi}(0)|^2 - \overline{\psi}(0)2\mathcal{F}^{-1}\{A(\mathbf{u}) \sin[\chi(\mathbf{u})]T(0,\mathbf{u})\}. \tag{3.22}$$

The term $\sin[\chi(\mathbf{u})]$ is called the linear contrast transfer function, and $T(0,u)$ the linear envelope term. The two govern the sign and magnitude with which the "$\mathbf{u}$" beam is passed on to the image. Thus the microscope acts like a linear filter. For a specimen about which there is no a-priori information, that is, no other diffraction information available, we would employ conditions where we have a broad pass band. This is generally close to the Scherzer defocus of $\Delta z = (C_s \lambda)^{1/2}$, and the width of the pass band is frequently used as a measure of the resolution of the microscope. (For further discussion of linear imaging theory, see [3.2,4,8]). With a periodic specimen most of the information is concentrated around the reciprocal lattice vectors $\boldsymbol{g}$ of the bulk material, as mentioned previously. Depending upon $\sin[\chi(\boldsymbol{g})]$ this lattice information is passed either with positive or negative sign, that is white or black contrast at the atomic columns. As a rule, the optimum choice of the defocus is one where $\sin[\chi(\boldsymbol{u})]$ is only slowly varying for spacings around $\boldsymbol{g}$, i.e. we choose a defocus where there is a broad pass band around the diffracted beam or beams of interest.

The additional effects arising from the more complete nonlinear imaging analysis are due to the envelope term $T(\boldsymbol{u} - \boldsymbol{v}, \boldsymbol{v})$. With the linear theory one can think of the envelope term as a soft aperture, cutting off the high-frequency components of the images. However, there are also high-frequency components due to the interference between the scattered beams, for example, $\boldsymbol{g}$ and $-\boldsymbol{g}$ beams leading to a $2\boldsymbol{g}$ frequency. The damping of them is very small with the nonlinear envelope, as against large with the linear analysis. These off-axis interference terms are also important for the background to the lattice fringes, for instance, the Fresnel-like fringes at a surface. In particular, a linear analysis is misleadingly optimistic [3.56,57]. Thus the full nonlinear analysis is obligatory for an accurate calculation.

3.2.4 Image Localisation

In the previous two subsections we have very briefly discussed some of the main aspects of the specimen scattering and the image formation process. For the case of a perfectly periodic specimen, for instance, when we are imaging the top surface of a periodic reconstruction or adlayer [3.24], the above equations can be used without any problems. However, with the profile imaging technique we are interested in a non-periodic object, almost a discontinuity. Because of the inherent localisation the diffraction from such a non-periodic object does not introduce any additional complications. However, we must carefully consider the localisation for the lens imaging, the question of where the information arrives in the image rather than the contrast. If we can obtain highly localised imaging conditions [3.58] where the information from one column of atoms in the object is present in only a very small area of the image, we have a faithful representation with monolayer sensitivity. Shifting one column in the object, or altering its ionicity substantially [3.59], would only shift the black or white disc representing this column in the image. If, however, the information is not localised, we may, with luck, still obtain an image which is sensitive to any surface relaxation, but it cannot be simply and directly interpreted as a rule. In this case, shifting one column of atoms in the object would alter the final image of a number of columns. Here we have a surface-sensitive signal, rather than a true, faithful, representation.

For this aspect of the imaging we need a form which shows the effects arising in the image plane. As mentioned above, most of the useful signal is concentrated around the reciprocal lattice g vectors, which implies an expansion of the electron wave leaving the specimen

$$\psi(\mathbf{r}) = \sum_{g} \varphi_g(\mathbf{r}) \exp\left(2\pi i \mathbf{g} \cdot \mathbf{r}\right), \tag{3.23}$$

which can also be written in a pseudo-Wannier form as

$$\psi(\mathbf{r}) = N^{-1/2} \sum_{lg} \varphi_g(l) w(\mathbf{r} - l) \exp\left(2\pi i \mathbf{g} \cdot \mathbf{r}\right), \tag{3.24}$$

for N lattice points. Here $w(\mathbf{r} - l)$ is given by

$$w(\mathbf{r} - l) = N^{-1/2} \int \exp\left(2\pi i \mathbf{u} \cdot \mathbf{r}\right) d^2 \mathbf{u} \tag{3.25}$$

(the integral being over the reciprocal-lattice unit cell). This is a function which is localised around the lattice site l. In this pseudo-Wannier form the effect of lens aberrations takes a very simple form. With, for simplicity, a kinematical analysis and using the linear imaging equation, the final image can be written

$$I(\mathbf{r}) = 1 - N^{-1/2} \sum_{g} \cos(2\pi \mathbf{g} \cdot \mathbf{r}) w(\mathbf{r} - l) \sum_{l'} [\varphi_g(l') L_g(l - l')] \tag{3.26}$$

and

$$L_g(l) = N^{-1/2} \int \{ \exp\left[i\chi(\mathbf{g} + \mathbf{u})\right] T(0, \mathbf{g} + \mathbf{u}) \dots$$

$$+ \exp\left[-i\chi(-\mathbf{g} - \mathbf{u})\right] T(0, -\mathbf{g} - \mathbf{u}) \} \exp\left(-2\pi i \mathbf{u} \cdot l\right) d^2 \mathbf{u}, \tag{3.27}$$

the integral again being over the reciprocal-lattice unit cell. (A more accurate form can be derived using a semi-linear analysis, as described in [3.57]).

The localisation function $L_g(l)$ parametises what we are interested in. If it is substantial only when $l = 0$, we have an image which can be simply interpreted. If, however, it is large over a number of different lattice points, we have an average image of the surface which, with luck, can still be interpreted as a surface-sensitive signal, but not simply interpreted. As a rule, the possible localised defoci will be those where $\sin[\chi(\mathbf{g})] = -1$, i.e. $\chi(\mathbf{g}) = (2n + 1)\pi/2$. Here the linear transfer function is stationary with respect to small angles on either side of the bulk g diffraction beam, i.e. we are centering a pass band over the region of interest. We note that this condition is also stationary with respect to variations in the operating conditions of the microscope such as lens defocus or spherical aberration. Hence it reduces the magnitude of any errors due to imprecise determination of any of these parameters.

An example of the role of localisation is shown in Fig. 3.5. This shows a region of a (100) surface at two defoci together with the localisation function. For the localised focus in Fig. 3.5b, one can clearly see a substantial displacement of one column of atoms. This is believed to be due to a surface Shockley partial dislocation [3.29]. The localisation function shown in Fig. 3.5d is only significant at the origin, so the image is a faithful representation of the object. For the defocus of Fig. 3.5a and c the localisation function is significant not

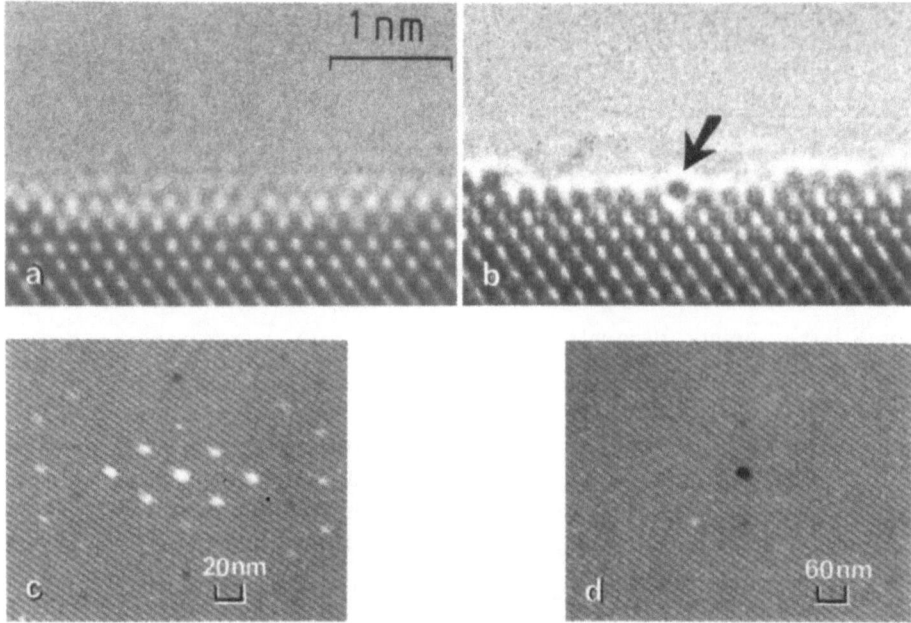

Fig. 3.5. Experimental images of a gold(100) surface for defoci of −20 nm and −60 nm in **(a)** and **(b)**, respectively, with the localisation functions shown in **(c)** and **(d)**, respectively. Note the behaviour with defocus of the displaced column of atoms (*arrowed*)

only at the origin, but also at the neighbouring lattice points. Thus the final image is an average, and the displaced column cannot be seen.

3.2.5 Image Simulation

To convert from a qualitative analysis to a quantitative one, it is necessary to calculate the images as a function of the specimen structure and compare these simulated images to the experimental results. Calculations are also the only scientifically rigorous way to check the degree of image localisation, and safeguard against any misleading artifacts in thick specimens or as a result of surface contaminants. As a rule we match a full through focal series of experimental images. In principle, this is sufficient to completely determine the amplitude and phase of the exit wave.

As in the previous subsections, the two different aspects of the image-formation procedure, namely the specimen scattering and the lens imaging, are dealt with separately. The specimen scattering is evaluated using a fast numerical technique called the multislice algorithm [3.2,60–62]. The basis of the technique is to employ the two analytical solutions

$$\psi(\varrho, z + \Delta z) = \psi(\varrho, z) \exp\left[-i\sigma \int_{z}^{z+\Delta z} V(\mathbf{r}) dz\right] \qquad (3.28)$$

for the propagation of the wave from z to $z + \Delta z$, and

$$\psi(\varrho, z + \Delta z) = (i/\lambda\Delta z)\psi(\varrho, z) \otimes \exp(-i\varrho^2/2\lambda\Delta z) \tag{3.29}$$

for propagation of a wave in vacuum, $\otimes$ representing a convolution. These two types of solutions are applied sequentially by a real-space multiplication for the potential and a reciprocal-space multiplication for the transverse spreading. The combined form

$$\psi(\varrho, z + \Delta z) = (i/\lambda\Delta z)\left\{\psi(\varrho, z) \exp\left[-i\varrho \int\limits_z^{z+\Delta z} V(\mathbf{r})dz\right]\right\}$$
$$\otimes \exp(-i\varrho^2/2\lambda\Delta z) \tag{3.30}$$

is a valid integration provided that the step Δz is sufficiently small and there is a sufficiently fine sampling of the potential $V(\mathbf{r})$ so as to include high-angle scattering effects. (The algorithm has been checked against analytical solutions or numerical results obtained by other techniques, see, for instance, [3.62].) Given the atomic co-ordinates, the potential $V(\mathbf{r})$ is generated from a sum of Gaussians matched to either the tabulated electron or x-ray atomic scattering factors (for the latter using the Mott formula to convert). The lens-imaging effects are numerically calculated using optical-imaging theory, as sketched in Sect. 3.2.4, either the reciprocal-space integration of (3.19) or by an incoherent integration in the image plane. The latter is often faster in practise.

Before dealing with specifics for surface-profile image simulations, a few comments on the limitations of the multislice diffraction calculations are in order. The algorithm is exact for *elastic* scattering, provided that the diffraction effects are included to a sufficiently high angle and the slice thickness is sufficiently small. In principle, one uses successively finer step lengths and higher-order beams until the calculations converge. Unfortunately, the convergence is frequently oscillatory in nature since we are using Fourier-transform techniques which can introduce Gibbs phenomena (ringing effects). Therefore it is possible to introduce quite serious artifacts unless substantial care is taken. Unavoidable errors are due to uncertainties in the exact form of the specimen potential and as a result of inelastic scattering. Fortunately, the electron scattering is dominated by the atomic core potentials, so the uncertainties in the potential due to the valence/conduction electrons are not particularly important. For instance, detailed refinements of the Fourier coefficients of the potential of silicon by convergent-beam diffraction by *Smith* and *Lehmpfuhl* [3.63] only resulted in a fraction of a percent correction.

More important, in principle, are the effects arising from inelastic scattering by plasmons [3.64,65] or phonons [3.2,64,66,67]. Fortunately these are weak for thin specimens with typical mean-free paths of several thousand Angstroms. Plasmon scattering, either from bulk or surface plasmons leads to energy losses of typically 15–25 eV with essentially no change in momentum normal to the beam. As pointed out by *Howie* [3.64], these can be considered as an additional energy spread, leading to further chromatic aberration effects. To a good approximation these electrons are sufficiently far out of focus that they only contribute to the background rather than the fringe structure of the image [3.68]. (It is worth noting that if microscopes continue to improve, this will no longer be true.)

In principle, phonon scattering is a substantially more important phenomenon. Although phonons themselves have long wavelenghts (small wavevectors), the actual electron scattering process is an Umklapp phenomenon [3.66]. Thus the phonons act as an additional diffraction source. However, the matrix elements depend upon $\nabla V(\mathbf{r})$, and therefore correspond to high-angle scattering typically maximum at about $20\,\mathrm{nm}^{-1}$. Such high-angle beams are too highly aberrated to contribute to the lattice structure of the final image, and can be crudely considered as a simple adsorption phenomenon.

Thus both types of inelastic effects contribute primarily to the background structure rather than the fringes which portray the atomic columns, and their neglect should not introduce substantial errors. There is a genuine experimental problem with the background level from high-order beams. These are substantially displaced, for instance, by $10\,\mathrm{nm}$ or so. Thus at a surface of thickness, say $5\,\mathrm{nm}$, the high-order background contribution may be due to a region of thickness $20\,\mathrm{nm}$. We must therefore accept the possibility of background errors. Fortunately, in practise these do not appear to be severe and, in any case, we are primarily interested in the intensity variations, not the background.

To summarise the last few paragraphs, the state-of-the-art at present is that it is possible to calculate the scattering from thin crystals excepting small absolute background errors arising from inelastic or high-angle scattering. Calculating the scattering from thicker crystals is, in principle, possible, but as yet no fast method which takes account of the inelastic scattering is available.

In practise, the procedure for surfaces is to use a rectangular array, as illustrated in Fig. 3.6. This is sampled in both directions using a mesh of about $0.005\,\mathrm{nm}$ spacing. This is fine enough for fairly accurate measurements of the apparent atomic locations. The reciprocal of this spacing is half the maximum

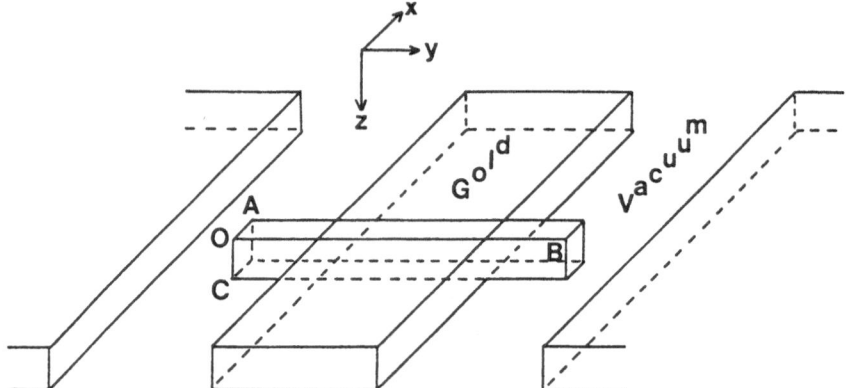

Fig. 3.6. Sketch of the configuration used for calculating a surface image. Shown here are both the periodically continued cell, and the unit cell for the calculation OABC. Typical dimensions would be OA, $0.2\,\mathrm{nm}$, OB, $4\,\mathrm{nm}$ with a slice thickness OC of $0.14\,\mathrm{nm}$

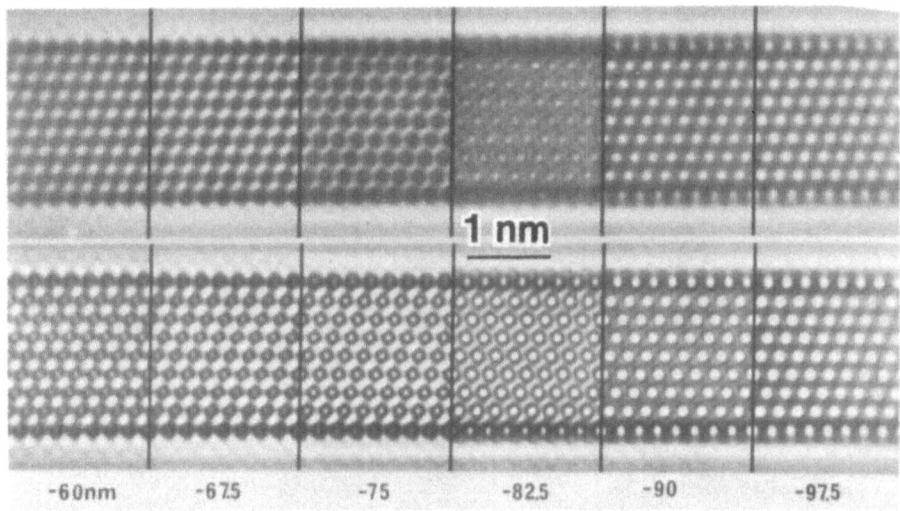

Fig. 3.7. Illustrative calculations for a gold(111) surface for a range of objective lens defoci, as marked on the figure, for thicknesses of 2.1 nm (*top*) and 4.3 nm (*bottom*)

distance in reciprocal space included in the calculation. However, due to aliasing effects the calculation is only applicable to the central two-thirds of the reciprocal space array. Thus 0.005 nm sampling is valid for diffraction out as far as 67 reciprocal nm, which is generally adequate for a calculation to converge at 500 keV.

The actual calculations use Fast Fourier Transforms [3.67] which inherently repeat the unit cell. Thus the true calculation is from an array of such cells. The continuation along the surface is useful, and it is only necessary to use a cell of width equal to the surface repeat. Normal to the surface we have to be careful. If either of the surfaces within the cell, or one of these surfaces and its periodically continued neighbour are too close we will obtain unphysical interference effects between the surfaces. This would invalidate the calculation. From our earlier discussion of the image localisation, this is not a particularly important problem as far as the specimen scattering is concerned (for a heavy material where the electrons are tightly bound). It is important for the imaging part of the calculation. If, however, we are only interested in localised defocus values we can use a relatively small separation. In practise, surface-to-surface separations of about 1 nm were found to be adequate for localised lens defoci, the larger value of about 2 nm being used to prevent interference at some of the more delocalised defoci. An example of the results of such a calculation is shown in Fig. 3.7.

3.3 Applications

There are two main uses of the profile imaging technique, namely qualitative and quantitative. By qualitative we mean applications where the overall fea-

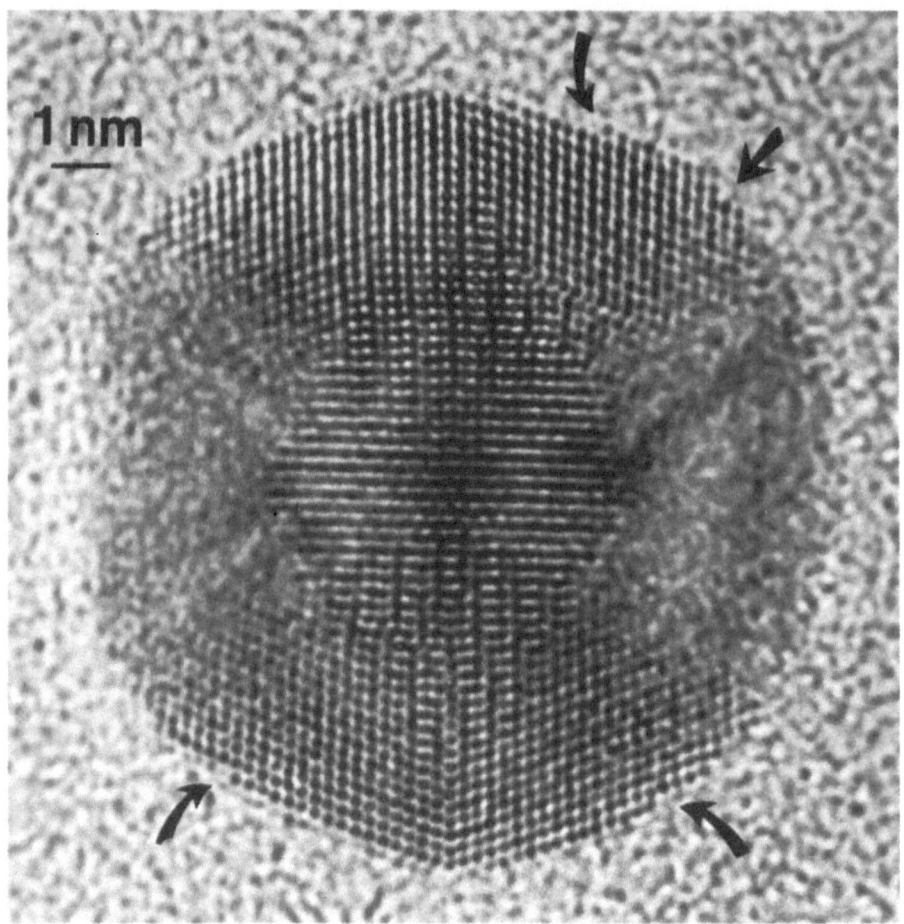

Fig. 3.8. Image of a small, silver multiply-twinned particle, the black dots representing the atomic columns. The background is amorphous carbon. Note the steps on the (111) facets as arrowed

tures such as the density of surface steps or the distribution of surface facets is the primary interest. Subject to checks that the imaging conditions are localised, the information can generally be directly read off the electron micrographs. Quantitative applications require substantially more detailed numerical analysis synergistic with digitised images. This is necessary for any evaluation of surface relaxation and, perhaps more importantly, when there are unusual surface features in order to rule out erroneous interpretations. For instance, careful analysis is required when there is some form of surface superstructure to exclude the possibility that this is due to surface contaminants.

3.3.1 Qualitative Imaging

Perhaps the most wide-spread use in the future will be appraisal of the general surface structure. For instance, information on the surfaces of heterogeneous

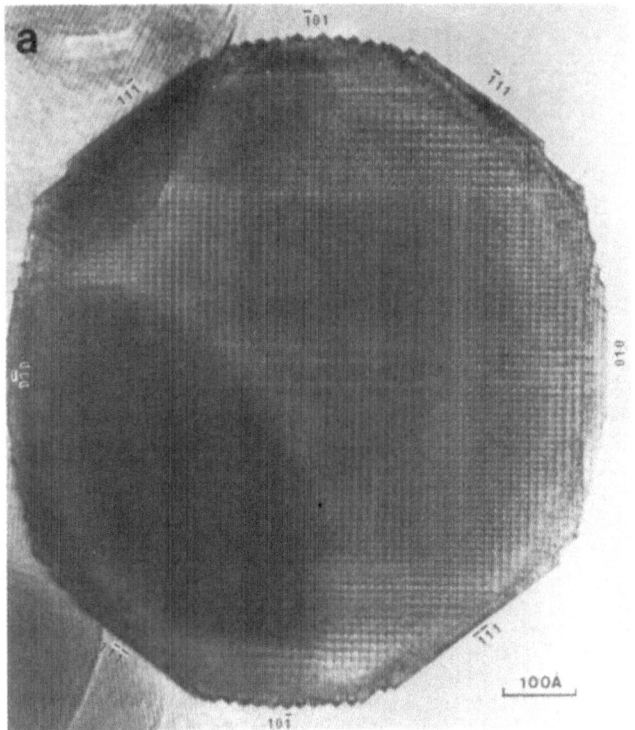

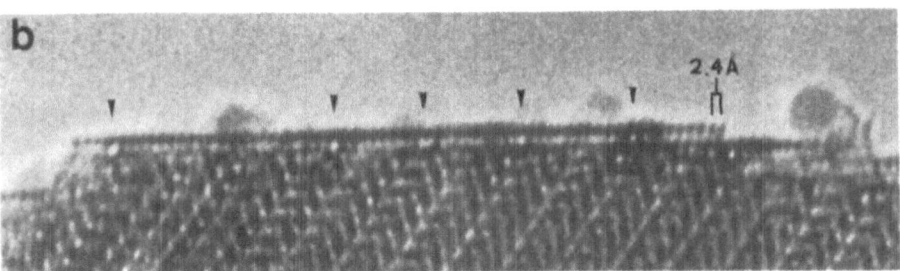

Fig. 3.9. Profile images of a γ'-alumina particle, courtesy of *Iijima* [3.33,34]. In (a) the overall facetting can be clearly seen, including a microfacetting reconstruction of the (101) surfaces. Shown in (b) is an enlargement of a region of (111) surface where it appears that an epitaxial layer of aluminium with 2.4 Å spacing has formed, probably due to desorption induced by electronic transitions (DIET) of oxygen

catalyst and, in particular, knowledge of what surface facets and steps are present is a very important application. At present precious little is known about this field due to the lack of any viable experimental technique to access the information. Some examples of the type of information available are shown in Fig. 3.7–11. These include surface steps, rafts, microtwins and facetting.

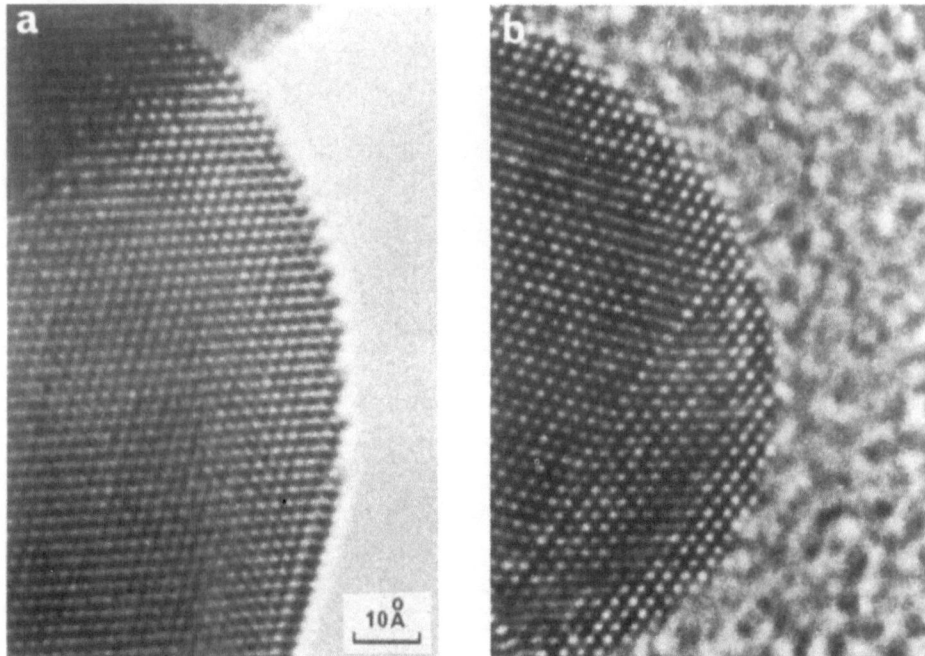

Fig. 3.10. Two images of gold particles showing the facetting, taken from [3.28]. In (a) the carbon has been cleaned away and the atomic columns show black contrast whilst in (b) the columns are white with carbon surrounding the surface

3.3.2 Quantitative Analysis

Quantitative analysis is a substantially more time consuming application when we are interested in refining the locations of the atomic columns or need to resolve ambiguities concerning any surface adlayers. Here, a sophisticated display system and some means of digitising the experimental micrographs is essential. Image processing of this type is a complete field in its own right: for some reviews the reader is referred to [3.69–71]. We will concentrate here upon two aspects of quantitative analysis, namely the measurement of surface relaxations and the detection of chemisorbed layers.

a) Surface Relaxations

At least at present the primary limitation for measurements of surface relaxations is shot noise. Taking a magnification for the final image of 1×10^6, a pixel size of 25 micrometers (corresponding to 0.025 nm), and an exposure of 1 electron per micrometer we have 625 electrons per pixel. Taking the contrast level as 10%, the signal-to-noise ratio is therefore 2.5:1. One can justifiably interpolate by a factor of two on to a finer mesh and then carry out a translational average to obtain a higher pixel accuracy. In practise, we have yet to find a specimen with a sufficiently repeatable structure to make more than a single

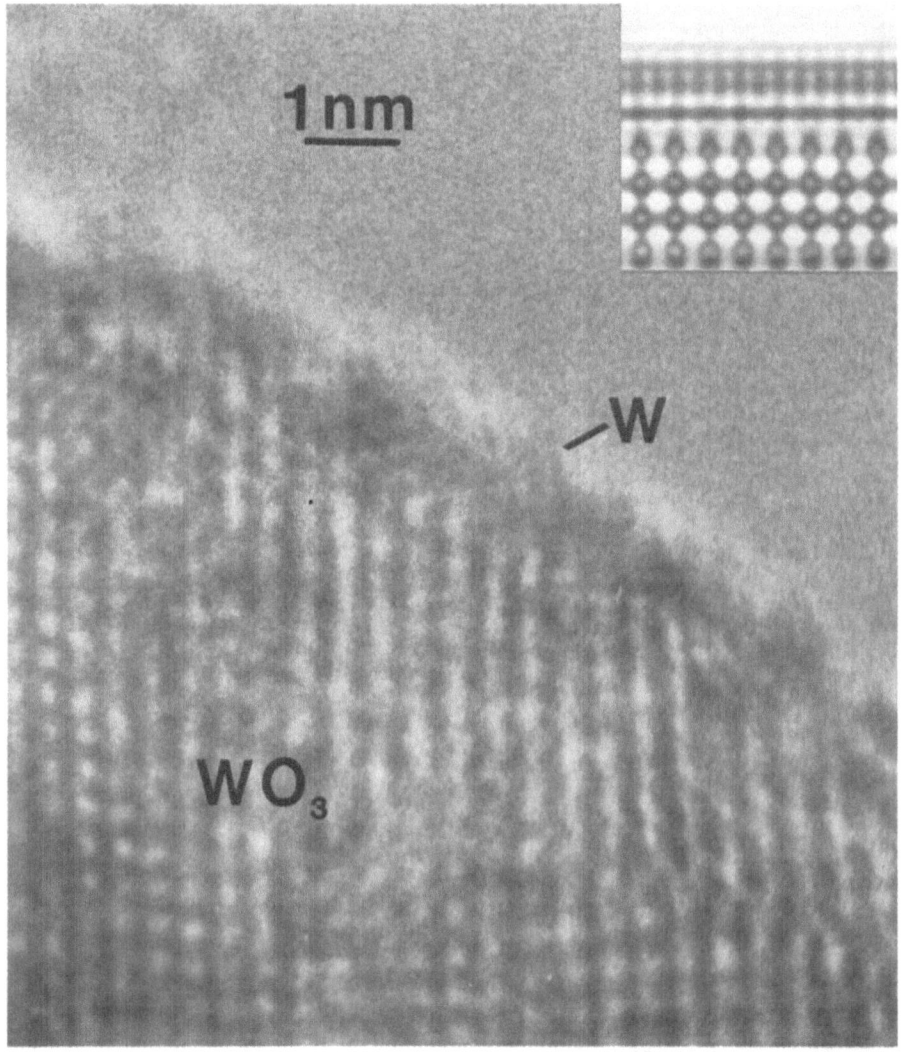

Fig. 3.11. Image of a WO_3 particle where a few layers of epitaxial tungsten have formed in the beam due to oxygen DIET [3.35,36]. Inset is a numerical calculation for three monolayers of tungsten on a $WO_3(100)$ surface

interpolation and a three-point translational average realistic. A simulation of the effects of shot noise is shown in Fig. 3.12. Note that the noise obscures the finer, second-order structure and the Fresnel-like fringes at the surface.

An illustration of the use of such averaging techniques is given in Fig. 3.13. Here the original image showed some uneven structure (a). By performing a translational three-point average we obtain the results shown in (b). Subtracting this from the original gives a clearer idea of the contrast due to the

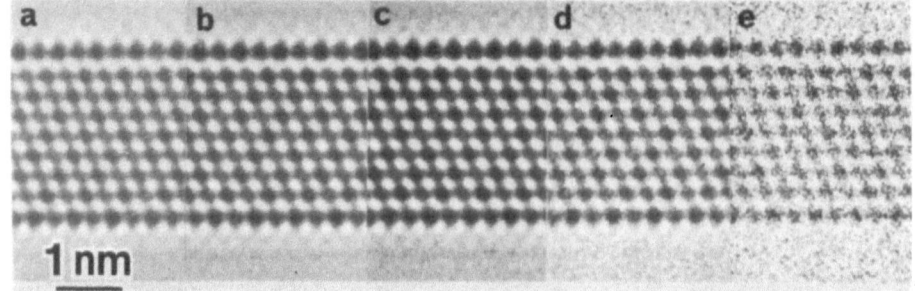

Fig. 3.12. Illustration of the effect of shot-noise upon images, taken from [3.24]. Shown is a region of calculated (111) surface which has been digitally treated to noise corresponding to total doses of ∞, 10^6, 10^5, 10^4 and 10^3 electrons per Angstrom squared. Note the obscuring of the Fresnel structure and a large (15%) expansion of the top surface. Typical experimental conditions correspond to doses of 10^4 electrons per Angstrom squared

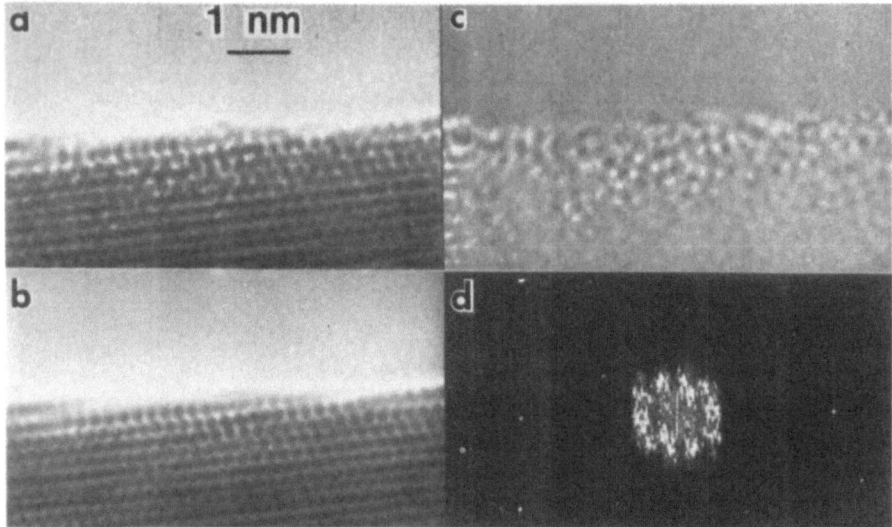

Fig. 3.13. Illustration of the role of averaging techniques. The experimental image in (a) shows some uneven features just inside the surface. Subtracting a three point translational average (b) gives the result shown in (c) ($\times 10$ contrast). The power spectrum of (c) is shown in (d), and indicates the presence of some residual amorphous carbon

structural perturbations. The speckle pattern of this image is characteristic of amorphous carbon, indicating that this surface is not completely clean.

To measure features such as surface relaxations it is first necessary to identify a characteristic feature. Here we are interested in some aspect of the image
that is sensitive, treating the image as a signal rather than a true representation of the object. What we use will depend very strongly upon the precise
spacings of the specimen and the characteristics of the microscope. For gold

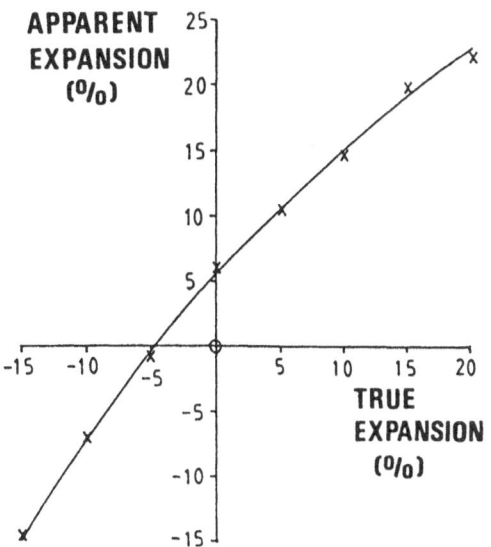

APPARENT EXPANSION (%)

TRUE EXPANSION (%)

Fig. 3.14. Plot of the apparent surface expansion, y axis, against the true expansion, x axis, for a gold(111) surface [3.24]. The relationship is approximately linear, excepting a static 5% shift. The data points do not fall exactly on to a smooth line due to the finite pixel size in the calculations

using the Cambridge microscope [3.31] the location of the first minimum at a defocus of about –60 nm (at 500 keV, –67 nm at 575 keV) was useful. This is the most localised defocus with black contrast at the atomic columns and $\sin[\chi(g)] = -1$ for the gold 0.235 nm (111) spacing. Here the normal surface spacing was faithfully represented in the final images, excepting a 5% static pseudo-expansion from the imaging optics. The calibration curve is shown in Fig. 3.14. As mentioned earlier, this choice of defocus also guarantees minimal changes as a function of the microscope parameters such as defocus and spherical aberration. For instance, there was essentially no change in the surface image for defoci between –55 and –65 nm. Outside this range there were noticeable changes in the overall features of the images, and the outermost minimum image of the column of atoms moved in towards the surface.

In addition to calibration with respect to one particular lens defocus, other features such as changes in the surface Fresnel structure and any characteristic effects at other defoci can be used. An example will be described later when we discuss the gold(110) surface.

b) Surface Adlayers

The other main use of quantitative analysis is when there are uncertainties concerning the chemical composition of a surface layer. Without any a-priori information concerning what elements are present, this is an insoluble problem. At present no high-resolution electron microscopes have any Auger capability, but this may change in the future. Frequently we may be able to limit the range of possibilities by other spectroscopies or from knowledge of what elements were initially present and the history of the specimen.

The most general technique for identifying foreign elements is to use a series of images taken at different values of the objective lens defocus. For a

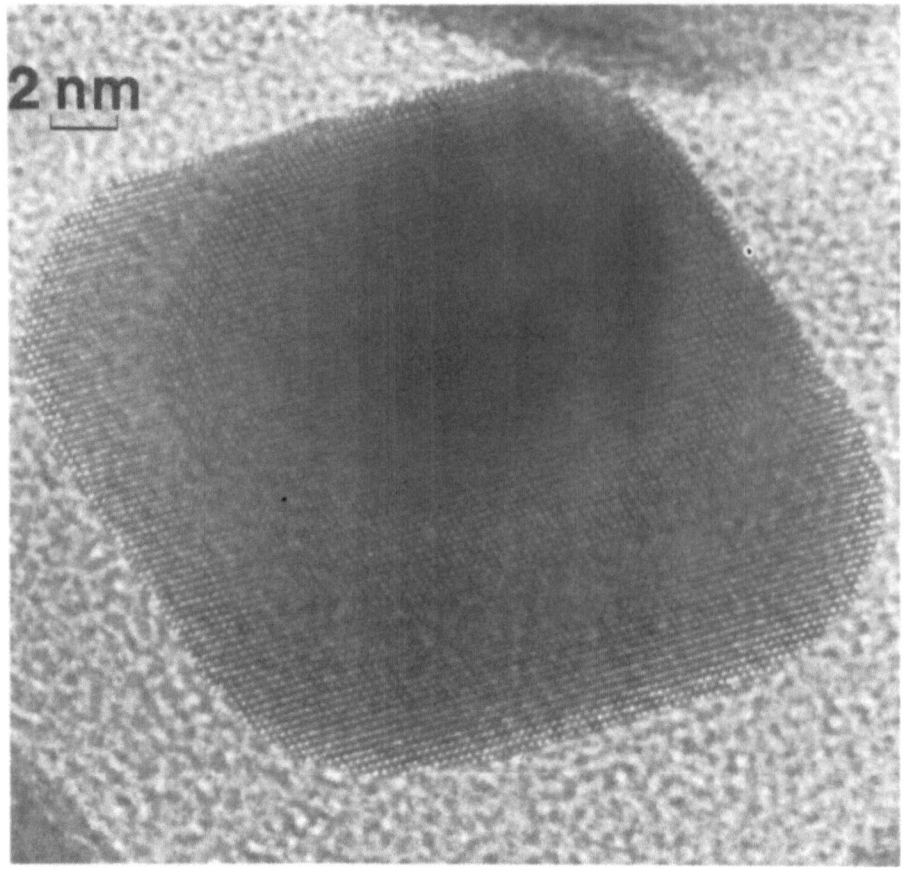

Fig. 3.15. Image of a small gold particle over which a small amount of silver was evaporated, the atomic columns displaying black dot contrast. No obvious differences due to the external silver layer can be seen at this or any other defocus

thin specimen the scattering in the presence of an ordered monolayer can be written

$$\psi(\mathbf{r}) = 1 - i\sigma t \sum_l v_a(\mathbf{r} - \mathbf{l}) - i\sigma t \sum_m v_b(\mathbf{r} - \mathbf{m}), \qquad (3.31)$$

where $\mathbf{l} \cdot \hat{\mathbf{x}} < 0$, $\mathbf{m} \cdot \hat{\mathbf{x}} = c$ for a separation c of the monolayer from the surface, with $v_a(\mathbf{r})$ and $v_b(\mathbf{r})$ the atomic potentials of the bulk and surface species, respectively. In the diffraction plane the wave is

$$\overline{\psi}(\mathbf{u}) = \delta(\mathbf{u}) - i\sigma t \left[\sum_g \frac{\tilde{v}_a(\mathbf{u})}{\mathbf{g} - \mathbf{u}_x} + \sum_q \tilde{v}_b(\mathbf{u} - \mathbf{q}) \right], \qquad (3.32)$$

where the different $\mathbf{q}$ values correspond to the periodic components of the superlattice along the surface. If the atomic scattering terms $\tilde{v}_a(\mathbf{u})$ and $\tilde{v}_b(\mathbf{u})$ for the two elements are similar it will be almost impossible to distinguish between them. For instance, silver and gold have very similar scattering potentials

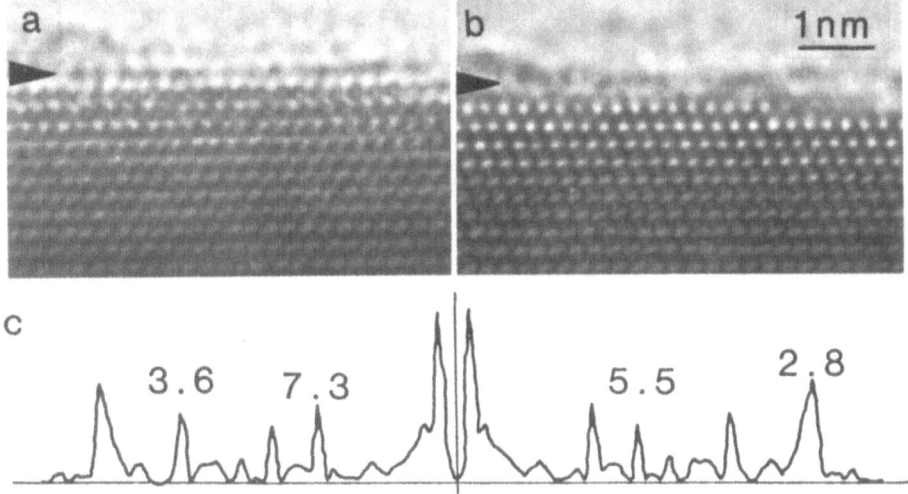

Fig. 3.16. Image of a region of gold(111) surface which has an ordered carbonaceous over-layer, arrowed, which is believed to be benzene [3.24]. At the focus of –60 nm in (a) both can be seen, whilst the carbon vanishes at the defocus in (b) of –100 nm. Shown in (c) is a digital power spectrum showing the spacings present in the carbonaceous layer. Note the 7.3 Å Van der Waals benzene periodicity

and, in practise, cannot be differentiated, as shown in the experimental image (Fig. 3.15). If there is a noticeable difference, the elements can be differentiated. A focal series is sufficient, in principle, to determine the amplitude and phase of the wave in real space, therefore it also determines $\tilde{v}_a(\boldsymbol{u})$ and $\tilde{v}_b(\boldsymbol{u})$. For instance, gold and carbon have different scattering potentials which is manifest by very low contrast for the carbon at a defocus of –100 nm, as shown in Fig. 3.16. As a rule, a different focal behaviour at the surface from that of the bulk crystal will be an indication of a change in chemical composition.

3.4 A Note of Warning

Through much of the analysis herein we have been dealing with the surface profile imaging results based upon our experience with gold surfaces. This is a particularly simple system, provided it is oriented correctly and relatively thin. To a good approximation the results are independent, or only very slowly dependent upon the thickness. Hence it is straightforward and simple to obtain faithful images.

It would be wrong to conclude that other systems will be just as simple. Materials with large unit cells are favourites of high-resolution electron micro-scopists, and these suffer from a number of problems associated with alignment of the microscopes [3.72]. It may prove difficult, or perhaps impossible, to simultaneously obtain localised conditions for the large number of different spacings in such a material. As yet no systematic study of surface imaging using a large-

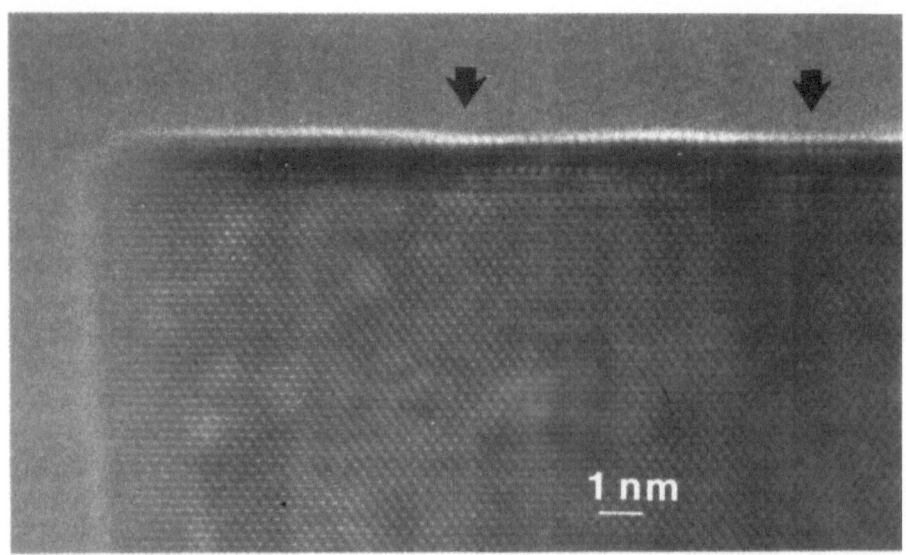

Fig. 3.17. Image of an MgO surface on a (110) zone, courtesy of K. Tanji. The crystal is cubic with (100) faces, so there is a wedge of half angle 45°. Near the extinction contours (*arrowed*) severe distortions of the surface fringe structure are evident

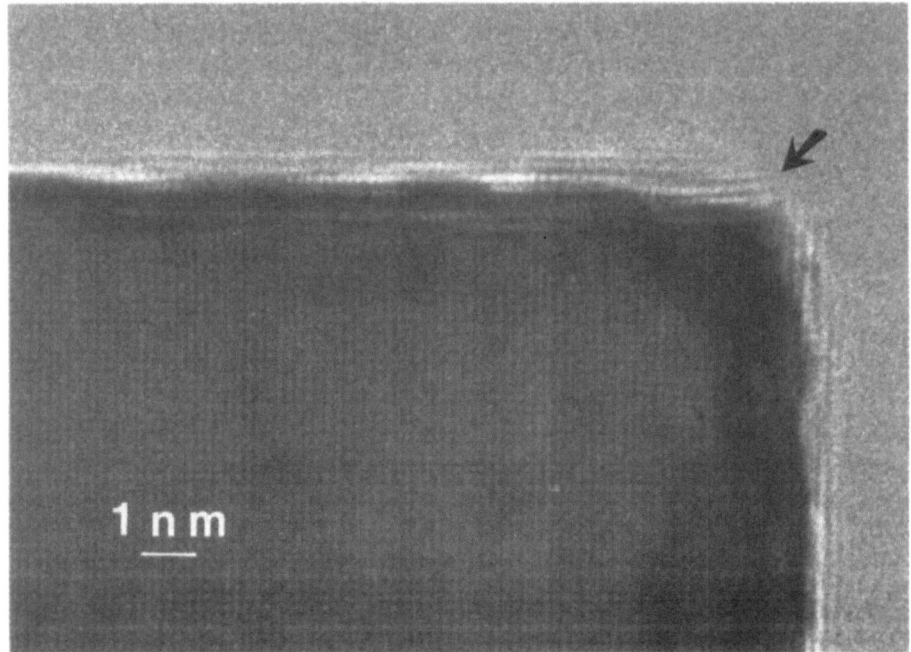

Fig. 3.18. Image of an MgO surface on a (100) zone, courtesy of K. Tanji. The fringes shown are the 0.2 nm (200) lattice fringes. For the microscope used, a JEOL 200CX, no well localised defocus exists for these spacings. Structure representing surface steps is apparent, as are severe artifacts (*arrowed*)

cell specimen has been carried out, so we must await future work to answer some of these questions.

In many materials the thickness dependence of the diffraction is also substantially stronger than it is for gold. Here, and for any thicker specimens or those incorrectly oriented the diffraction localisation may break down. One particular artifact is worth specific mention, namely the effect of extinction contours. These are due to the oscillatory nature of the scattering; the intensity of any diffracted beam reaches a maximum at some thickness and then decreases. At the low point there is often a 90°phase change which leads to a half-spacing shift in the lattice fringes. At the surface this leads to unusual effects, an example being shown in Fig. 3.17.

Another effect is also noteworthy. Fringes representing the atomic structure can be obtained for a non-localised defocus. For instance, very fine fringes can often be obtained which are not localised at any choice of lens defocus. Experimentally these still seem to represent some of the surface features such as surface steps, an example being shown in Fig. 3.18. There are, however, substantial distortions to the fringes, primarily around the steps. These are almost certainly an artifact of the imaging system.

3.5 Gold(110)

In this subsection we illustrate the use of the surface profile imaging technique in both qualitative and quantitative modes by discussing the results for gold(110) surfaces.

When carbon covered, the (110) surface displays few remarkable features (for instance, Fig. 3.19). Digital analyses imply the possibility of a marginal normal contraction, although this is below the minimal error bar of 5% for any quantitative measurements. The only interesting phenomenon is the presence of strong (100) fringes, particularly for specimens tilted slightly away from the true (110) zone-axis orientation (Fig. 3.20). The (100) fringes are forbidden in the bulk of an fcc material, but become allowed at a surface.

During the cleaning procedure by electron-beam-induced etching of carbonaceous contaminants by water vapor [3.25], there is comparatively little change, exceedingly small compared to the very large changes observed on (111) surfaces [3.23,25]. One feature similar to the (111) surface is the persistence of approximately a monolayer of carbonaceous material, probably some complicated hydrocarbons. These can be distinguished by their focal dependence, being almost invisible for a defocus of –100 nm but clearly visible at a defocus of –60 nm. An example is shown in Fig. 3.21. (Gold is evident at both defoci [3.24,25]).

When clean, the behaviour of the surface is fairly complicated. Locally one can find areas which show the 2×1 missing row reconstruction (Fig. 3.22). Digital line traces (Fig. 3.23) normal to the surface at different locations show some interesting features. The first intensity minimum at approximately the location

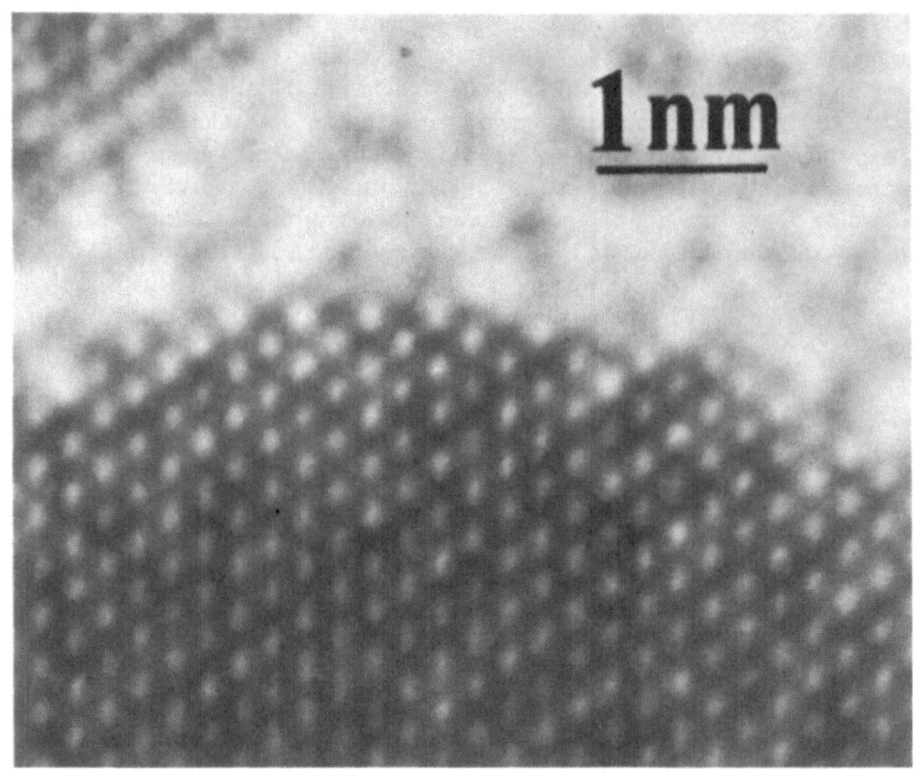

Fig. 3.19. Area of gold(110) surface on a small particle covered in amorphous carbon, taken from [3.20]. The atomic columns are white, with perhaps a minor surface contraction

(001)

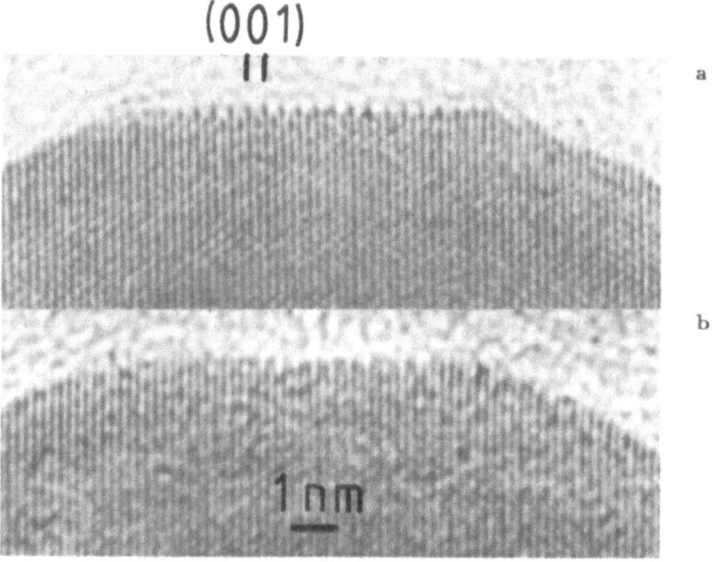

a

b

Fig. 3.20. Area of gold(110) surface for defoci of −60 and −100 nm in (a) and (b), respectively, tilted slightly off the (110) zone axis orientation. Strong (001) fringes can be seen at the surface

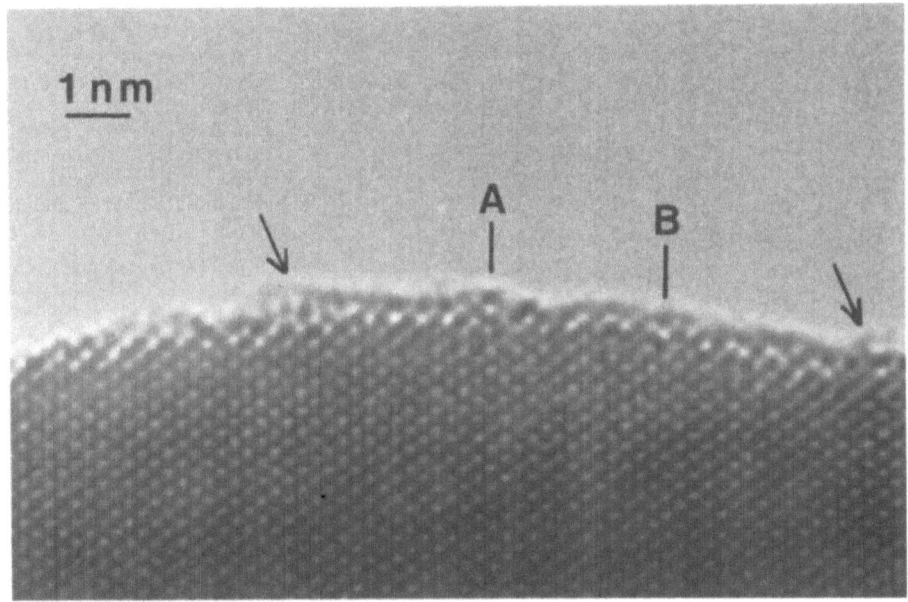

Fig. 3.21. Area of gold(110) surface with about a monolayer of carbonaceous covering (*arrowed*). Atomic columns are black. Two atomic columns labelled *A* and *B* can also be seen in Fig. 3.25 when the surface is clean

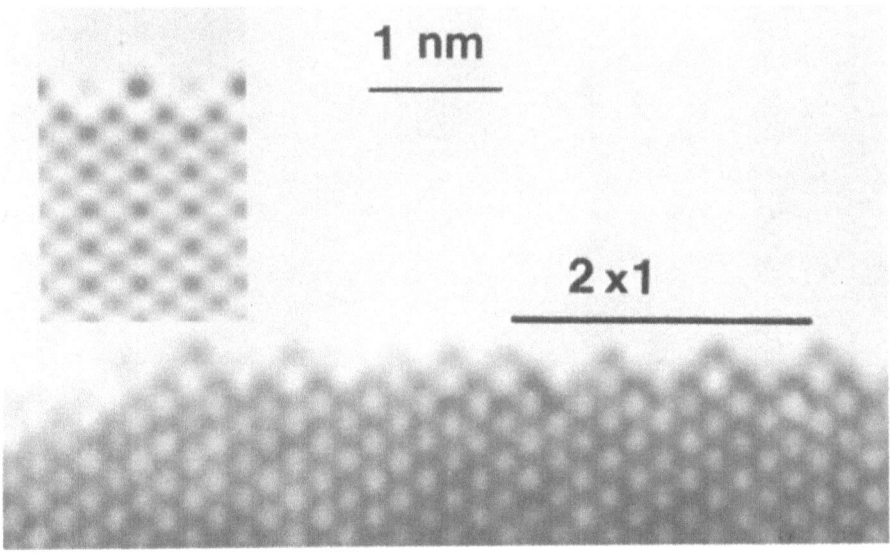

Fig. 3.22. Area of gold(110) surface which locally shows the 2×1 missing row structure, with an image simulation inset [3.20]

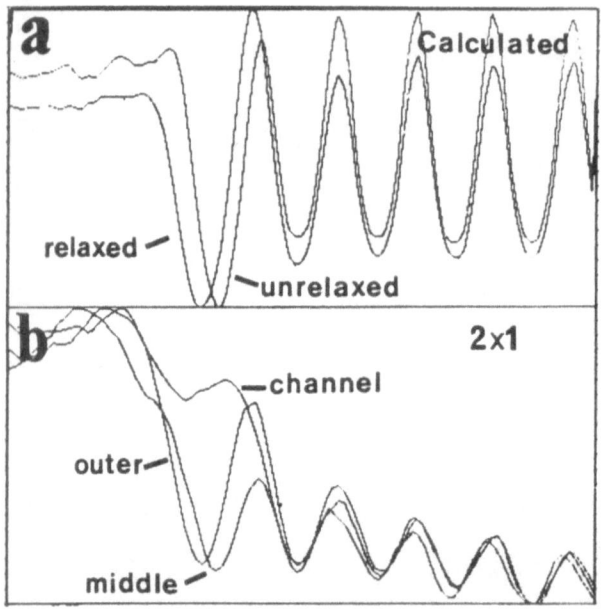

Fig. 3.23. Digital line trace analysis normal to the surface. In (**a**), calculated results for zero, and a 23.5% outwards expansion. In (**b**), experimental results from the outermost atoms of the corrugated surface, middle and the channels [3.20]

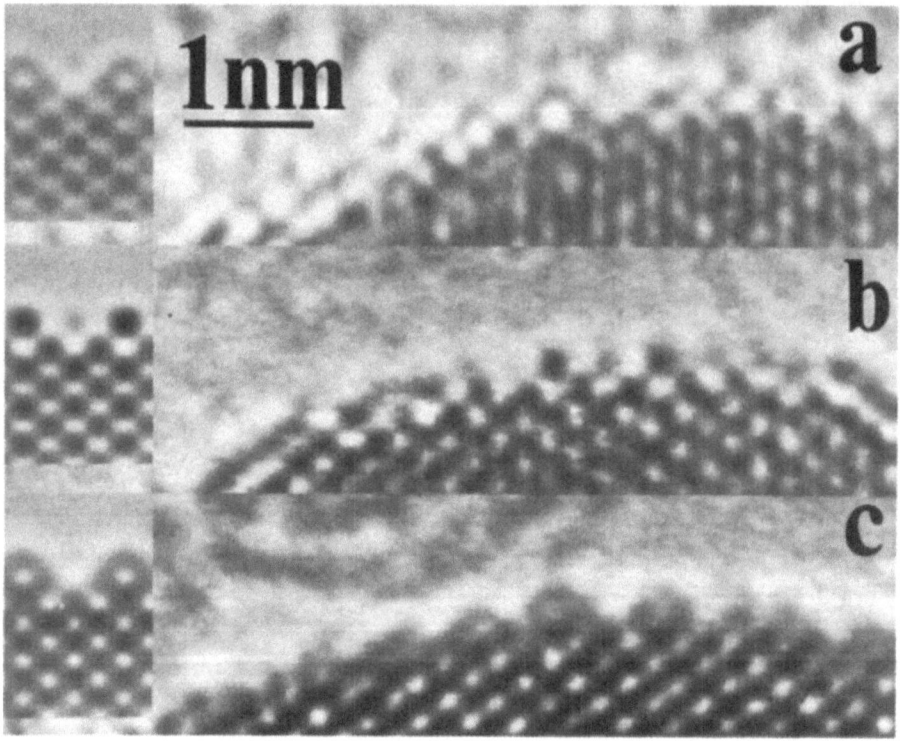

Fig. 3.24. Match between the experimental and calculated results, the latter inset on the left for defoci of −20, −60 and −100 nm in (**a–c**) [3.20]

of the outermost column of atoms is displaced normal to the surface by 0.06 nm. There is also a substantial rise in the strength of the first maximum just inside this. The two are characteristics of a large, normal expansion. However, it is not valid to simply read the lattice location from the images because of Fresnel-like effects at a surface which lead to an essentially structure-dependent static displacement [3.24]. Rigorously, it is necessary to compare the experiments with calculated images. These indicate that the true displacement is 0.057 nm, with an error of ±0.014 nm. The source of these error bars is primarily shot-noise since there are comparatively few electrons within each pixel when the image is digitised. In principle, with an extensive homogeneous surface it would be possible to remove shot-noise effects. Then other, secondary criteria such as the precise lens defocus would become important.

However, it is not a good idea simply to fit to one defocus value. A more general criterion is to fit a focal series. An example is shown in Fig. 3.24. One feature to note is the strong grey halo around the white central image of the atomic columns for the defoci of about −20 nm and about 100 nm in both the experimental and theoretical images. These are another characteristic of a large outward expansion.

Along similar lines we can also determine the normal expansion from local regions of 1×1 surface. From the line traces this is 0.029 nm±0.014 nm.

Turning next to the overall picture for the (110) surface this is substantially more complicated. Large areas of such surface are not flat, but quite rough and inhomogeneous (Fig. 3.25). This is not particularly surprising. On fcc crystals large stable (111) and (100) facets exit, but only relatively small (311) and

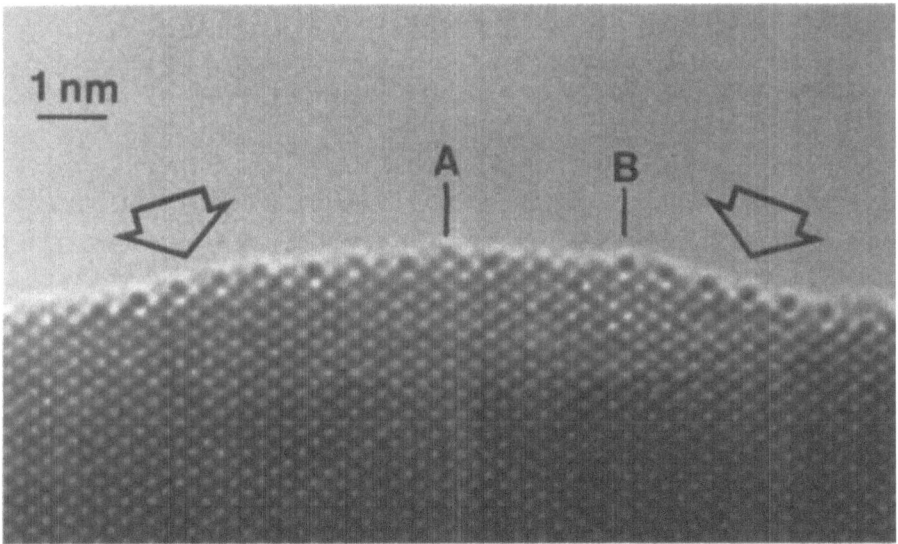

Fig. 3.25. Area of clean gold(110) surface, in fact the same as that shown in Fig. 3.21 at a later stage, for a defocus of −60 nm with the atomic columns black. Columns labelled A and B as in Fig. 3.21. Note the (311) microfacets arrowed [3.27]

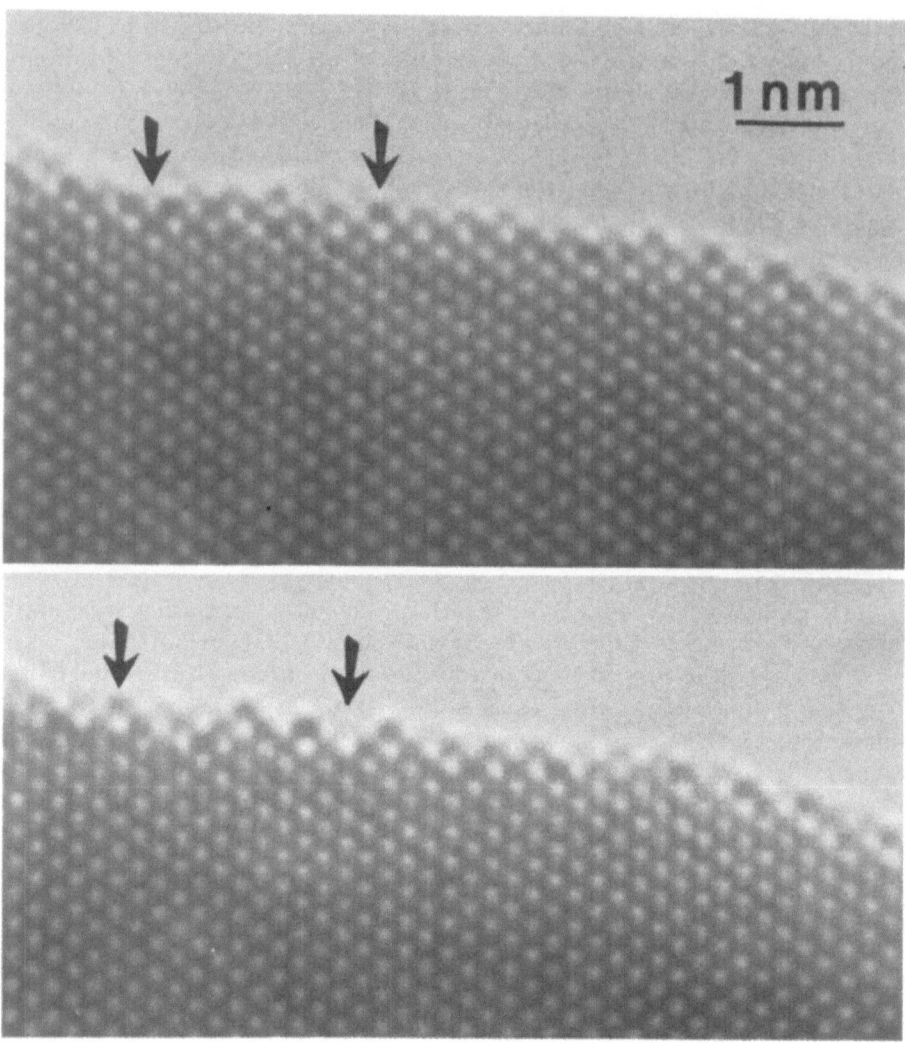

Fig. 3.26. Images of a region of a gold(110) surface taken about 10 seconds apart for black contrast of the atomic columns. Some of the atomic columns (*arrowed*) have moved

(110) facets [3.73,74]. The remainder of the surface in an equilibriated crystal is made up of vicinal, essentially curved surfaces. Indeed, the surfaces of small gold particles can be fitted quite well using only (111) and (100) facets with a little additional step rounding [3.75,76].

Thus the existence of an exceedingly inhomogeneous surface is not surprising. Furthermore, it is not clear if flatter surfaces can be obtained by annealing such surfaces in UHV. At, say, 300°C there is substantial sublimation of gold favouring the kinetically stable (111) faces. Thermodynamic stability is only attained by annealing in equilibrium with the vapor [3.77].

On this surface one can see local areas of 2×1 structure and local regions of 1×1. These generally show evidence for large normal expansions. It seems these normal expansions are more a characteristic of the (111)-like steps on the surface [3.78], and they are also clearly evident at local regions of (311) surface (for instance Fig. 3.25). However, the surface is also quite mobile with atomic columns changing in position frequently between micrographs, as illustrated in Fig. 3.26. This poor order along the (001) direction is also apparent from x-ray work [3.79,80], which indicates that the order along the beam direction of the electron micrographs is exceedingly good. What appears to be happening is column hopping along the open (001) direction. A recent model of Au and Pt(110) surfaces as essentially a two-dimensional gas which will display 2×1 ordering [3.81,82] seems to be an excellent description of the system.

Summarising, the (110) gold surface is rough along (001) with large normal expansions. It does not seem to be particularly stable, tending to decompose into (111) microfacets [3.83], consistent with the marginal stability of (110) surfaces [3.73,74]. This comes from qualitative analyses. Quantitative work indicates that these are large normal expansions associated with the (111)-like steps on the surface.

3.6 Other High-Resolution Techniques

Whilst the profile method of high-resolution surface imaging has already proven its worth, it would be wrong to neglect alternative approaches. As yet these have not achieved so much impact, but this is likely to change over the next few years. Two techniques would appear to be promising, namely high-resolution imaging with the beam normal to the surface and transmitted through the specimen, and high-resolution imaging in a reflection incidence configuration. We will briefly discuss these techniques in this section.

3.6.1 Transmission Geometry

The essence of this approach is to use the surface diffraction at the top and bottom surfaces of a thin film to produce an image, as illustrated in Fig. 3.27. The image is formed by interfering the surface superlattice reflections to produce what is best thought of as a representative signal, rather than a true image of the surface. The big gain over simply collecting the diffraction pattern is that the signal is sensitive to the relative phases of the different superlattice beams. Knowledge of the superlattice spacing only gives the size of the surface unit cell, not a-priori any information upon its contents. This phase information refines to some extent at least the contents of the cell.

Unfortunately there are two important problems with this approach, and two major limitations. The problem is that the signal levels are small, and the images are insensitive to any surface contaminants. A layer of low atomic weight adatoms would produce only minimal changes. In principle, this could be overcome by quantitative image analysis including checks that there were no

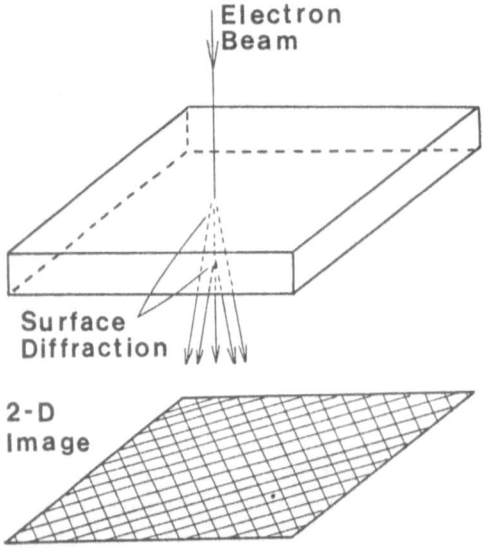

Electron Beam

Surface Diffraction

2-D Image

Fig. 3.27. Diagram to illustrate the transmission approach to surface imaging

Fig. 3.28. High-resolution image representing the structure of the Au(111) surface, courtesy of *Nihoul* [3.84]

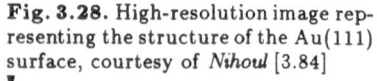

1 n̲m

additional features present in the images. This would be difficult. In particular, the devil's advocate case when the adsorbed layer had the same structure as that of the surface would be almost impossible to detect. This could be overcome with an Auger spectrometer to monitor the surface cleanliness, which may be practicable in a few years time.

The second problem is that the image will be sensitive to both the top and bottom surfaces. This ambiguity may lead to some awkward questions in the image interpretation.

The two limitations are uncertainties in the registry of the surface with the bulk, and any normal relaxations. To determine the registry one needs images which show both the bulk and the surface superlattice. Except for heavy atoms on a light substrate, the low surface signal level will make this almost impossible. The normal spacing poses a problem which is essentially insoluble. Even a change as large as 0.05 nm will almost certainly not affect the experimental results.

At present the most detailed study using this technique has been the work of *Nihoul* et al. [3.84]. They succeeded in obtaining representative images of the Au(111) 23×1 surface (Fig. 3.28). Although the results indicate a small contraction and a surface shear, the agreement between the experimental and calculated images was not perfect. However, the prospects look good for the future.

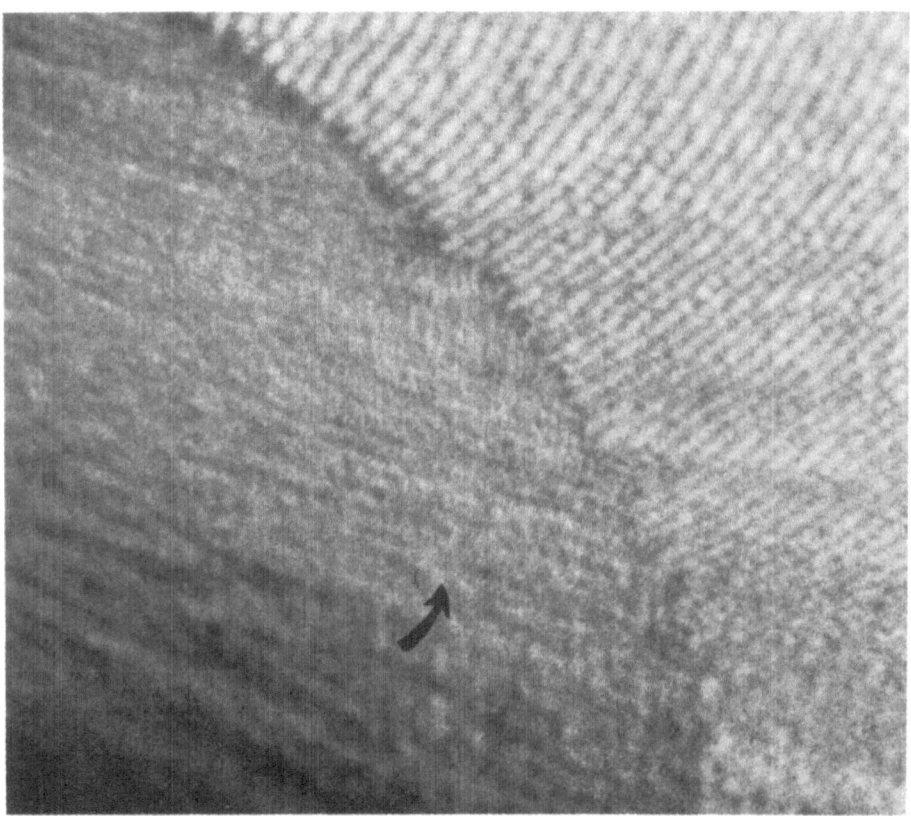

Fig. 3.29. Reflection electron microscopical image from a vicinal gold surface showing 0.9 nm steps (*arrowed*). Courtesy of *Tung Hsu* [3.85]

3.6.2 Reflection Geometry

The second geometry is the same as that used for the profile technique, except that it employs bulk specimens and uses the reflected, RHEED, electrons to produce an image. Until very recently it was thought that chromatic aberrations, primarily due to surface plasmon excitations, would prevent high-resolution images from being obtained. The experiments have at least partially disproved this. *Tung Hsu* [3.85,86] has succeeded in obtaining reflection images of steps on a gold surface separated by about 1 nm (Fig. 3.29). More recently *Yagi* et al. [3.87] have obtained reflection images using the 2.35 nm 7×7 superlattice beams on Si(111). Although, in principle, the theory exists for interpreting such images, how to use it in practise has yet to be worked out. This, and perhaps images at better resolutions, are subjects for future research.

3.7 Conclusions and the Future

We have tried herein to give a feel of high-resolution electron microscopy, primarily the profile imaging technique. Whilst there are still a number of unsolved problems in, for instance, the specimen preparation, the technique holds considerable promise for the future. Perhaps the most important impact will not be the determination of the atomic structures of homogeneous surfaces to tenths of an Angstrom precision, but rather as a bridge between structural surface science on macroscopic crystals and real system such as heterogeneous catalysts. High-resolution imaging, like any imaging technique, is at its best for inhomogeneous surfaces when averaging, diffraction methods run into problems. It will be interesting to see over the next few years how well the assumption inherent in so many surface science techniques that the average structure is a valid representation will hold up.

Looking towards the future, generalisations except that progress will be rapid are somewhat suspect. A number of electron microscope manufacture are at present tooling up to produce UHV high-resolution microscopes. Thus the more chemically reactive materials should become accessible over the next few years. There is also quite a substantial growth in techniques at present, for instance, surface convergent beam diffraction [3.88] and reflection lattice imaging [3.85–87]. A technique which may prove useful is to monitor the energy loss at a surface [3.89,90], perhaps eventually to obtain information on the surface states (both their average and their spatially varying nature). An electron microscope is a versatile multitechnique instrument, and we can hope in the future to be able to cross-check conclusions using a variety of methods. Furthermore, at least in principle, one can hang other spectroscopical techniques on to the microscope, for instance, an Auger spectrometer. The future will tell.

Acknowledgements. This work would have been impossible without extensive collaboration with Dr. David J. Smith. The author would also like to thank, in particular, Prof. J.M. Cowley, Prof. V. Heine, Drs. A. Howie, R.F. Willis, M.A. O'Keefe, W.O. Saxton and G.J. Wood for numerous discussions and suggestions and Robyn Santucci for her patient typing. The author would also like to acknowledge SERC (UK) for financial support of the Cambridge High-Resolution Electron Microscope.

References

3.1   P. Hirsch, A. Howie, R.B. Nicholson, D.W. Pashley, M.J. Whelan: *Electron Microscopy of Thin Crystals* (R.E. Krieger, New York 1977)
3.2   J.M. Cowley: *Diffraction Physics* (North-Holland, Oxford 1981)
3.3   U. Valdre, R. Ruedl (eds.): *Electron Microscopy in Materials Sciences* (Commission of the European Communities, Luxembourg 1976) Vols. I–III
3.4   L. Reimer: *Transmission Electron Microscopy*, Springer Ser. Opt. Sci., Vol. 36 (Springer Berlin, Heidelberg 1984);
        L. Reimer: *Scanning Electron Microscopy*, Springer Ser. Opt. Sci., Vol. 45 (Springer, Berlin, Heidelberg 1985)
3.5   R.H. Geiss (ed.): *Analytical Electron Microscopy* (Plenum, New York 1981)
3.6   J.I. Goldstein, J. Hren, D. Joy (eds.): *Introduction to Analytical Electron Microscopy* (Plenum, New York 1979)
3.7   C.J. Humphreys: Rept. Progr. Phys. **42**, 122 (1979)
3.8   J.C.H. Spence: *Experimental High-Resolution Electron Microscopy* (Clarendon, Oxford 1981)
3.9   K. Yagi, K. Takayanagi, K. Kobayashi, N. Osakabe, Y. Tanishiro, G. Honjo: Surf. Sci. **86**, 174 (1979)
3.10   N. Osakabe, Y. Tanishiro, K. Yagi, G. Honjo: Surf. Sci. **97**, 393 (1980)
3.11   K. Takayanagi: J. Microsc. **136**, 287 (1984)
3.12   K. Takayanagi, K. Yagi: Trans. Japan Inst. Metals **24**, 337 (1983)
3.13   A.J. Forty: Contemp. Phys. **24**, 271 (1983)
3.14   J. Venables: Ultramicrosc. **7**, 81 (1981)
3.15   A. Howie: In *Electron Microscopy and Analysis 1981*, ed. by M.J. Goringe (Institute of Physics, London 1982)
3.16   The articles by a variety of authors in Chemica Scripta **41** (1978/79)
3.17   W. Neumann, M. Pasemann, J. Heydenreich: In *Crystals, Growth Properties and Applications*, Vol. 7 (Springer, Berlin, Heidelberg 1982) p. 1
3.18   D.J. Smith: Helvetica Physica Acta **56**, 464 (1983)
3.19   U. Gonser (ed.): Topics Current Phys., Vol. 40 (Springer, Berlin, Heidelberg 1986)
3.20   L.D. Marks, D.J. Smith: Nature **303**, 316 (1983)
3.21   L.D. Marks: Phys. Rev. Lett. **51**, 1000 (1983)
3.22   D.J. Smith, L.D. Marks: Proc. 7th Intern. Conf. High Voltage Electron Microscopy, ed. by R.M. Fisher, R. Gronsky, K.H. Westmacott (Lawrence Berkeley Lab., Berkeley CA 1983) p. 53
3.23   L.D. Marks, V. Heine, D.J. Smith: Phys. Rev. Lett. **52**, 656 (1984)
3.24   L.D. Marks: Surf. Sci. **139**, 281 (1984)
3.25   L.D. Marks, D.J. Smith: Surf. Sci. **143**, 495 (1984)
3.26   D.J. Smith, L.D. Marks: 12th European Congr. on Electron Microscopy, Budapest, Hungary (1984) p. 599
3.27   D.J. Smith, L.D. Marks: Ultramicrosc. **16**, 101 (1985)
3.28   L.D. Marks, D.J. Smith: In *Catalyst Characterization Science*, ed. by M. Deviney, J. Gland, ACS Symposium Series 288 (1985) p. 341
3.29   D.J. Smith, L.D. Marks: Surf. Sci. **157**, 2367 (1985)
3.30   D.J. Smith, L.D. Marks: Mat. Res. Soc. Symp. Proc. **41**, 129 (1985)
3.31   D.J. Smith, R.A. Camps, V.E. Cosslett, L.A. Freeman, W.O. Saxton, W.C. Nixon, H. Ahmed, C.J.D. Catto, J.R.A. Cleaver, K.C.A. Smith, A.E. Timbs: Ultramicroscopy **9**, 203 (1982);
        D.J. Smith, R.A. Camps, L.A. Freeman, R. Hill, W.C. Nixon, K.C.A. Smith: J. Microscopy **130**, 127 (1983)

3.32   J.L. Hutchinson, N.A. Briscoe: ULtramicrosc. **18**, 435 (1985)
3.33   S. Iijima: Japan J. Appl. Phys. **23**, L347 (1984)
3.34   S. Iijima: Surf. Sci. **156**, 1003 (1985)
3.35   A.K. Petford, L.D. Marks, M. O'Keeffe: Surf. Sci., in press
3.36   L.D. Marks, A.K. Petford, M. O'Keeffe: In Proc. 43, EMSA Meeting, Louisville, Kentucky (San Francisco Press, CA 1985) p. 266
3.37   D.J. Smith, L.A. Bursill, L.D. Marks, P.J. Lin: In Proc. 43, EMSA Meeting, Louisville, Kentucky (San Francisco Press, CA 1985) p. 256
3.38   D.J. Smith: Private communication
3.39   P.W. Hawkes (ed.): *Magnetic Electron Lenses*, Topics Current Phys., Vol. 18 (Springer, Berlin, Heidelberg 1982)
3.40   K.H. Herrman, D. Krahl, H.P. Rust: Ultramicrosc. **3**, 227 (1978)
        C.J.D. Catto, K.C.A. Smith, W.C. Nixon, S.R. Erasmus, D.J. Smith: In *Electron Microscopy and Analysis 1981*, ed. by M.J. Goringe (Inst. Physics, London 1982) p. 123
3.41   H.M. Smith (ed.): *Holographic Recording Materials*, Topics Appl. Phys., Vol. 20 (Springer, Berlin, Heidelberg 1977)
3.42   A.M. Glauert (ed.): *Practical Methods in Electron Microscopy* (North-Holland, Amsterdam 1974)
3.43   N.H. Tolk, M.M. Traum, J.C. Tully, T.E. Madley (eds.): *Desorption Induced by Electronic Transitions DIET I*, Springer Ser. Chem. Phys., Vol. 24 (Springer, Berlin, Heidelberg 1983);
        W. Brenig, D. Menzel (eds.): *Desorption Induced by Electronic Transitions DIET II*, Springer Ser. Surf. Sci., Vol. 4 (Springer, Berlin, Heidelberg 1985);
        M.L. Knotek: Phys. Today **37**, 24 (1984)
3.44   K. Fujiwara: J. Phys. Soc. Japan **17** (Suppl. BII,118 (1962)), 618
3.45   A. Howie: In the discussion following Ref. [3.44]
3.46   M.V. Berry: J. Phys. C**4**, 697 (1971);
        K. Kambe, G. Lehmpfuhl, F. Fujimoto: Z. Naturforsch. **29a**, 1034 (1974);
        A.M. Ozori de Almeida: Acta Crysta. A**31**, 435 (1973)
3.47   G. Kurizki: Phys. Rev. B**33**, 49 (1986)
3.48   J. Van Bladel: *Relativity and Engineering*, Springer Ser. Electrophys., Vol. 15 (Springer, Berlin, Heidelberg 1984)
3.49   H.L. Skriver: *The LMTO Method*, Springer Ser. Solid-State Sci., Vol. 41 (Springer, Berlin, Heidelberg 1984)
3.50   L.I. Schiff: *Quantum Mechanics* (McGraw-Hill Kogakusha Ltd., Tokyo 1968)
3.51   D. Gratias, R. Portier: Acta Crysta. A**39**, 576 (1983)
3.52   L.D. Marks: Ultramicrosc. **14**, 351 (1984)
3.53   M.A. O'Keefe: In Proc. 37th. Ann. EMSA Meeting, San Antonio, TX (Claitors, Baton Rouge, LA 1979) p. 556
3.54   W.O. Saxton: J. Microsc. Spectrosc. Electron. **5**, 55 (1980)
3.55   K. Ishizuka: Ultramicrosc. **5**, 55 (1980)
3.56   M.A. O'Keefe, W.O. Saxton: In Proc. 41st Ann. EMSA Meeting, Phoenix, Az. 1983 (Claitors, Baton Rouge, LA 1983) p. 288
3.57   L.D. Marks: Ultramicrosc. **12**, 237 (1984)
3.58   L.D. Marks: Ultramicrosc. **18**, 33 (1985)
3.59   J.C.H. Spence: In Proc. 42nd Ann. EMSA Meeting, San Francisco, CA 1984 (San Francisco Press, San Francisco, CA 1984) p. 364
3.60   J.M. Cowley, A.F. Moodie: Acta Cryst. **10**, 609 (1967)
3.61   P. Goodman, A.F. Moodie: Acta Cryst. A**30**, 280 (1974)
3.62   P.G. Self, M.A. O'Keefe, P.R. Buseck, A.E.C. Spargo: Ultramicrosc. **11**, 35 (1983)
3.63   P.J. Smith, G. Lehmpfuhl: Acta. Cryst. A**31**, 5220 (1975)
3.64   A. Howie: Proc. Roy. Soc. A**271**, 268 (1963)
3.65   H. Raether: *Excitation of Plasmons and Interband Transitions by Electrons*, Springer Tracts Mod. Phys. (Springer, Berlin, Heidelberg 1980)
3.66   P. Rez, C.J. Humphreys, M.J. Whelan: Phil. Mag. **35**, 81 (1977);
        Y.H. Otsuki: *Charged Beam Interactions with Solids* (Taylor and Francis, Philadelphia 1983)
3.67   H.J. Nussbaumer: *Fast Fourier Transform and Convolution Algorithms* , 2nd ed., Springer Ser. Inf. Sci., Vol. 2 (Springer, Berlin, Heidelberg 1982)

3.68   P. Rose: Ultramicrosc. **1,** 167 (1975)

3.69   W.O. Saxton: *Computer Techniques for Image Processing in Electron Microscopy* (Academic, New York 1978)

3.70   D.L. Misell: *Image Analysis, Enhancement and Interpretation* (North Holland, Amsterdam 1978)

3.71   P.W. Hawkes (eds.): *Computer Processing of Electron Microscope Images*, Topics Current Phys., Vol. 13 (Springer, Berlin, Heidelberg 1980).

3.72   D.J. Smith, L.A. Bursill, G.J. Wood: Ultramicrosc. **16,** 19 (1985)

3.73   M. Flytzani-Stephanopoulos, L.D. Schmidt: Prog. Surf. Sci. **9,** 83 (1979)

3.74   M. Drechsler: In *Surface Mobilities on Solid Materials*, ed. by V.T. Binh (Plenum, New York 1983) p. 405

3.75   L.D. Marks: J. Crystal Growth **61,** 556 (1983)

3.76   L.D. Marks: Phil. Mag. A**49,** 81 (1984)

3.76   J.J. Metois, G.D. Spiller, J.A. Venables: Phil. Mag. A**46,** 1015 (1982)

3.78   V. Heine, L.D. Marks: Surf. Sci. **165,** 65 (1986)

3.79   I.K. Robinson: Phys. Rev. Letts. **50,** 1145 (1983)

3.80   I.K. Robinson, Y. Kuk, L.C. Feldman: Phys. Rev. B**29,** 4762 (1984)

3.81   D. Wolf, H. Jagodzinski, W. Moritz: Surf. Sci. **77,** 283 (1978)

3.82   J.C. Campuzano, A.M. Lahee, G. Jennings: Surf. Sci. **152/153,** 68 (1985)

3.83   G. Binnig, H. Rohrer, Ch. Gerber, E. Weibel: Surf. Sci. **131,** L379 (1983)

3.84   G. Nihoul, K. Abdelmoula, J.J. Metois: Ultramicrosc. **12,** 353 (1983–1984)

3.85   T. Hsu: Ultramicrosc. **11,** 167 (1983)

3.86   T. Hsu, S. Iijima, J.M. Cowley: Surf. Sci. **137,** 551 (1984)

3.87   N. Shimizu, Y. Tanishiro, K. Takayanagi, K. Yagi: Ultramicrosc. **18,** 453 (1985)

3.88   J.A. Eades, M.D. Shannon, M.E. Meichle: Proc. 42nd Ann. EMSA Meeting, San Francisco, CA 1984 (San Francisco Press, San Francisco, CA 1984) p. 516

3.89   A. Howie: Ultramicrosc. **11,** 141 (1983)

3.90   R.B. Milne, A. Howie: J. Microsc. **136,** 279 (1984)

4. Surface Channeling and Its Application to Surface Structures and Location of Adsorbates

C. Varelas

With 31 Figures

In surface channeling swift light ions incident, at a grazing angle, on the crystal surface along a low-index direction are utilized as versatile probes for surface analysis. We review the development of this field and the application of the surface channeling technique in surface structure analysis, surface reconstruction, and location of adsorbed atoms.

4.1 Overview

A beam of monoenergetic, charged particles like protons or heavy ions ("projectiles"), incident upon a crystalline target penetrates much deeper into the bulk if aligned with density packed (i.e., low index) lattice directions, than at random incidence. This phenomena is well known as channeling [4.1–4]. The basic principle of channeling rests on the *steering action* by the regular array of lattice atoms. Fast ions incident towards a low-index lattice direction experience strong steering by the potential of the atom rows ("axial channeling"). Yet, this effect is not limited to the bulk. As well, axial channeling is possible for fast ions incident at a grazing angle on the crystal surface along a low-index surface direction. The scattering of the ions arises now from successive, correlated collisions with many atoms along the low-index surface direction ("surface axial channeling" or "surface channeling" [4.5]).

In the following, the basic ideas of lattice steering, as far as they are relevant for surface axial channeling, are outlined in Sect. 4.2. Section 4.3 is devoted to calculation of the Auger-electron yields induced by surface channeled ions. Section 4.4 discusses the realization of experiments. The application of surface channeling in surface structure analysis and location of adsorbed atoms is reported by a number of selected examples in Sect. 4.5. Finally, in Sect. 4.6 the surface sensitivity and the applicability of axial channeling are examined and the effect of thermal vibrations and surface steps are treated.

4.2 Principles

4.2.1 Continuum Potential

The important point by axial channeling is that the successive repulsive actions of a row of target atoms on the fast moving ion can be approximated by a smooth (continuum) potential U(r) provided the distance of closest approach towards any individual atom is larger than a certain critical distance r_c. This can be understood from the following arguments.

It is sufficiently accurate to use binary interaction potentials between projectile and target atoms, conveniently of the Thomas-Fermi type

$$V(R) = (Z_1 Z_2 e^2/R)\phi(R/a_{TF}). \tag{4.1}$$

Z_1 and Z_2 are the atomic numbers of projectile and target atom, respectively,

$$a_{TF} = 0.8853\, a_0/(Z_1^{1/2} + Z_2^{1/2})^{2/3} \quad \text{for heavy ions}, \tag{4.2a}$$

$$a_{TF} = 0.8853\, a_0/Z_2^{1/3} \quad \text{for protons} \tag{4.2b}$$

is the Thomas-Fermi screening radius, $a_0 = 0.0529\,nm$ the Bohr-radius, and $\phi(R/a_{TF})$ the Thomas-Fermi screening function [4.6]. The total interaction between a projectile at position $\boldsymbol{R}$ and the target surface atoms is taken as the sum over all individual binary interactions

$$U_{surf}(\boldsymbol{R}) = \sum_{\boldsymbol{T}} V(\boldsymbol{R} - \boldsymbol{T}), \tag{4.3}$$

where the $\boldsymbol{T}$'s are the translation vectors of the surface structure. In axial channeling we are interested in the behavior of atomic rows of the lattice. Therefore, we split $U_{surf}(\boldsymbol{R})$ up in an appropriate sum over atomic rows [hkl] which are parallel to the projective direction. Such an axial potential is given by

$$U_{[hkl]}(r, z) = \sum_{n=-\infty}^{+\infty} V\left(\sqrt{r^2 + (z - nd_{[hkl]})^2}\right), \tag{4.4}$$

where r and z are cylinder coordinates, with r being the radial distance from the row axis and z the axial direction (Fig. 4.1). $d_{[hkl]}$ is the interatomic spacing along the atomic row [hkl]. We assume, for simplicity, a primitive row with all its atoms identical and with an uniform interatomic spacing. For non primitive structures, see [4.7].

$U_{[hkl]}$ is a periodic function in z and, therefore, can be expanded in a Fourier series with the propagation constant $k = 2\pi/d_{[hkl]}$

$$U_{[hkl]}(r, z) = (1/d_{[hkl]}) \int_{-\infty}^{+\infty} V(\sqrt{r^2 + \xi^2})\,d\xi$$

$$\times \left(1 + 2\sum_{n=1}^{\infty} \cos nk\xi \cdot \cos nkz\right). \tag{4.5}$$

The terms containing $\cos nkz$ vary between +1 and −1 along z. Their contribution practically cancels if the projectile is only slightly deflected $(dr/dz \ll 1)$ over the periodicity length $d_{[hkl]}$. This will be fulfilled only for $r > r_c$, where the critical value of approach r_c between projectile and target atoms comes out to

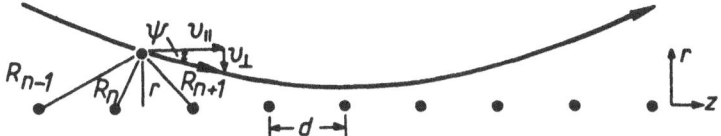

Fig. 4.1. Trajectory of a projectile scattered at a rigid atomic row

be for monotonously decreasing potentials [4.4,7]:

$$r_c = (Z_1 Z_2 e^2 d_{[hkl]}/E)^{1/2} \{1 + \log[0.2 E a_{TF}^2/(Z_1 Z_2 e^2 d_{[hkl]})]\}, \qquad (4.6)$$

where E is the kinetic energy of the projectile. Thus, for $r > r_c$ the only term contributing on the average in (4.5) is the zero term of the Fourier series

$$U_{[hkl]}(r) = (1/d_{[hkl]}) \int_{-\infty}^{+\infty} V(\sqrt{r^2 + \xi^2}) d\xi \qquad (4.7)$$

which is independent of the z coordinate. That means for $r > r_c$ the discreteness of the lattice structure along the channeling direction can be disregarded. This is physically justified by the fact that the steering action of the crystal comes by a large number of correlated collisions with a row of atoms along the z direction. The approximation is the same as in the transition from the lattice theory of crystal elasticity to a continuum theory. For this reason $U_{[hkl]}(r)$ is generally denoted as a *continuum row potential*. It must be emphasized that in the directions perpendicular to the channeling direction the discreteness of the crystal structure has been retained. In the case of $r < r_c$, the z dependence cannot be neglected; then the projectile "feels" the individual atomic potentials.

So far the target atoms are considered static and exactly placed at lattice sites. Taking into account their thermal vibrations, r_c becomes essentially the value of the thermal displacements. In lattice steering experiments the fast ions see essentially a static arrangement of thermally displaced atoms, even at high temperatures. This fact permits to incorporate rather simply the effects of thermal displacements on r_c [4.7]

$$r_c^2 (\text{thermal}) = r_c^2 + r_{th}^2, \qquad (4.8)$$

where r_{th}^2 is the mean-square thermal displacement of the atoms perpendicular to the row. In the following $r_c(\text{thermal})$ is simply indicated as r_c. As a rule of thumb, r_c turns out to be in the most cases $\lesssim a_{TF}$ [4.4].

In order to proceed further we have to specify the atomic potential $V(R)$ in (4.1). *Lindhard* et al. [4.8] have proposed the following approximation to the Thomas-Fermi potential

$$V(R) = (Z_1 Z_2 e^2/R)[1 - R/(R^2 + 3a_{TF}^2)^{1/2}]. \qquad (4.9)$$

Insertion of (4.9) into (4.7) gives

$$U_{[hkl]}(r) = (Z_1 Z_2 e^2 / d_{[hkl]}) \ln (3a_{TF}^2 / r^2 + 1). \qquad (4.10)$$

This equation is particularly convenient in analytical calculations.

The most widely used analytical expression of the Thomas-Fermi screening function $\phi(R/a_{TF})$ in (4.1) is the Molière's approximation [4.9]. The so-called Thomas-Fermi-Molière potential is

$$V(R) = (Z_1 Z_2 e^2 / R) \sum_{i=1}^{3} \alpha_i \exp(-\beta_i R / a_{TF}) \qquad (4.11)$$

with $\alpha_i = (0.1, 0.55, 0.35)$ and $\beta_i = (6.0, 1.2, 0.3)$. The expression of the corresponding continuum-row potential reads

$$U_{[hkl]}(r) = (2Z_1 Z_2 e^2 / d_{[hkl]}) \sum_{i=1}^{3} \alpha_i K_0(\beta_i r / a_{TF}), \qquad (4.12)$$

where $K_0(x)$ is the zero-order modified Bessel-function. Equation (4.12) has been shown to give good agreement between theory and experiment in axial channeling studies.

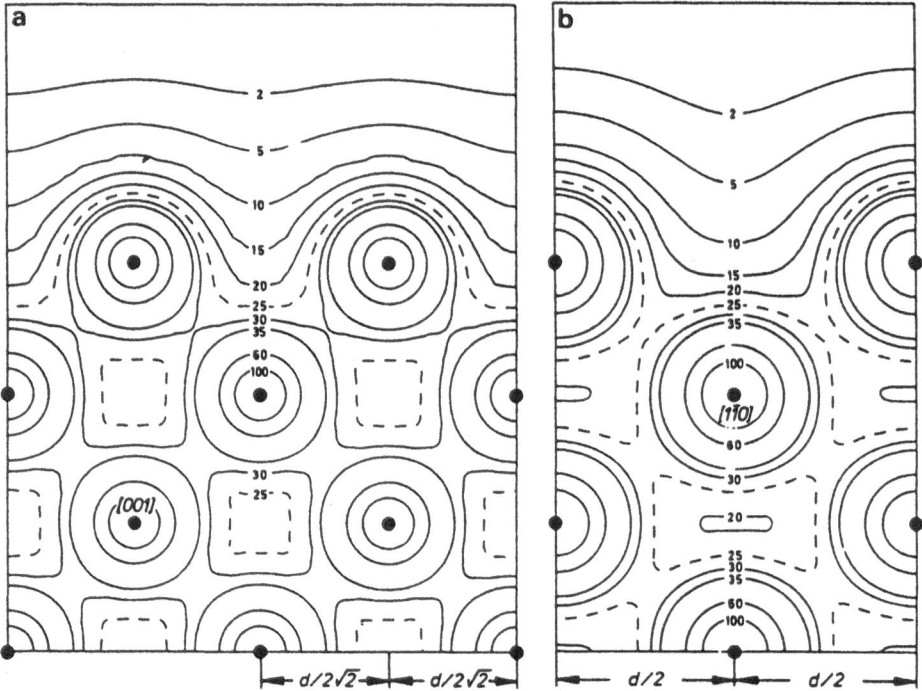

Fig. 4.2. Contours of constant potential energy [eV] near the Ni(110) surface in the plane transverse to (**a**) the [001] and (**b**) the [1$\bar{1}$0] direction. The calculations are based on the superposition of continuum row potentials, (4.12) for interaction with protons. The position of the [001] and the [1$\bar{1}$0] rows of nickel atoms are indicated by (•). d = 0.352 nm

Figure 4.2 illustrates, as an example, contours of constant potential energy near the Ni(110) surface in the plane transverse to the [001] and [1$\bar{1}$0] direction. Such contour lines result from the superposition of continuum row potentials as derived in (4.12) for interaction with protons.

4.2.2 Transverse Energy

The motion of a projectile in continuum row potentials

$$U(\mathbf{r}) = \sum_j U(\mathbf{r} - \mathbf{r}_j),$$

where j runs over all atomic rows, is independent of the coordinate along the axial direction. We can describe, therefore, the projectile motion in a plane transverse to the axial direction, in which for a conservative potential $U(\mathbf{r})$ the total energy is constant:

$$U(\mathbf{r}) + E \, \sin^2 \psi = E_\perp = \text{const.} \tag{4.13}$$

$$E \, \sin^2 \psi = (M_1/2)v^2 \sin^2 \psi = M_1 v_\perp^2/2 \quad \text{(Fig. 4.1)}$$

is the kinetic energy in the transverse motion, $E = M_1 v^2/2$ being the entire kinetic energy of the projectile and ψ its momentary angle towards the row direction. Since in typical channeling experiments the angles ψ are small (a few degrees) we may replace, to a good approximation, $\sin^2 \psi$ by ψ^2.

Equation (4.13) can then rewritten as

$$U(\mathbf{r}) + E\psi^2 \equiv E_\perp = \text{const.} \tag{4.14}$$

It is hence appropriate to define $E_\perp$ as the *transverse energy* of the projectile.

The conservation of transverse energy along a given projectile trajectory in a continuum potential is a fundamental law of the channeling theory [4.1]. Whenever (4.14) obtains the direction of the projectile motion, $E_\perp$ is preserved within small angular fluctuations of at most $\psi = (E_\perp/E)^{1/2}$. The transverse energy of channeled projectiles is determined by the potential energy $U(\mathbf{r}_0)$ at the starting position $\mathbf{r}_0$ and their starting transverse kinetic energy $E\psi_0^2$, i.e.,

$$E_\perp = E\psi_0^2 + U(\mathbf{r}_0). \tag{4.15}$$

If we set $U(\mathbf{r}_0)$ equal to zero, at the starting position $\mathbf{r}_0$ far away above the surface we obtain from (4.15) $E_\perp = E\psi_0^2$. Projectiles of transverse energy $E_\perp$ can approach the atomic row only until its transverse kinetic energy has been completely converted into potential energy, $E\psi_0^2 = U(\mathbf{r}_m)$, $\mathbf{r}_m$ being the minimum distance of approach to the atomic row. Regardless of the validity of the continuum potential, $\mathbf{r}_m$ is restricted to be larger than the critical distance $\mathbf{r}_c$. The maximum transverse energy a row of atoms can withstand, acting as a continuum potential line source, is $E_\perp(\max) = U_{[hkl]}(\mathbf{r}_c)$. Thus, the angle towards the row direction cannot exceed a maximum value ψ_c, the *critical angle*, given by

$$\psi_c = [U_{[hkl]}(r_c)/E]^{1/2}. \tag{4.16}$$

Using (4.6,8,12,16) for 150 keV protons channeled along the [001] nickel atomic rows at room temperature (T = 300 K, r_{th} = 5.4 pm) we obtain r_c = 10.7 pm. The maximum transverse energy is $E_\perp$(max) = $U_{[hkl]}(r_c)$ $\cong$210 eV and ψ_c = 2.1°.

One can derive a simple estimation of ψ_c from Lindhard's row potential (4.10) setting r_c $\cong$0.7 a_{TF} in (4.16)

$$\psi_c \cong (2Z_1Z_2e^2/E\,d_{[hkl]})^{1/2}. \tag{4.17}$$

For the example above, the critical angle comes out to be 2.2°, which agrees well with the exact calculated ψ_c value.

Note in Fig. 4.2 the shallow energy barrier of a few electron volts between neighboring atomic rows as compared to the maximum transverse energy of 210 eV the row allows for 150 keV protons being channeled. This means that "channeled" projecties are generally *not* confined to a lattice channel flanked by atomic rows. In effect, they can house in large areas in the transverse plane.

Relationships for r_c, $E_\perp$(max) = $U_{[hkl]}(r_c)$, and ψ_c may also be derived for compounds (e.g., GaAs, NaCl) and for monoatomic crystal structures with a basis (e.g., Si), starting from the Fourier expression of the total interaction potential between a projectile and the target atoms as in Sect. 4.2.1. For details the reader is referred to [4.7].

4.2.3 Flux Distribution

It has already been pointed out that a projectile with transverse energy $E_\perp$ cannot cross the equipotential contour line $E_\perp = U(r)$. Thus, the contour lines shown, e.g., in Fig. 4.2, are the border lines of the area in which the transverse projectile motion is confined. We now assume that the concepts of statistical mechanics are applicable to such a transverse motion. The approximation is that statistical equilibrium is attained rapidly during projectile-target inter-action. Then the two-dimensional transverse motion during axial channeling leads to an uniform probability of finding a projectile with a given $E_\perp$ any-where inside the area $A(E_\perp)$ bounded by $U(r) = E_\perp$. Since experimentally the starting positions r_0 of the projectile trajectories with the sample angle ψ_0 towards the row direction are randomly distributed over the full transverse plane, not a single but a distribution of $E_\perp$ will be present, see (4.14). The relative weights of the various $E_\perp$ depend on the topography of the contours $U(r)$ and, of course, on the incidence angle ψ by which the distribution can be altered. Statistical equilibrium causes each group around a given $E_\perp$ to become uniformly distributed over an area $A(E_\perp)$, as discussed before. These uniform distributions over the r's inside the various $A(E_\perp)$ add up in their common re-gion, i.e., at the positions with lowest potential energy $U(r)$. Thus, high peaks in flux density can be obtained in these regions. At well-aligned incidence such

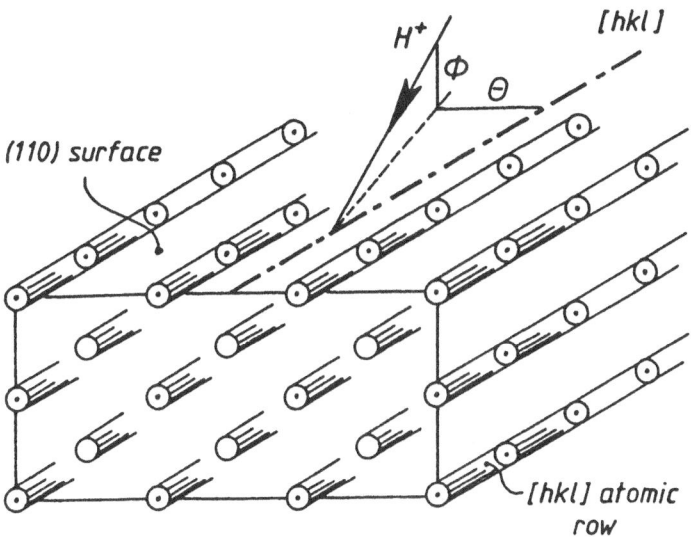

Fig. 4.3. Illustration of the angles of incidence used in the surface channeling experiments

peaks can become as high as ten times the original flux density of the incident beam. This phenomenon is called *flux peaking*. The opposite effect, a *flux depression* occurs near the atomic rows. In fact, the spatial average over flux peaking and flux depression must yield the original incident ion flux density. Flux peaking and flux depression are most important features in channeling.

The assumptions made so far, i.e., validity of a continuum potential in the transverse motion, transverse energy conservation, statistical equilibrium, were introduced merely for the sake of simplicity and to elucidate the fundamental principles of channeling. It is very difficult to calculate a realistic ion-flux density distribution in an analytically closed form. Furthermore, regarding the incident conditions of surface channeling, the ion-flux density distribution is strongly influenced by two angles of incidence: (i) the angle of incidence between the ion beam and the surface plane, denoted here by ϕ, and (ii) the lateral angle of incidence between the ion beam and a low-index surface direction, denoted here by θ. These angles are illustrated in Fig. 4.3, where

$$\psi^2 = \theta^2 + \phi^2 \tag{4.18}$$

for the small angles as they arise in channeling. That is the reason, why ion-flux density distributions with the respective dependence on the angles of incidense, on the lattice structure and temperature, including scattering and energy loss to target atoms, can be achieved with computer simulations. Such calculations are performed on the basis of binary collisions between projectile and target atoms, as described in Sect. 4.3.2. A typical result is shown in Fig. 4.4 for various lateral angles of incidence θ and a constant angle $\phi = 1°$.

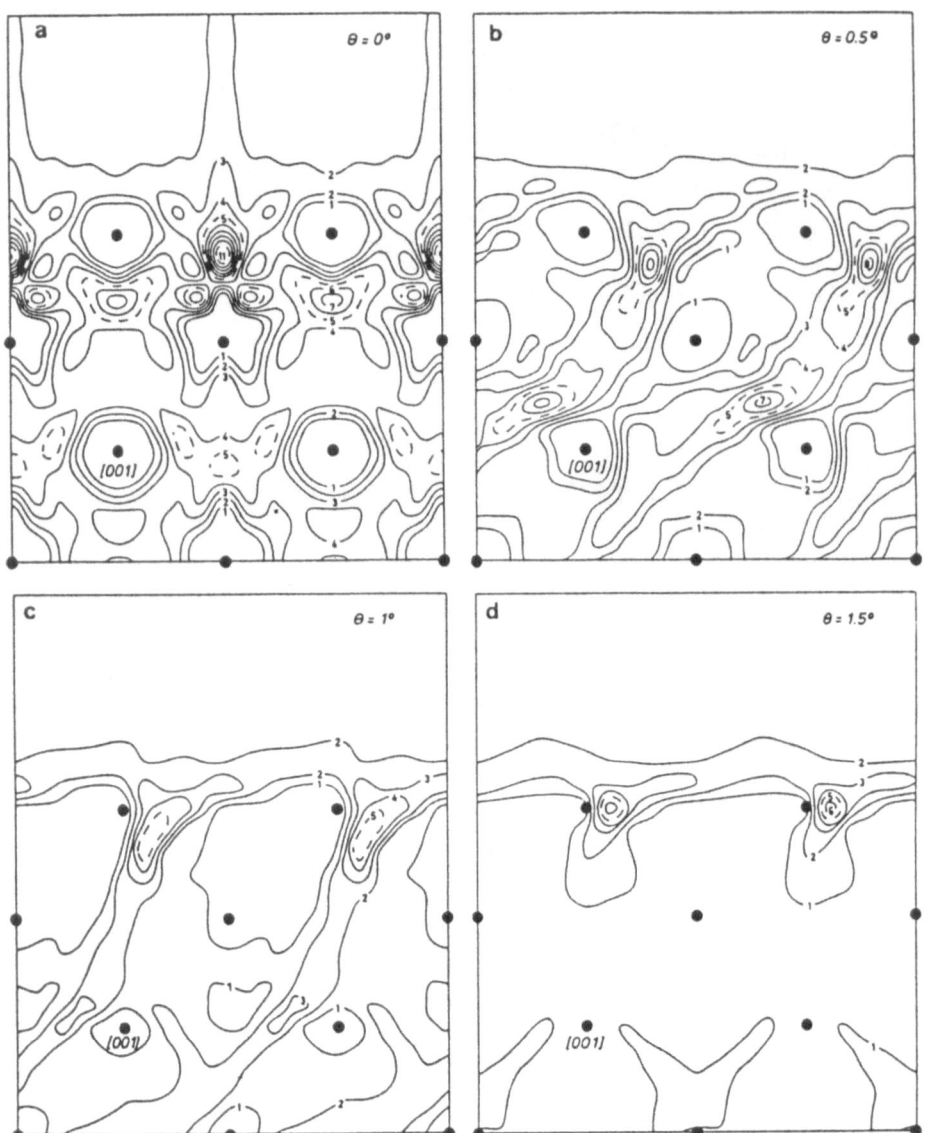

Fig. 4.4a–d. Flux density contours in the transverse plane (normalized to the incident flux density) of protons channeled along the [001] direction in the (110) surface region of a nickel crystal for various lateral angles of incidence, $\theta = 0°$, $0.5°$, $1.0°$, and $1.5°$ calculated by computer simulations [4.10]. All the other parameters are constant: $\phi = 1°$, $E = 150\,keV$, $T = 300\,K$. The nickel atomic rows are indicated by (•)

4.2.4 Dechanneling

The consideration of thermal vibrations and scattering at target cores and electrons inevitably breaks the strict validity of the fundamental rule of conservation of transverse energy, see (4.14). We must admit that the transverse energy of a channeled projectile can change with interaction length. For instance, if we start a bunch of projectiles with a sharply defined transverse energy $E_\perp$, then we shall find a broadening of $E_\perp$ to smaller and higher values with increasing interaction length along the target. On the average, the centroid of the distribution of $E_\perp$ drifts to higher $E_\perp$. The drift rate depends on the magnitude of $E_\perp$ and is slow as long as $E_\perp$ is noticeable smaller then the maximum transverse energy $E_\perp(\max)$. Then, (4.14) is still a good approximation. The drift rate increases steeply if the transverse energy of the projectile comes close to $E_\perp(\max)$ [4.11,12]. This is understandable, because a large $E_\perp$ means close approach of the projectile trajectory to the target atoms and hence a strong scattering interaction. Eventually the projectile will suffer a large angle deflection, thereby loosing its original direction completely.

Another obvious perturbation of $E_\perp$ must result from lattice defects: displaced atoms, foreign atoms, dislocations, subgrain and other boundaries, regions damaged by radiation or even only deformed by elastic strains. A channeled projectile can experience a single large-angle deflection on such defects, which causes instantaneous dechanneling. There is, however, an even higher probability that a lattice defect merely triggers a subsequent dechanneling process ([4.2,3] and references therein). Primarily, the lattice defect causes a slight deflection which, on the average, is likely to increase the transverse energy of the projectile. Although its $E_\perp$ might be still smaller than $E_\perp(\max)$, it has come closer to this threshold and thereby the "normal" dechanneling mechanism we mentioned before is drastically enhanced.

4.2.5 Classical Scattering Versus Quantum Diffraction

We have so far implied that the projectiles scatter according to the laws of classical mechanics. In the present subsection we shall show that such an approximation is reasonably well justified in usual surface channeling experiments. We employ the Bohr condition for the scattering angle ϑ to be well defined in an atom-atom collision with impact parameter b, namely $k_\perp \cdot b \gg 1$.

The transverse wave vector $k_\perp$ is given by $k_\perp = k \cdot \vartheta$ with $\hbar k = (2M_1 E)^{1/2}$. Hence, taking ϑ equal to ψ_c,

$$k_\perp b \cong (4M_1 Z_1 Z_2 e^2 / d_{[hkl]})^{1/2} b / \hbar \tag{4.19}$$

according to the estimate of ψ_c given in (4.17). The typical impact parameter b of channeled projectiles with $E_\perp(\max)$ is, as stated before, of the order of $a_{TF} = a_0 / Z_2^{1/3}$, see (4.2b). These relations combine to a condition for the validity of classical scattering

$$[(M_1/m_e) Z_1 Z_2^{1/3} (4a_0 / d_{[hkl]})]^{1/2} \gg 1, \tag{4.20}$$

where m_e is the electron mass and originates in (4.20) from $a_0 = \hbar^2/m_e e^2$. For protons and typical target values, e.g., $Z_2 \cong 20$, $4a_0/d_{[hkl]} \cong 1$, the left-hand side of (4.20) yields a characteristic value of about 70 and, therefore, classical scattering comes out to be a good approximation.

4.3 Surface Channeling Angular Yield Profiles

The main feature of channeled projectiles is their unique impact parameter distribution (flux-density distribution). The small impact parameters towards atomic rows are less populated. On the average, the channeled projectiles have a large impact parameter to the atoms in the rows. Inelastic reactions with decreasing cross section for larger impact parameters between the incoming projectile and the target atoms exhibit a reduction of the reaction yield if the projectiles are incident parallel to the major crystallographic directions. The production of Auger electrons by inner-shell ionization is an example for such an impact-parameter-dependent process. Other impact-parameter-dependent processes are, e.g., energy loss, backscattering yield, nuclear reactions, and x-ray excitation. For the application of surface channeling (grazing ion incidence to the surface) as a surface-analysis method it is advantageous to use the ion-induced Auger-electron emission, which will be discussed in detail in Sect. 4.3.1.

If the angle θ between the ion beam and the low-index direction [hkl] with fixed angle ϕ (Fig. 4.3) is increased the steering effect becomes smaller, because $E_\perp$ is increased, and the ion-flux density becomes more homogeneous (Fig. 4.4). Thus, the average impact parameter between the projectiles and the atoms in the [hkl] atomic rows is reduced. Therefore, the emission yield of Auger electrons is increased, and for even larger anlges θ it reaches the value which is obtained in a random distribution of impact parameters. The yield of emitted electrons is usually normalized to the random value, which corresponds to the yield that would be measured from a hypothetical amorphous target surface. For incidence, in random directions, the projectiles experience no steering and therefore the flux density distribution shows no lateral dependence. The schematic θ-angular dependence of the Auger-electron yield measured in a surface channeling experiment is illustrated in Fig. 4.5 for a constant angle ϕ between the ion beam and the surface plane. It stands to reason that ϕ always must be smaller than the critical angle ψ_c.

The quantity $\theta_{1/2}$ shown in Fig. 4.5 is defined to be the half-width at half minimum (HWHM) of the yield. The corresponding total declination of the ion beam to the low-index direction [hkl] is

$$\psi_{1/2}^2 = \theta_{1/2}^2 + \phi^2. \tag{4.21}$$

It was found, to a good approximation [4.2,7,13] within 5 or 10%, that $\psi_{1/2} = \psi_c$. Then, we obtain from (4.21)

$$\theta_{1/2} = \psi_c(1 - \phi^2/\psi_c^2)^{1/2} \tag{4.22}$$

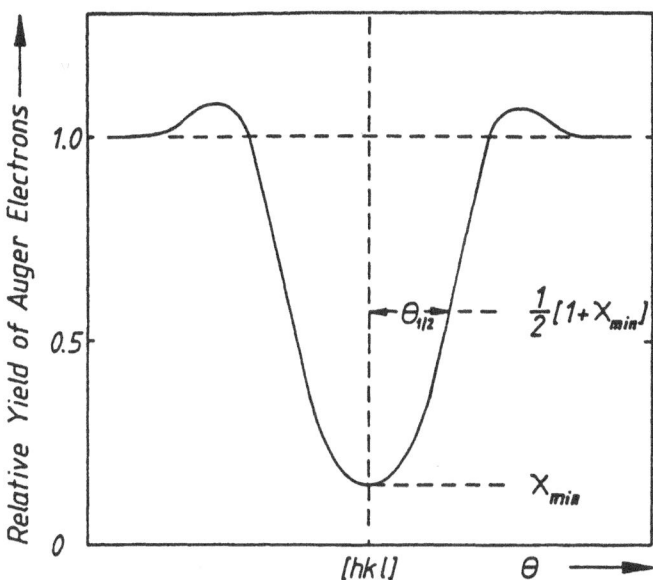

Fig. 4.5. Normalized yield of ion-induced Auger electrons in dependence on the lateral angle of incidence θ for a fixed angle of incidence ϕ to the surface plane, typically measured in surface channeling experiments

and for $\phi \ll \psi_c$ it follows $\theta_{1/2} \cong (2Z_1 Z_2 e^2 / E d_{[hkl]})^{1/2}$ according to (4.17). If Z_1, Z_2, E, and ϕ are fixed by the experimental conditions, $\theta_{1/2}$ is a measure of the interatomic distance $d_{[hkl]}$ in the steering rows.

On the other hand, the unique flux density distribution of channeled projectiles (Fig. 4.4) leads itself to the location of adsorbed atoms on the crystal surface. The Auger-electron emission from the adsorbed atoms is enhanced, if the adsorbates are sitting between the atomic rows, and reduced, if the adsorbates are sitting in or near the lattice rows.

Thus, the comparison of Auger-electron yields measured by ion incidence along different low-index directions (simultaneously for substrate and adsorbate atoms) allows a straightforward estimation of the surface structure and of the adsorbate position (Sect. 4.4). Calculations, however, are a prerequisite for the quantitative interpretation of the experimental data and the planning of experiments designed to test specific models.

4.3.1 General Considerations

The yield of Auger electrons emitted by ion impact on a perfect rigid crystal can be given by

$$Y(E^{(0)}, \phi, \theta) = \sum_{n_z} \int_y \int_x \int_E F(r, E; E^{(0)}, \phi, \theta)$$

$$\cdot I(E, b(r)) \cdot P_A \cdot P_E(y) dE \, dx \, dy \qquad (4.23)$$

$F(r, E; E^{(0)}, \phi, \theta)$:   ion flux density (projectiles per unit area and per energy interval),

$I(E, b(r))$:   Ionization probability of an inner-shell by ion impact with an energy E and an impact parameter $b(r)$.

Hereby, the ionization probability is connected with the ionization cross section via

$$\sigma_I(E) = 2\pi \int_0^\infty I(E, b) b \, db \qquad (4.24)$$

$\sigma_I(E)$:   total ionization cross section induced by a homogeneous ion flux,

P_A: Auger efficiency (probability for filling up the innershell vacancy by an Auger transition),

$P_E(y)$:   escape probability for an Auger electron produced at the depth y,

ϕ, θ: angles of incidence (Fig. 4.3),

$r = (x, y, z)$:   coordinates of the projectile relative to the atomic lattice position

n_z: number of transverse planes to the low-index direction [hkl]. $n_z x$ (distance of the transverse planes)$= z$,

E: momentary projectile energy,

$E^{(0)}$:   projectile energy at the incidence, and

$b(r)$:   impact parameter of the projectile to the nearest target atom.

For fixed ion incidence with ϕ and $E^{(0)}$ constant, (4.23) can be simplified to

$$Y(\theta) = \sum_{n_z} \int_y \int_x \int_E F(r, E; \theta) \cdot I(E, b(r)) \cdot P_A \cdot P_E(y) dE \, dx \, dy. \qquad (4.25)$$

Equations (4.23 and 25) are based on the assumption of a rigid lattice, where the position of the target atoms is a δ-shaped distribution. In a real lattice, thermal vibrations cause a Gaussian-shaped distribution of the target atom positions. This thermal distribution $A(r_0)$ of the target atom positions also affects the Auger-electron yield, since in the ionization-probability function the impact parameter between the projectile and the thermally elongated target atom, has to be considered. The probability of ionizing a target atom, whose position is distributed according to the function $A(r_0)$, by the impact of the projectile at r, is the convolution of the distribution of the target atoms with the ionization probability function

$$I^*(E, r) = \int_{r_0} I(E, b(r - r_0)) A(r_0) dr_0. \qquad (4.26)$$

$I^*(E, r)$ can be regarded as the thermally broadened ionization probability function with reference to the coordinates of the rigid lattice. Therefore, in (4.23

and 25) $I(E, b(r))$ has to be replaced by $I^*(E, r)$, if thermal vibrations are to be taken into account.

a) Ionization Probability

In general, the ionization of inner-shells can occur in collisions initiated by the projectile itself or by the action of an energetic recoil atom, electron or photon, produced by the incident ion. In the first case, one distinguishes between the potential electron emission, which involves the potential energy of the incoming ion, e.g., the Auger neutralization, and the kinetic electron emission, where the energy required to liberate electrons into vacuum is provided by the kinetic energy of the projectile. These processes have extensively been discussed in [4.14].

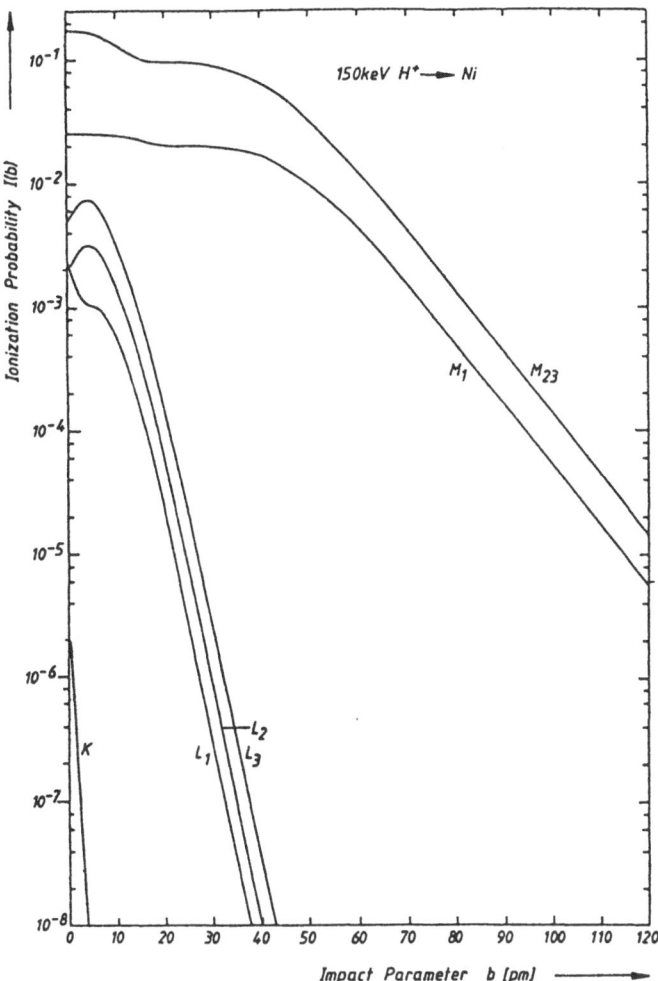

Fig. 4.6. Ionization probability versus the impact parameter for various nickel subshells by impact of 150 keV protons [4.16]

In fact, for fast, bare, charged projectiles like protons under surface channeling experimental conditions, the only ionization mechanism is the Coulomb ionization [4.14]. For heavier ions, also quasimolecular electron promotion [4.15] and ionization processes by recoil target atoms are important.

Besides the dependence on the impact parameter b and on the projectile energy E, the ionization probability is also dependent on the orbital of the target atom to be ionized. The values of the Coulomb ionization probability I(E, b,shell) can be taken from tables calculated, e.g., in the straight-line version of the semiclassical approximation model by *Hansteen* et al. [4.16]. Figure 4.6 shows as an example the ionization probability I(b) of some subshells of nickel atoms by impact of 150 keV protons. The functions $I_W(b)$, with $W = K, L_1, L_2, L_3, M_1, M_{23}$ in Fig. 4.6 represent the primary ionization by direct ion impact. Primarily created vacancies can be shifted to lower energetic levels by x-ray, Auger and Coster-Kroning transitions. If an Auger electron transition WXY is observed in the experiment, not the primary vacancy distribution by direct ion impact I_W, but the vacancy distribution after shifting I_W^{total} has to be considered. An exception are the KXY-Auger transitions, because all the K-shell vancancies are primary vacancies, $I_K^{total} = I_K$. The total ionization probability of the nickel L_3-subshell, for example, is given by [4.17]

$$I_{L_3}^{total} = I_K(f_{KL_1}f_{L_1L_3} + f_{KL_1}f_{L_1L_2}f_{L_2L_3} + f_{KL}f_{L_2L_3} + f_{KL_3})$$
$$+ I_{L_1}(f_{L_1L_2}f_{L_2L_3} + f_{L_1L_3})$$
$$+ I_{L_2}f_{L_2L_3}$$
$$+ I_{L_3}, \tag{4.27}$$

$$I_{L_3}^{total} = 0.85I_K + 0.70I_{L_1} + 0.10I_{L_2} + I_{L_3}, \tag{4.27a}$$

I_W^{total}: ionization probability of the W-shell by primary ion impact and subsequent vacancy shifting,

I_W: ionization probability of the W-shell only by primary ion impact, and

f_{WX}: transition probability W→X-shell (Auger, Coster-Kronig, x-ray transitions).

A reduction of expressions similar to (4.27a) is possible, because the ionization probabilities of the various W-shells (W = K, L, M, etc.) by ion impact differ by some orders of magnitude: As an example compare Fig. 4.6. A larger contribution is provided by Coster-Kroning transitions between the subshells, e.g. L_1, L_2, L_3, which cause a difference between I_W^{total} and I_W, e.g., of up to 15% from $I_{L_3}^{total}$ to I_{L_3} of nickel by 150 keV proton impact.

The final vacancies described by I_W^{total} can be deexcited via two competing processes, by x-ray emission or by Auger excitation. The probability for filling the W shell with an X shell electron by an x-ray transition is the fluorescence yield ω_{WX} and thus the probability for a WXY-Auger transition is $P_{A_{WX}} = (1 - \omega_{WX})$. In the energy range of Auger electrons of interest, $E_{WXY} < 1000\,eV$, the fluorescence yield ω_{WX} is less than 0.05 [4.17]. Therefore, the probability for the production of an ion-induced WXY-Auger electron, $P_{A_{WX}} I_W(b, E)$, is essentially determined by the W-shell ionization probability $I_W(b, E)$.

b) Escape Probability for Auger Electrons

The electrons produced in the crystal must penetrate through the solid and overcome the surface potential barrier to get into the electron-energy ana-lyzer. For the transport of excited electrons the mean-free path of the electrons before being inelastically scattered is relevant. After inelastic scattering the Auger electron still exists, but its energy has diminished and it only appears in the background of the emitted electron energy spectrum, not contributing to the relevant line strength [4.18]. For the electron escape depth inelastic scat-tering processes with electron-hole-creation and electron-plasmon-interactions are important [4.19,20]. In most materials the elastic scattering cross sections for electrons above 200 eV is much smaller than the inelastic scattering cross section. It can, therefore, be assumed that an Auger electron emitted by a deexciting atom, will travel in a straight line until it is either emitted from the surface or inelastically scattered. If the upper condition is not fulfilled, the escape mean-free path becomes a function of both the elastic and the inelastic mean-free path [4.21,22]. The probability dP_E of an electron with the energy E_e originating from an Auger process at the depth y, to escape from the crystal without inelastic scattering in a solid angle $d\Omega$ at an angle α with respect to the surface normal is given by

$$dP_E(y) = \exp[-y/\lambda(E_e) \cos\alpha](d\Omega/4\pi), \qquad (4.28)$$

where $\lambda(E_e)$ is the escape mean free path of an electron with the energy E_e.

The diffraction of electrons at the surface potential barrier can be neglected as long as the work function is much smaller than the electron energy.

c) Ion-Flux Density Distribution

As mentioned in Sect. 4.2.3, the ion-flux density gets its inhomogeneous form by the scattering at the atoms of the low-index rows. It is, however, very diffi-cult to calculate the ion-flux density distribution $F(\mathbf{r}, E; E^{(0)}, \phi, \theta)$, see (4.23), in closed form. Unfortunately, only in fairly simple models the numerical eval-uation of the ion-flux density distribution, as calculated within the frame-work of the continuum model (Sect. 4.2), is possible. An alternative way is to turn to numerical computation right away and to estimate the interaction of the projectiles with the target by computer simulation.

4.3.2 Computer Simulation

Generally, the ion-flux density distribution and the whole Auger-electron yield $Y(\theta)$, see (4.23), will be determined in a computer simulation. The basic con-cepts of such calculations were developed in [4.10,14,23]. From the following outline it will become apparent that the method allows projectile trajectories

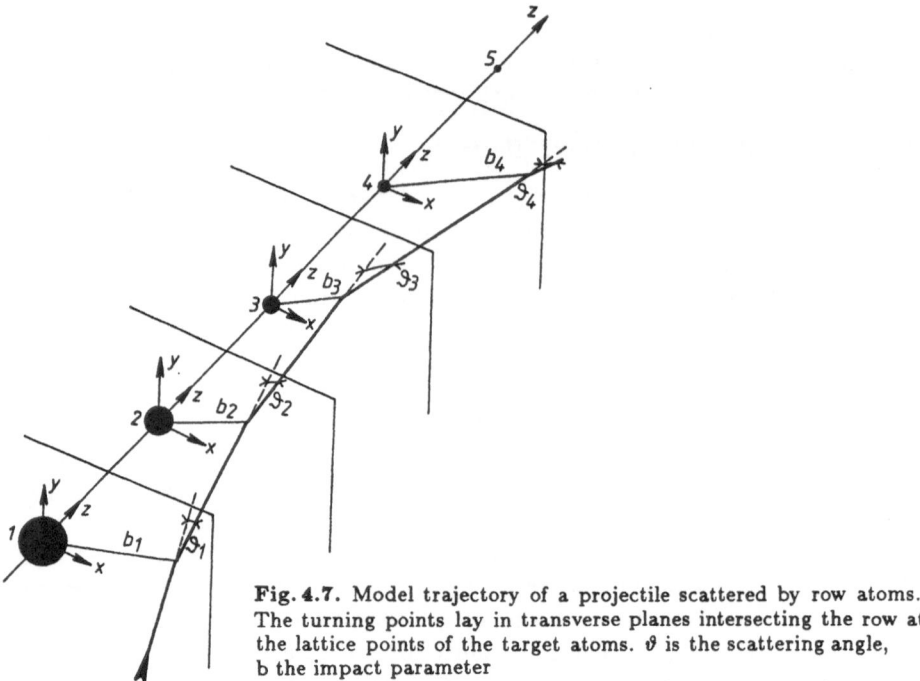

Fig. 4.7. Model trajectory of a projectile scattered by row atoms. The turning points lay in transverse planes intersecting the row at the lattice points of the target atoms. ϑ is the scattering angle, b the impact parameter

in crystal and at crystal surfaces to be calculated in a rather simple way. In the simulation the trajectories of the projectiles are calculated one by one. At the start, the projectile energy $E^{(0)}$, the angles of incidence $\phi^{(0)}$ and $\theta^{(0)}$ (till now simply denoted by ϕ and θ) and the starting coordinates $(x^{(0)}, y^{(0)}, z^{(0)})$ relative to the lattice points of the target atoms are given. In order to simulate realistic incident beams, the starting points of the trajectories are varied so that all started projectiles are spatially distributed homogeneously over one unit cell of the crystal surface (x-z plane). The distance $y^{(0)}$ of the projectiles from the surface at the start is to choose large enough, twice or more of the lattice constant. The scattering is treated as a sequence of binary collisions between the projectile and the individual target atoms (Fig. 4.7). This is justified by the short range of the atomic interaction potential. Since under channeling conditions the scattering angles ϑ are small, they may be calculated by means of the momentum approximation [4.24], which is adequate for $\vartheta \lesssim \pi/10$ [4.25]. Using the Thomas-Fermi-Molière potential (4.11) for the projectile-target atom interaction one obtains in laboratory systems

$$\vartheta = (Z_1 Z_2 e^2 / Ea_{TF}) \sum_{i=1}^{3} \alpha_i \beta_i K_1 (\beta_i b / a_{TF}) \tag{4.29}$$

where $K_1(\chi)$ denotes the modified Bessel function of order one, and b the impact parameter of the binary collision.

For the purpose of fast computation, (4.29) can be fairly well approximated by [4.25]

$$\vartheta = (Z_1 Z_2 e^2 / E a_{TF}) \begin{cases} 1.01(1/s)e^{-0.5s} & \text{for } s \leq 1.5 \\ 0.03|s - 3.8| + 0.64(1/s)e^{-0.3s} & \text{for } 1.5 \leq s \leq 30 \end{cases}$$

(4.30)

with $s = b/a_{TF}$. Other available options are given in [4.23].

Under channeling conditions b may be replaced by the distance of the projectile from an atom in the transverse plane

$$b \cong b' = (x^2 + y^2)^{1/2}.$$

(4.31)

Thermal vibrations of the lattice atoms, including anisotropies at the surface, are simulated by Gaussian-distributed displacements of the atoms, whose width is determined by the Debye model. This changes the impact parameter into

$$b^* = [(x - x_{th})^2 + (y - y_{th})^2]^{1/2}.$$

(4.32)

The new angles $\theta^{(1)}$ and $\phi^{(1)}$ of the projectile after the first binary collision read

$$\theta^{(1)} = \theta^{(0)} + (x - x_{th})\vartheta/b^*,$$
$$\phi^{(1)} = \phi^{(0)} + (y - y_{th})\vartheta/b^*,$$

(4.33)

where ϑ is to be taken from (4.29).

In the momentum approximation the trajectories are sequences of rectilinear motions with break-points being situated in the plane transverse to the particular low-index direction, in which the projectiles are channeled, at the lattice points of the target atoms (Fig. 4.7). Thereafter, the position of the projectile impact in the next transverse plane is given by

$$x^{(1)} \cong x^{(0)} + a \cdot \theta^{(1)},$$
$$y^{(1)} \cong y^{(0)} + a \cdot \phi^{(1)}\vartheta,$$

(4.34)

where a is the separation between successive transverse planes. Here again, the scattering angle in dependence on the projectile energy and the impact parameter is computed, and so on. The program performs transformations of the coordinates taking into account the periodical arrangement of the lattice atoms in the bulk and on the surface and also the position of adsorbate atoms.

Surface steps are simulated according to [4.26]. The height of one step corresponds to the periodicity length perpendicular to the surface. The distance between the steps can be varied and is given by the number of collisions, S, from step to step. Consecutive upward or downward steps and alternating series of upward and downward steps can be simulated. The influence of steps on surface channeling is discussed in Sect. 4.6.4.

In transversing a transverse plane, the projectiles loose the energy $\Delta E(b^*)$. It suffices to substract this amount from the kinetic energy of the projectile after each collision, since under channeling conditions only the electronic (inelastic) energy loss (e.g., excitation of electrons and of plasma oscillations) should contribute to slowing down, leaving the (4.33) unchanged. The nuclear (elastic) energy loss is negligably small (Sect. 4.6.1) and will thus not be calculated.

There are various models available to account for the electronic energy loss of the projectile. For projectile energies of interest here, the model proposed by *Oen* and *Robinson* [4.27] leads to adequate results. It includes an exponential dependence of energy loss by electron excitation on the distance of closest approach which, in the momentum approximation, is set equal to the impact parameter

$$\Delta E(b^*) \cong (0.045 k E^{1/2}/\pi a_{TF}^2) \exp(-0.3 b^*/a_{TF}). \qquad (4.35)$$

The stopping cross-section derived from (4.35) by integration over all impact parameters gives *Lindhard*'s estimation [4.28] and the parameter k can then be taken from this theory.

After each collision the position of the projectile $x^{(i)}, y^{(i)}$ in the transverse plane is stored. The projection of the interaction points of all the trajectories with each transverse plane onto one plane gives the spatial distribution of the projectiles. The density of the interaction points in this presentation is the ion-flux density distribution, as it is given in Fig. 4.4. This result corresponds with the ion-flux density distribution that will be obtained, if not only one unit cell is bombarded with projectiles, but the full surface plane, like in the experiment, and the flux is measured at only one intersection plane. Therefore, the ion-flux density distribution normalized to the incoming ion-flux density in simulation and experiment are the same. This fact allows the comparison of the computed and experimental results, which is, e.g., necessary for the exact location of adsorbates (Sect. 4.4.4).

For the calculation of the Auger-electron yields at each collision the ionization probability $I(E^{(i)}, b^{*(i)}, \text{shell})$ is evaluated with the impact parameter $b^{*(i)}$ given by (4.32), and the momentary projectile energy $E^{(i)}$. As mentioned above, the values of the Coulomb ionization probability can be taken from tables [4.16]. The ionization probabilities found for each collision are multiplied by the escape probability P_E of the produced Auger electron according to their escape depth $y^{(i)}$ and their mean-free path $\lambda(E_e)$, see (4.28). The resulting Auger-electron emission probabilities are summed up over all the collisions. This sum is the Auger-electron yield $Y(\theta)$ for a certain lateral angle of incidence $\theta^{(0)}$ for $\phi^{(0)}$ and $E^{(0)}$ constant.

4.4 Applications

In this section the application of the surface channeling technique, combined with ion induced-Auger-electron spectroscopy, to surface analysis will be illustrated. As an example serves a Ni(110) surface bombarded with 150 keV protons under axial surface channeling conditions. For investigations making use of the surface channeling technique the reader is referred to [4.10,14,23,26,29–32].

4.4.1 Experimental

For the experimental set-up and adjustment, the electron energy analysis, the recording of ion-induced Auger-electron spectra, we refer to [4.10,14]. In Fig. 4.8 the set-up of the experiments is shown schematically.

A well-collimated monoenergetic beam of 150 keV protons is incident on the Ni(110) surface with a grazing angle $\phi = 1°$. The electrons emitted from the crystal by proton bombardment are energy analyzed by a cylindrical mirror analyzer (CMA). An ion-induced Auger spectrum $dN(E_e)/dE_e$, originating from a nickel surface which was not been cleaned before, is shown as an example in Fig. 4.9. The ion-induced spectrum is similar to spectra obtained by electron bombardment. The decrease of the ionization cross section with the binding energy causes a rapid fall-off of the Auger peak intensities. The spectrum exhibits the Auger peaks of nickel $M_{23}VV$ at 61 eV and M_1VV at 102 eV, of sulfur $L_3M_{23}M_{23}$ at 152 eV, of chlorine $L_3M_{23}M_{23}$ at 181 eV and of carbon $KL_{23}L_{23}$ at 273 eV. By an amplification of a factor of 40, the Auger peaks of oxygen $KL_{23}L_{23}$ at 510 eV and of nickel $L_3M_{23}M_{23}$ at 716 eV, $L_3M_{23}V$ at 783 eV, and L_3VV at 848 eV become visible. In all the following channeling experiments the yield of Auger electrons induced by protons has been studied as a function of the lateral angle of incidence θ. As a quantity for the yield of Auger-electron

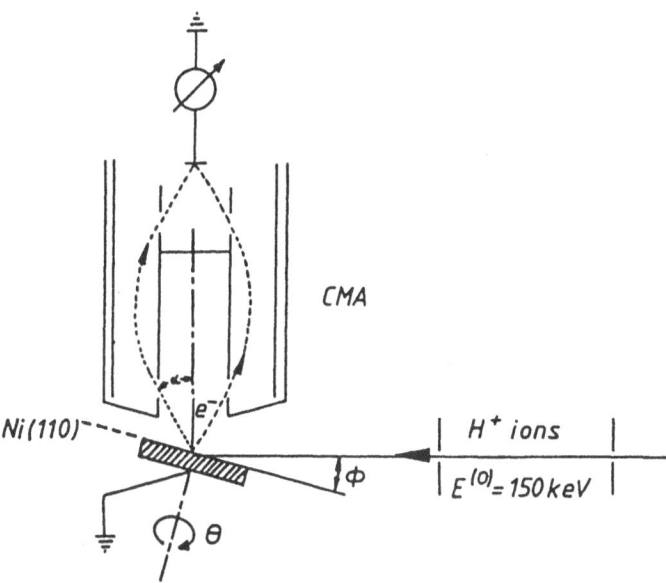

Fig. 4.8. Experimental arrangement with assignment of the angles used. (ϕ: angle of incidence to the surface plane. θ: lateral angle of incidence to the low index direction [hkl] in the surface)

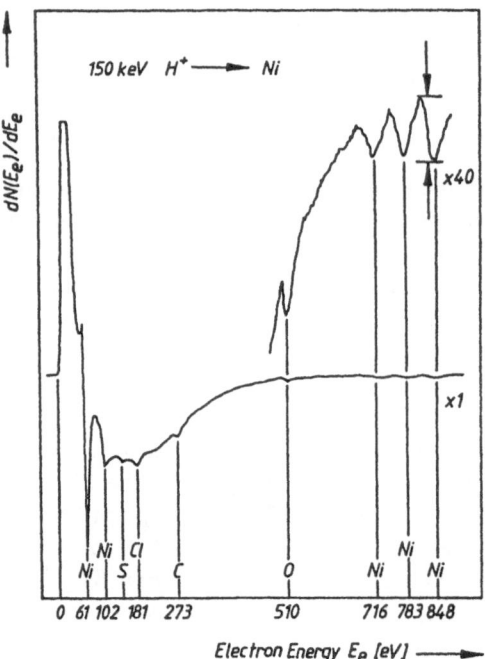

Fig. 4.9. Spectrum of Auger electrons induced by 150 keV protons incident on a not cleaned nickel surface with $\phi = 5°$ and θ=random [4.10]

emission the peak-to-peak height of the observed Auger line, e.g. nickel L_3VV marked by two arrows in Fig. 4.9, in the electron energy spectrum has been taken.

4.4.2 Surface Structure and Reconstruction

For proton incidence along the [001] direction on a clean Ni(110) surface, the angular scan of the Auger-electron yield at 848 eV (nickel L_3VV transition) is shown in Fig. 4.10. The yield of the Auger electrons exhibits a distinct minimum, as it is discussed in Sect. 4.3 (Fig. 4.5). The channeling dip has a half width at half minimum (HWHM) of $\theta_{1/2} = 1.7°\pm0.2°$ corresponding to $\psi_{1/2} = 2.0°$, using $\phi = 1°$ in (4.21). This value is in good agreement with the value of 2.2°calculated from (4.17) with an interatomic distance $d_{[001]}$ equal to the bulk value. A similar result is obtained for proton incidence along the [1$\bar{1}$0] direction, Fig. 4.11. The value of $\psi_{1/2} = 2.8°$ corresponding to the HWHM of the angular yield profile $\theta_{1/2} = 2.6°\pm0.2°$ agrees again with the calculated value 2.7° using in (4.17) the interactomic distance of the bulk along the [1$\bar{1}$0] lattice direction. Thus, we can conclude that the clean Ni(110) surface has the same structure as the (110) lattice planes in the bulk.

A more reliable evidence is given by the comparison of the whole angular yield profile obtained by experiment and calculation. The solid lines in Figs. 4.10 and 4.11 present the result of computer simulations taken a bulk-like atom arrangement on the clean Ni(110) surface. The computation is performed for a

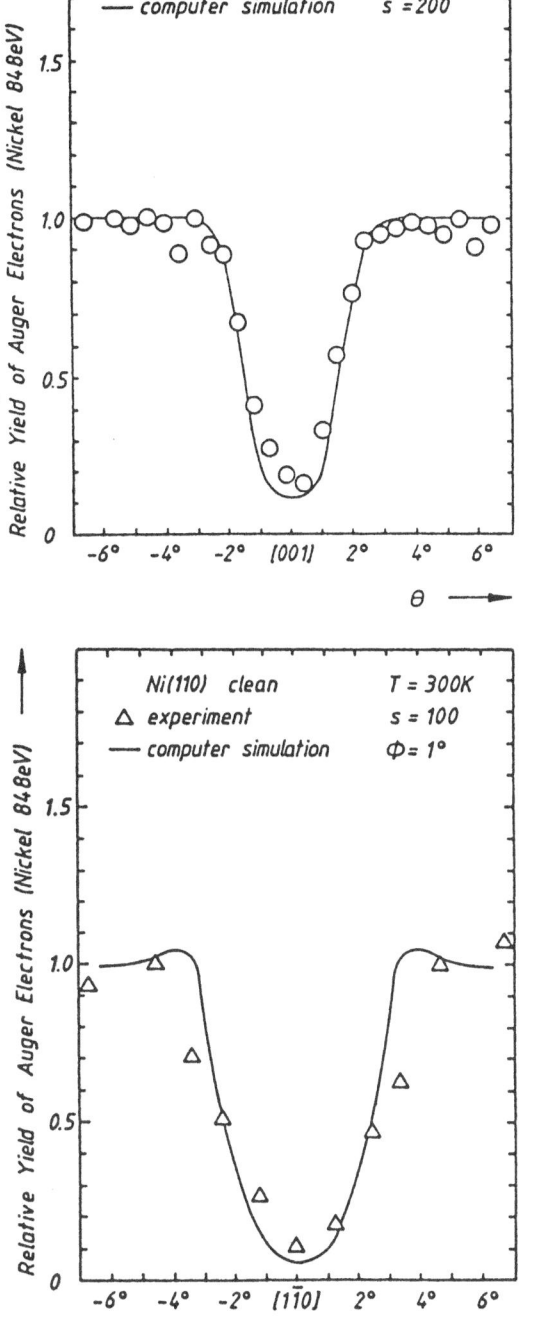

Fig. 4.10. Yield of 848 eV Auger electrons of Ni target atoms induced by 150 keV protons as a function of the lateral angle of incidence θ with respect to the [001] direction in the Ni(110) surface for $\phi = 1°$. The full line is the result of a computer simulation for a clean stepped Ni(110) surface with a step distance, s, of 200 atomic distances [s = $200 \cdot (g/2\sqrt{2})$] and a step height of $g/\sqrt{2}$, g: lattice constant. The escape probability of the Auger electrons is included in the calculation using a mean free path of $\lambda = 1.5\,\text{nm}$ [4.14]

Fig. 4.11. Yield of 848 eV Auger electrons of Ni target atoms induced by 150 keV protons as a function of the lateral angle of incidence θ with respect to the [1$\bar{1}$0] direction in the Ni(110) surface for $\phi = 1°$. The full line is the result of a computer simulation for a clean stepped Ni(110) surface with a step distance of s = 100 atomic distances = $100 \cdot (g/2)$ and a step height of $g/\sqrt{2}$, g: lattice constant. The escape probability of the Auger electrons is included in the calculation using a mean free path of $\lambda = 1.5\,\text{nm}$ [4.10]

stepped surface and a temperature of 300 K. The influence of the temperature and of the steps is discussed in Sects. 4.6.2 and 4. The good agreement with the experimental angular yield profiles indicates a bulk-like surface structure.

4.4.3 Surface Reconstruction

The example discussed so far concerned the clean Ni(110) surface. However, the surface channeling technique is applicable to covered surfaces and consequently to Sect. 4.4.2 it can be used to study the reconstruction, e.g. adsorption induced reconstruction, of the surface.

The clean Ni(110) has been now covered by half a monolayer of oxygen and the experiments have been repeated under otherwise identical conditions as for the clean surface (Z_1, Z_2, E, and ϕ remain constant). The angular-yield profiles of the induced L_3VV Auger electrons of nickel for the oxygen-covered surface are shown in Figs. 4.12 and 13. For proton incidence along the [001] direction the obtained channeling dip indicated by open circles in Fig. 4.12, coincidences

Fig. 4.12   Fig. 4.13

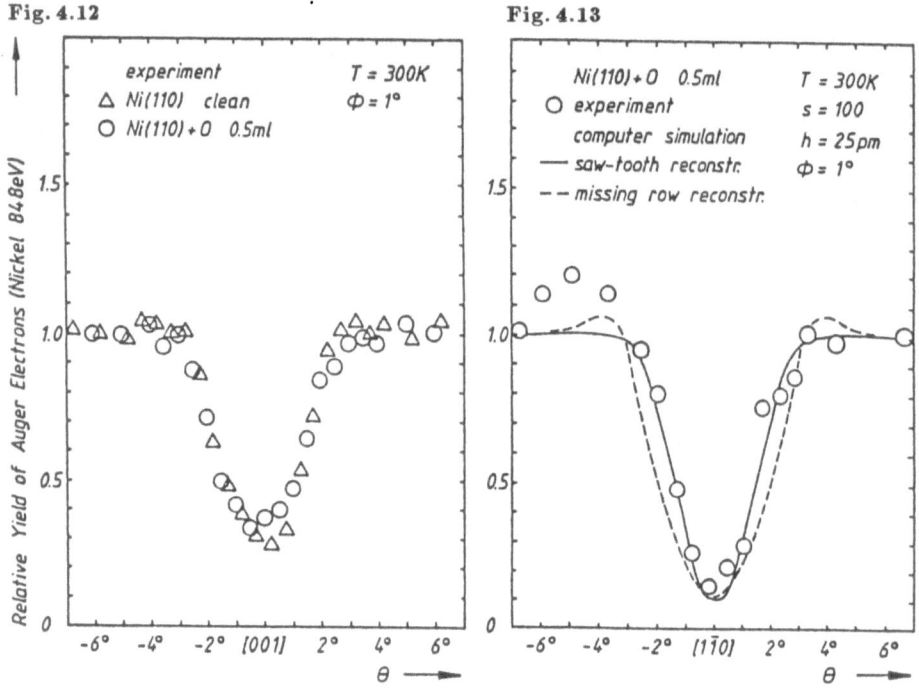

Fig. 4.12. Yield of 848 eV Auger electrons of Ni target atoms induced by 150 keV protons as a function of the lateral angle of incidence θ with respect to the [001] direction for a clean ($\triangle$) and for an oxygen coverd (O) Ni(110) surface [4.10]

Fig. 4.13. Yield of 848 eV Auger electrons of Ni target atoms induced by 150 keV protons as a function of the lateral angle of incidence θ with respect to the [1$\bar{1}$0] direction for a Ni(110) surface covered with 0.5 ML oxygen. The *solid* and the *dashed line* are the results of the computer simulation of a reconstructed (2×1) Ni(110)+0 surface for two reconstruction models, the saw-tooth and the missing row, respectively [4.10]. h is the distance of the oxygen atoms above the topmost nickel layer

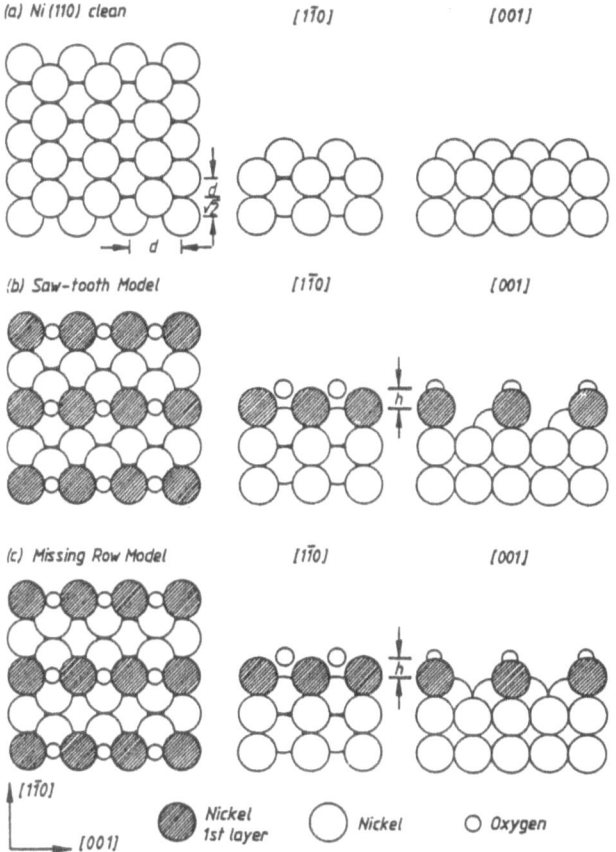

(a) Ni (110) clean   [1$\bar{1}$0]   [001]

(b) Saw-tooth Model   [1$\bar{1}$0]   [001]

(c) Missing Row Model   [1$\bar{1}$0]   [001]

[1$\bar{1}$0]

[001]

⬤ Nickel 1st layer   ◯ Nickel   ○ Oxygen

Fig. 4.14a–c. Top view of the Ni(110) surface: (a) clean (1×1) surface, (b) saw-tooth reconstructed (2×1) Ni(110)+0 surface, and (c) missing row reconstructed (2×1) Ni(110)+0 surface. On the right the profiles of the surface region to the [1$\bar{1}$0] and [001] directions

with the dip measured for the clean surface, indicated by triangles. In this direction, therefore, the distance of the nickel atoms is the same for the clean and the oxygen covered surface. In contrast, the angular-yield profile of the nickel Auger electrons for proton incidence along the [1$\bar{1}$0] direction has changed drastically (Fig. 4.13). HWHM has decreased to $\theta_{1/2} = 1.6°$ corresponding to $\psi_{1/2} = 1.9°$. For the [1$\bar{1}$0] direction the angle $\psi_{1/2}$ of the clean Ni(110) surface relates to $\psi_{1/2}$ of the oxygen covered nickel surface as $\psi_{1/2}(\text{Ni})/\psi_{1/2}(\text{Ni} + 0) = 2^{1/2}$, and according to (4.17), $d_{[1\bar{1}0]}(\text{Ni} + 0)/d_{[1\bar{1}0]}(\text{Ni}) = 2$. Therefore, the distance of the nickel atoms in the [1$\bar{1}$0] surface direction must have become twice the value of the interatomic distance in the [1$\bar{1}$0] direction of the bulk. From these simple straightforward arguments it follows already that the oxygen covered Ni(110) surface is reconstructed in a (2×1) structure. Computer simulation for different reconstruction models [4.10] have shown that the (2×1) saw-tooth

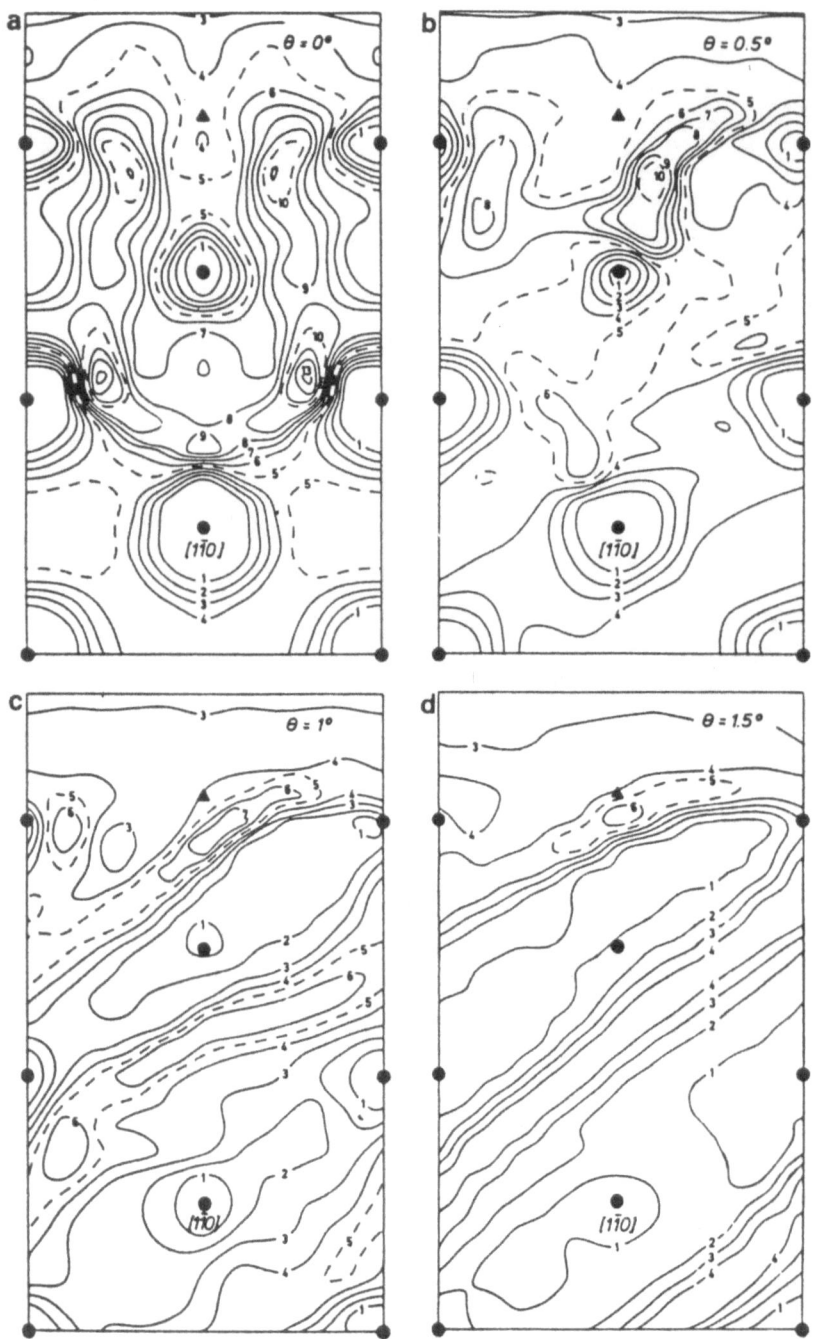

Fig. 4.15. Flux density contours in the transverse plane (normalized to half the incident flux density) of 150 keV protons channeled along the $[1\bar{1}0]$ direction in the region of a saw-tooth reconstructed (2×1) Ni(110) surface covered by 0.5 ML oxygen for various lateral angles of incidence; calculated by comptuer simulations for $\phi = 1°$; (●) nickel rows; (▲) oxygen rows [4.10]

reconstruction model can best explain the experimental results (solid line in Fig. 4.13). The saw-tooth model in which every second [001] row is shifted into the position of the [001] row immediately above exhibit a (2×1) reconstruction of the two top layers of the surface, as it is demonstrated in Fig. 4.14b. A competative reconstruction model, the missing-row model [4.33–35], does not give a good agreement with the experiment results, because the [1̄10] Ni rows below the reconstructed first layer are as densely packed as in the bulk (Fig. 4.14c) and still have enough influence on the steering that the channeling dip does not become as narrow as in the experiment (dashed line in Fig. 4.13). Besides the surface structure, all the other parameters in the computer simulation, as temperature and step distance, are the same for the clean (1×1) surface and the both (2×1) reconstruction models. The distance of the oxygen atoms above the topmost nickel layer has turned out to have negligible influence on the form of the computed angular-yield profile of nickel Auger electrons.

4.4.4 Location of Adsorbed Atoms

Another application of the surface channeling technique is the location of adsorbed atoms. As mentioned above, the unique flux density distribution of channeled projectiles (Figs. 4.4 and 15) leads itself to the location of adsorbed atoms on the crystal surface. The Auger-electron emission from the adsorbed atoms is enhanced, if the adsorbates are located between the atomic rows; it is reduced, if the adsorbates are sitting in or near the lattice rows, because in this case the adsorbates are shadowed by the repulsive potential of the lattice rows, as it is demonstrated in Fig. 4.16. A comparison of the angular-yield profiles of the Auger electrons of the adsorbates measured by ion incidence along various low-index directions gives a straightforward estimation of the adsorbate positions. Following, as an example, the location of adsorbed oxygen atoms on the Ni(110) surface covered by 0.5 monolayers oxygen is discussed.

Simultaneously with the Auger electrons of nickel, the Auger electrons of oxygen $KL_{23}L_{23}$ at 510 eV are measured. Figure 4.17 exhibits the angular-yield profile of the oxygen Auger electrons observed for proton incidence along the [001] direction. The dip has a minimum yield and a width similar to that obtained using the nickel Auger-electron emission by proton incidence along the same direction (Fig. 4.10). Because the potential of the oxygen atoms is not strong enough to cause such a wide and deep channeling dip (Fig. 4.16), the oxygen atoms must be shadowed by the repulsive potential of the [001] nickel atomic rows. The oxygen atoms are only excited if the incident protons break through the potential wall of the [001] nickel rows. Therefore, it can be concluded that the oxygen atoms must be located in the [001] nickel rows or near to them.

For proton incidence along the [1̄10] direction the angular yield profile of the 510 eV oxygen Auger electrons is quite different (Fig. 4.18). The minimum yield is enhanced by a factor of 5 compared with that for proton incidence along the [001] direction. That means the oxygen atoms are located in a region with

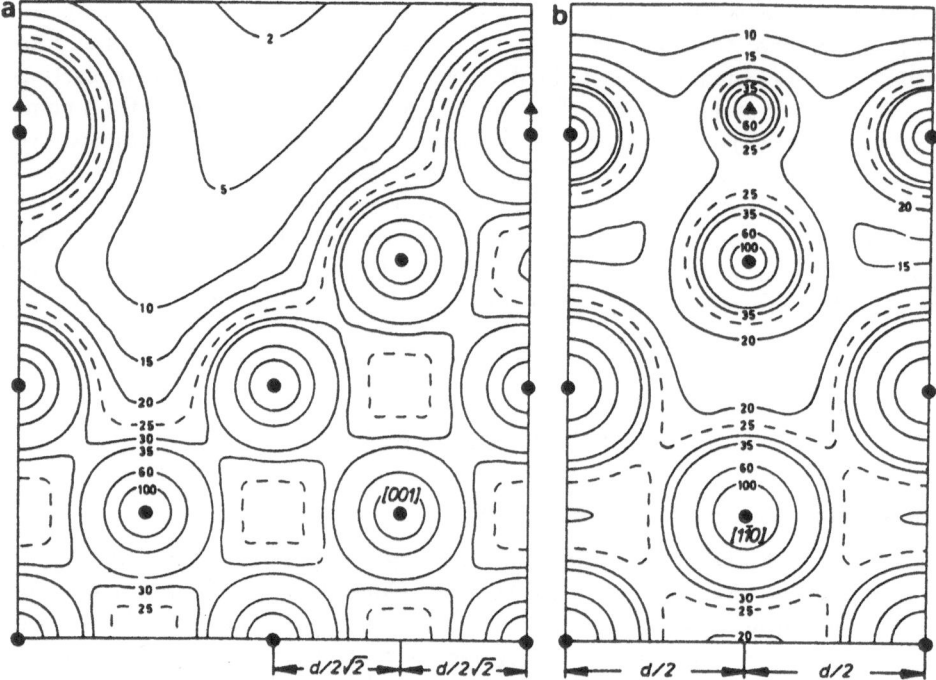

Fig. 4.16. Contours of constant potential energy [eV] of the saw-tooth reconstructed (2×1) Ni(110) surface covered by 0.5 ML oxygen in the plane transverse to (**a**) the [001] and (**b**) the [1$\bar{1}$0] direction. The contours are calculated by the superposition of continuum row potentials, (4.12), of the nickel (•) and oxygen rows (▲); d = 0.352 nm

enhanced flux density, therefore between the [1$\bar{1}$0] nickel rows. By combining the results of the two directions, it can be concluded that the oxygen atoms are located in the long bridge position between the nickel atoms in the topmost [001] rows of the reconstructed surface (Fig. 4.14b).

The comparison between measured and computed angular-yield profiles allows more quantitative statements of the positions of adsorbed atoms. Figures 4.17 and 18 exhibit the calculated angular yield profiles of the oxygen Auger electrons for proton incidence along the two lattice directions discussed here. In both directions the bridge position in the [001] nickel rows with a distance h = 25 pm of the oxygen atoms above the topmost nickel layer delivers the best fit.

With the spectroscopy of Auger electrons induced by surface channeled ions one is able to identify foreign atoms with a sensitivity depending on the ionization cross section of, e.g. 0.05 monolayers of oxygen, and simultaneously to estimate their position with an accuracy of about 10 pm. With this technique it seems to be possible to study all combinations of substrates and adsorbates, except, of course, hydrogen and helium. For the identification and the determination of the position of these adsorbates nuclear reactions induced by surface channeled ions are convenient [4.29,31].

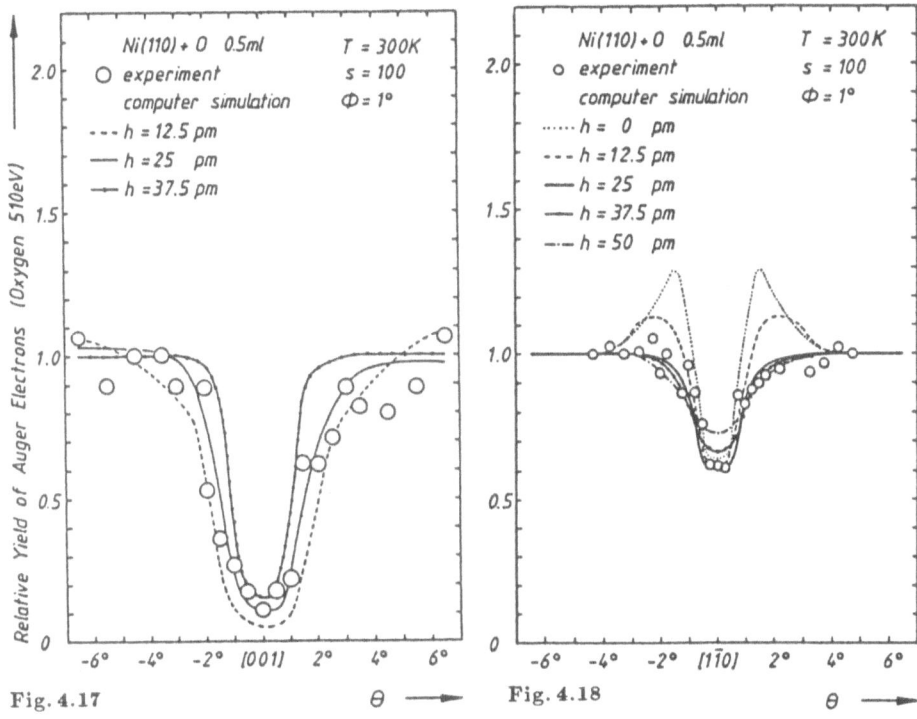

Fig. 4.17

Fig. 4.18

θ ⟶

θ ⟶

Fig. 4.17. Yield of 510 eV Auger electrons of the adsorbed oxygen atoms induced by 150 keV protons as a function of the lateral angle of incidence θ with respect to the [001] direction for a Ni(110) surface covered with 0.5 ML oxygen. The lines represent the results of computer simulations for various distances of the oxygen atoms above the topmost nickel layer [4.10]. The calculations are performed for a saw-tooth reconstructed surface

Fig. 4.18. As in Fig. 4.17, however, for proton incidence along the [1$\bar{1}$0] direction

4.5 Applicability of Surface Channeling to Surface Studies

The interest in applying surface channeling combined with ion-induced Auger-electron emission requires a high surface sensitivity of the method. Essentially, the selection of the most suitable Auger transition and of the most suitable angle of incidence to the surface is significant for the surface sensitivity and applicability of the surface channeling technique to surface studies.

4.5.1 Surface Sensitivity

The angular-yield profile of nickel L_3VV Auger electrons (Fig. 4.10) exhibits a deep and distinct minimum which indicates that the ion flux is already modulated – flux depression at the atomic rows and flux enhancement between the rows –, when the ion beam impinges on the first atomic layer. Projectiles incident at a very small angle to the surface, e.g. $\phi = 1°$ to the Ni(110) sur-

137

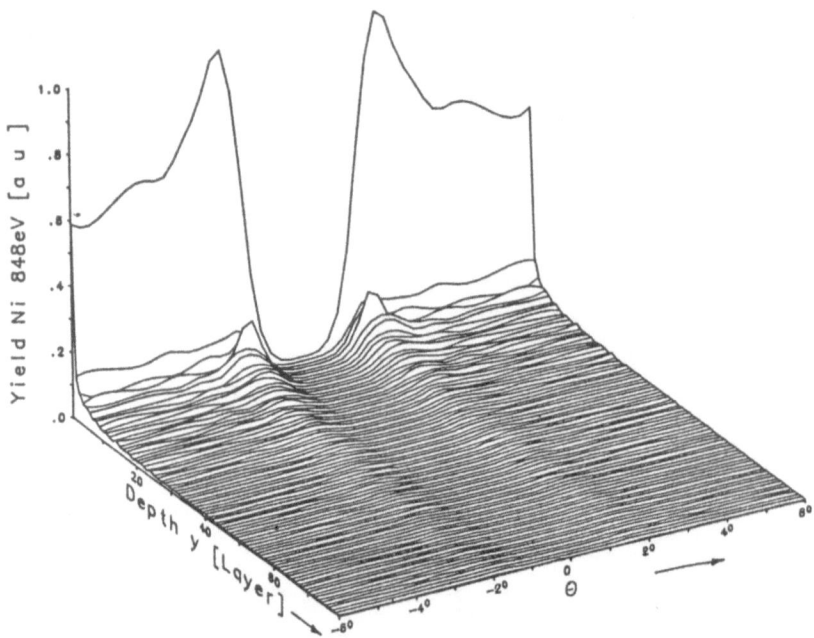

Fig. 4.19. Computed production profile of 848 eV Auger electrons (nickel L_3VV) induced by 150 keV protons as a function of the lateral angle of incidence θ with respect to the [001] direction and the depth beneath the surface plane for $\phi = 1°$. The depth extends from first layer to the 80th atomic layer corresponding with a depth y of 0 to 20 nm [4.14]

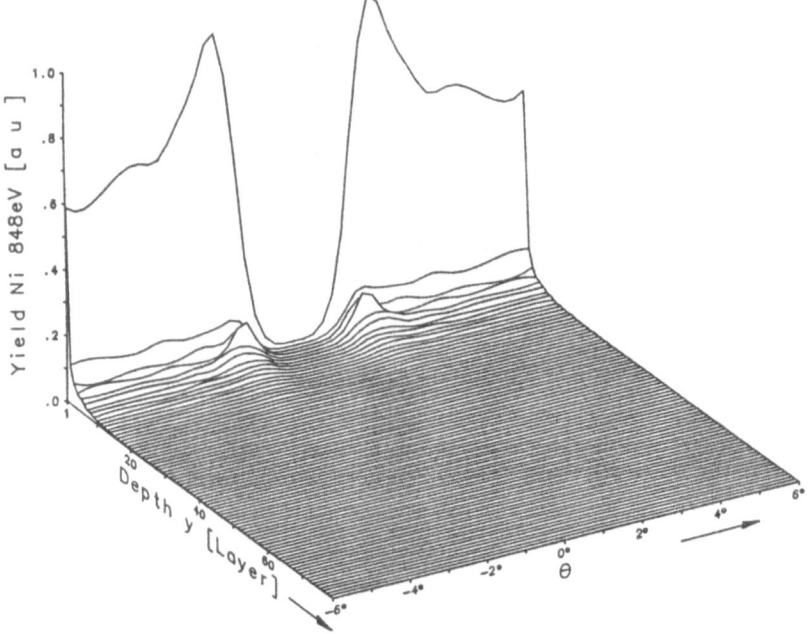

Fig. 4.20. Computed angular emission yield profile of the 848 eV Auger electrons as a function of the depth. Here, the production probability of the Auger electrons (Fig. 4.19) is multiplied by their escape probability using a mean free path λ of 1.5 nm [4.14]

138

face, are "prechanneled" by successive correlated collisions with many atoms of the surface rows, before they touch the surface. At their encounter with the topmost atomic layer, their impact parameters to the target atoms are already inhomogeneously distributed with an enhancement of projectiles with large impact parameters. The prechanneling effect can be visualized by the computer simulation. Figure 4.19 gives the angular production profile of the L_3VV nickel Auger electrons as a function of the depth y, where the Auger electrons are created. This presentation does intensionally not regard the escape probability P_E. The first atomic layer reveals a very strong dip of the angular production profile, which confirms the consideration above. The next deeper atomic layers in the surface region give only a minor contribution to the Auger-electron yield, the rest can be neglected at all. This comes from only a small number of protons penetrating deeper than a few atomic layers into the bulk. Additionally, these protons have smaller energies than the protons at the surface, because of energy loss. This reduces the Auger-electron production since the cross section of inner-shell ionization decreases approximately exponentially with the proton energy. Therefore, there are two reasons for the strong decrease of the Auger-electron production with the depth: first the decrease of the flux density and second the decrease of the kinetic energy of the projectiles with the depth. Furthermore, the small mean-free path of the produced Auger electrons ($\lambda = 1.5\,\text{nm}$ for $E_e = 848\,\text{eV}$) amplifies the sensitivity of axial surface channeling as a surface analytical method (Fig. 4.20).

4.5.2 Selection of the Auger Transition

The cross section for the Auger excitation depends on the projectile energy and the binding energy of the inner-shell electrons. Coulomb ionization yields the largest cross section if the projectile velocity is comparable to the velocity of the inner-shell electrons of the target atoms. For example, for proton incidence, the maximum cross section for the ionization of the nickel L_3 subshell lies at $1.25\,\text{MeV}$ far above the used proton energy of $150\,\text{keV}$ in the presented experiments. However, one has to compromise on the value of the cross section and an easily measurable angular-yield profile, half width $\theta_{1/2}$ of which is inverse proportional to the square-root of the projectile energy, see (4.17 and 22). On the other hand, for constant projectile energy, the ionization cross section strongly increases with decreasing electron binding energy (Fig. 4.6). Therefore, Auger transitions with low-energy-emitted Auger electrons would promise higher ionization cross sections. Moreover, electron energies of the order of $60\,\text{eV}$ guarantee a small mean-free path and, therefore, a higher surface sensitivity. By means of two different Auger transitions, nickel $M_{23}VV$ and L_3VV, the selection of the most suitable Auger transition for the application of the surface channeling technique will be examined in this section. The electron binding energies, $E_{M_{23}} = 68\,\text{eV}$ and $E_{L_3} = 855\,\text{eV}$, and the ionization cross sections for $150\,\text{keV}$ proton incidence, $\sigma_{M_{23}} = 6.7 \times 10^{-18}\,\text{cm}^2$ and $\sigma_{L_3} = 2.7 \times 10^{-20}\,\text{cm}^2$, for the M_{23}- and L_3-subshell, respectively, are quite different.

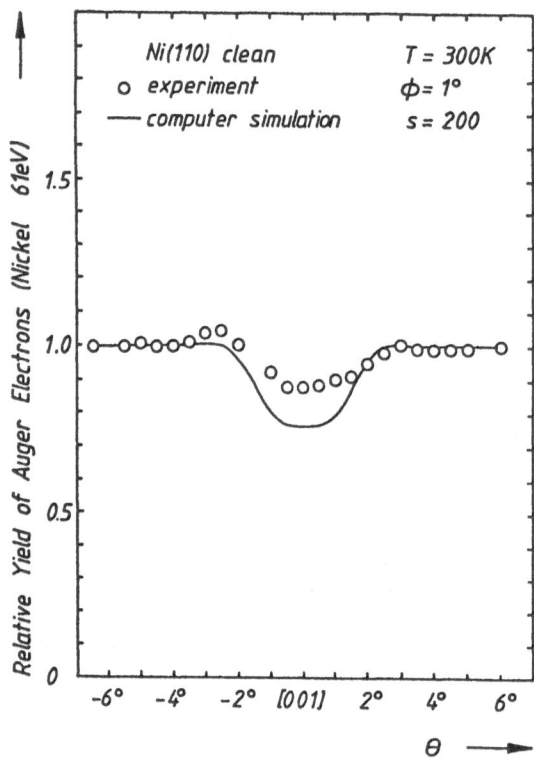

Fig. 4.21. Angular yield profile of the nickel $M_{23}VV$ (61 eV) Auger electrons [4.14]. The other conditions are identical as in Fig. 4.10, however, with $\lambda = 0.4\,nm$

Ni(110) clean   *T = 300K*
o *experiment* $\phi = 1°$
— *computer simulation*   *s = 200*

Relative Yield of Auger Electrons (Nickel 61eV)

$\theta \longrightarrow$

The angular-yield profile of the nickel $M_{23}VV$ Auger electrons at 61 eV presented in Fig. 4.21 is simultaneously measured with the angular-yield profile of the nickel L_3VV Auger electrons which is shown in Fig. 4.10. The minimum of the angular-yield profile of the $M_{23}VV$ Auger electrons is less pronounced and the minimum yield only slightly differs from random value. The difference of the two angular-yield profiles can easily be understood, if we look at the ionization probability of the L_3 and M_{23} subshell. The flux density distribution of the protons is the same for both angular yield profiles, but there is a difference in the close encounter process leading to the Auger-electron yield. Neglecting the shifting of vacancies (Sect. 4.3.1) one can proceed from the fact that the angular yield profile of L_3VV and $M_{23}VV$ Auger electrons is governed by the dependence of the ionization probability function of the L_3- and M_{23}-subshell on the impact parameter of the impinging protons. The ionization probability of the L_3-subshell has nonvanishing values only near the core (Fig. 4.6). The ionization probability of the M_{23}-subshell, however, is more extended and stretches far into the channel formed by the atomic rows. Even well channeled protons excite the branch of the M_{23}-subshell and, therefore, produce $M_{23}VV$ Auger electrons, as the computer result in Fig. 4.22 shows. If the escape probability of the produced 61 eV Auger electrons with $\lambda = 0.4\,nm$ is include in the calculation the contribution of the channeled protons, which have penetrated into the bulk, is still clearly visible (Fig. 4.23).

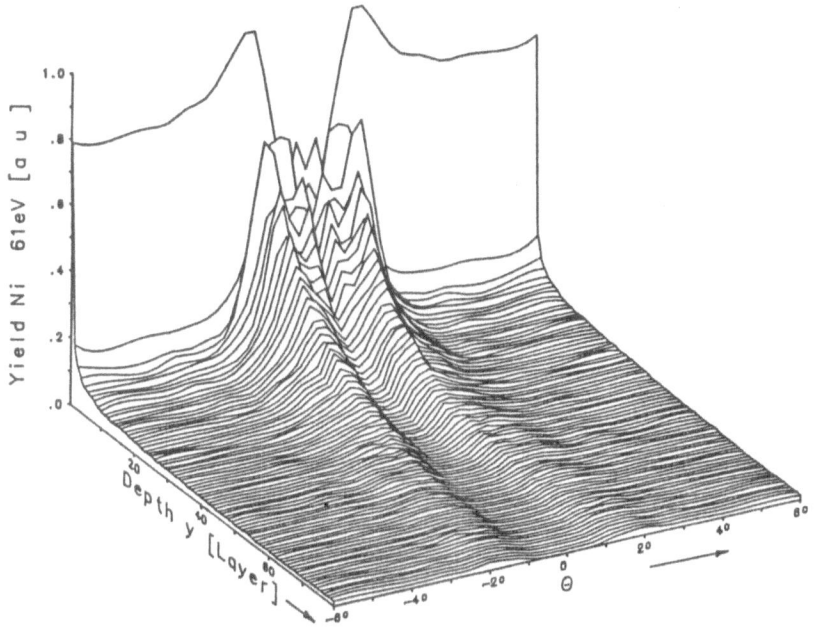

Fig. 4.22. Computed angular production profile of the nickel $M_{23}VV$ (61 eV) Auger electrons as a function of the depth [4.14]. The other conditions are identical as in Fig. 4.19

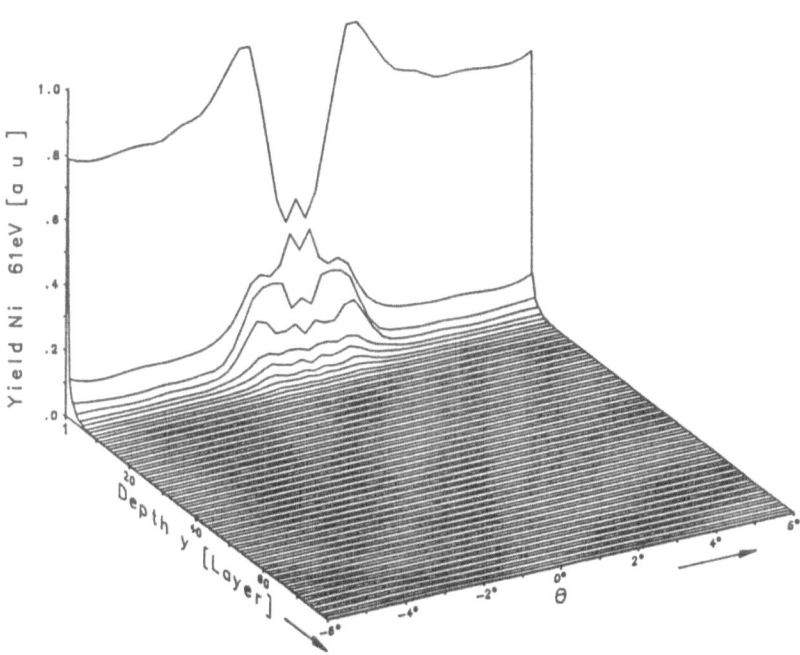

Fig. 4.23. Computed angular emission yield profile of the nickel $M_{23}VV$ (61 eV) Auger electrons as a function of the depth [4.14]. The other conditions are identical as in Fig. 4.20 but with $\lambda = 0.4\,\text{nm}$

The difference of the two Auger-electron angular-yield profiles can also be explained by the adiabatic distance of the ionization process given by

$$r_{ad} = \hbar v / E_W \qquad (4.36)$$

where v is the projectile velocity, and E_W the electron binding energy of the ionized shell W. The adiabatic distance for the M_{23}-subshell, $r_{ad} = 52\,pm$, is much larger than the minimum distance $r_c = 9.2\,pm$ for 150 keV proton channeling along the [001] nickel direction, see (4.6). This means that even well channeled protons with distances to the atomic rows $r_c < r < r_{ad}$ can excite $M_{23}VV$ Auger electrons. The L_3-subshell can significantly be excited only by dechanneled protons because the adiabatic distance $r_{ad} = 4.1\,pm$ is smaller than r_c. Therefore, the condition for a suitable Auger transition is that the adiabatic distance of the ionization process, (4.36) must be smaller than the critical distance for axial channeling, (4.6).

4.5.3 Selection of the Angle of Incidence to the Surface

At very small angles of incidence ϕ the transverse energy of the projectiles to the surface, $E\phi^2$, is very small and the reflection appears to be nearly specular (reflection at plane potential energy contours, see Fig. 4.2), independent of the lateral angle of incidence θ and thus of the atomic arrangement in the surface. By increasing ϕ the projectiles are scattered by successive correlated collisions with many atoms of the surface atomic rows, axial surface channeling, as is already discussed above. By further increasing the angle ϕ a considerable fraction of the projectiles can penetrate through the surface plane and be channeled between the lattice planes extending into the crystal. The transition from axial to planar channeling is governed by the angle ϕ to the surface. Planar channeling is developed for angles of incidence ϕ larger than a critical angle ϕ_c given by [4.36–38]:

$$\phi_c^2 = d_{ap}^2 U_p''(d_p/2)/8E, \qquad (4.37)$$

where d_{ap} is the distance between the atomic rows in the lattice plane, and d_p is the separation of the planes. U_p'' is the second derivative of the planar potential [4.1–4]. Numerical estimates can be obtained with, e.g., the planar continuum potential in the Thomas-Fermi-Molière approximation

$$\phi_c^2 = (\pi/4)(d_{ap}/a_{TF})(2Z_1 Z_2 e^2 / E d_{[hkl]}) \sum_{i=1}^{3} \alpha_i \exp\left(-\beta_i d_p / 2 a_{TF}\right) \qquad (4.38)$$

with $\alpha_i = (0.6; 0.66; 0.105)$, $\beta_i = (6; 1.2; 0.3)$, and $d_{[hkl]}$ the interatomic distance in the [hkl] rows in the surface. For 150 keV protons, for example, the critical angle is $\phi_c = 1.5°$ for the transition from axial channeling along the [001] direction to planar channeling between the $(1\bar{1}0)$ planes of the nickel crystal.

Figure 4.24 presents the measured yield of the nickel L_3VV Auger electrons induced by 150 keV protons with an angle of incidence to the surface $\phi = 4.5°$. The lateral angle of incidence θ is again counted to the [001] direction as in

142

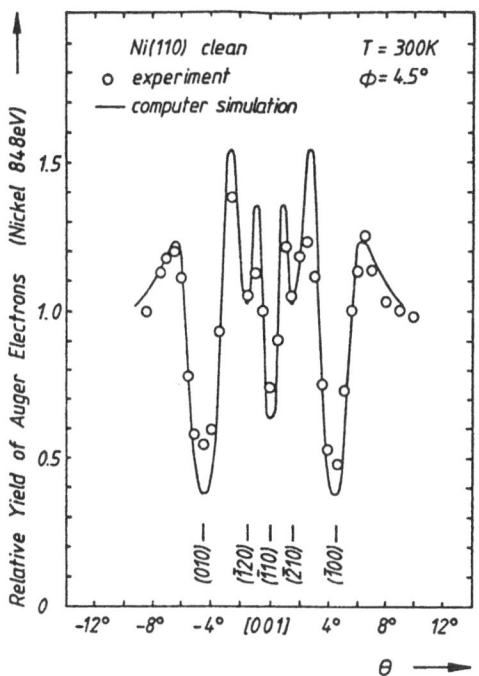

The figure shows a plot with:
- Y-axis: Relative Yield of Auger Electrons (Nickel 848 eV), ranging from 0 to 1.5
- X-axis: θ, ranging from $-12°$ to $12°$, with $[001]$ marked at center

Legend/labels in plot:
Ni(110) clean   T = 300K
o experiment $\phi = 4.5°$
— computer simulation

Lattice plane markers along x-axis: (010), (120), (110), (210), (100)

Fig. 4.24. Yield profile of nickel L_3VV (848 eV) Auger electrons induced by 150 keV protons as a function of the lateral angle of incidence θ with respect to the [001] direction in the Ni(110) surface for $\phi = 4.5°$ (planar channeling). The escape probability of the Auger electrons is included in the calculation using the mean free path λ of 1.5 nm [4.14]

axial surface channeling (Fig. 4.10). The estimated position of the low-index lattice planes (uwv) is given by

$$\theta_{(uvw)} = \phi \cot\varphi_{(uvw)}, \tag{4.39}$$

$$\cos\varphi_{(uvw)} = (u, v, w)(m, n, o)/[(u^2 + v^2 + w^2)(m^2 + n^2 + o^2)]^{1/2}, \tag{4.40}$$

where $\varphi_{(uvw)}$ is the angle between the (uvw) plane and the (mno) surface [in the experiments here: (mno) = (110)]. The position of the minima in the Auger-electron yield is in accordance with the low-index lattice planes. The result of the computer simulation gives by the solid line in Fig. 4.24 confirms the experimental result. The calculated Auger-electron production and emission yield for planar channeling as a function of the lateral angle of incidence θ and the depth y is shown in Fig. 4.25 and 26, respectively. In planar channeling the effect of prechanneling is less pronounced as in axial surface channeling. The Auger-electron yield of the first atomic layer has indistinct minima. Only in some layers below the surface the channeling dips become clearly visible. That means that the strong inhomogeneity of the ion-flux density, caused by channeling, is developed within the crystal in deeper atomic layers and, therefore, the Auger-electron production in dependence on the angle θ is formed below the surface. Thus, planar channeling is not sensitive for surface analysis. As an example, Fig. 4.27 shows the measured nickel L_3VV Auger-electron yields induced by 150 keV protons incident along the $[1\bar{1}0]$ direction with $\phi = 4.5°$ on clean and an oxygen covered Ni(110) surface. Besides the experimental uncer-

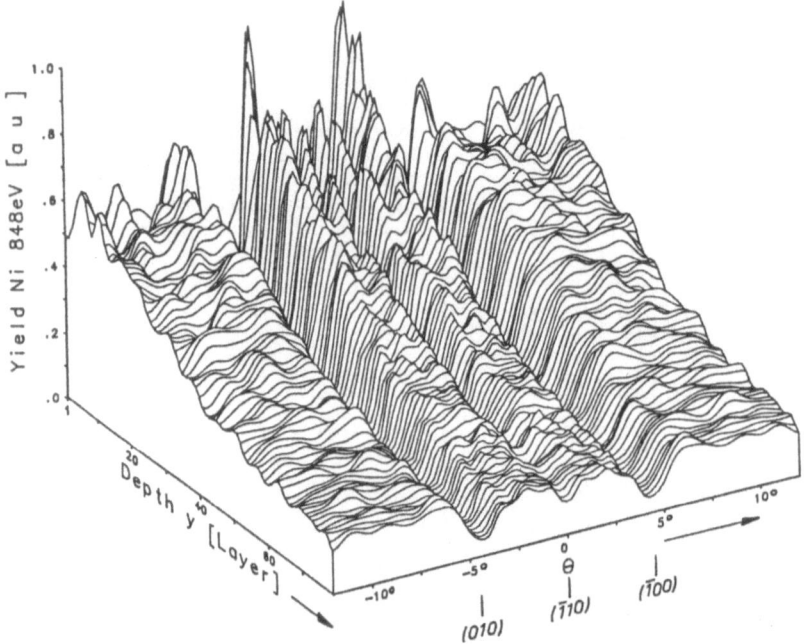

Fig. 4.25. Computed production profile of the nickel L_3VV (848 eV) Auger electrons induced by 150 keV protons as a function of the lateral angle of incidence θ with respect to the [001] direction and the depth y beneath the (110) surface plane for $\phi = 4.5°$ (planar channeling) [4.14]

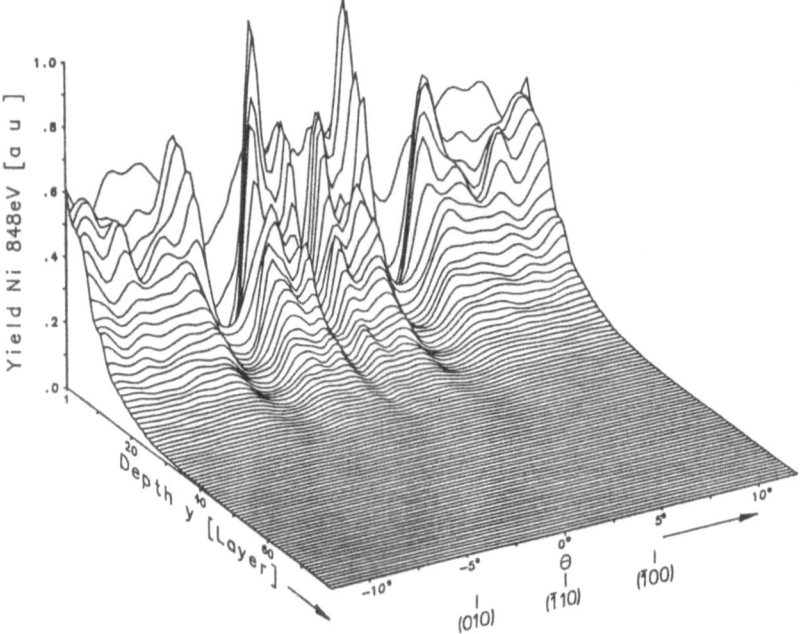

Fig. 4.26. Computed emission yield profile of the 848 eV Auger electrons for the same conditions of incidence as in Fig. 4.25 including in the calculation the escape probability of the electron with $\lambda = 1.5$ nm [4.14]

144

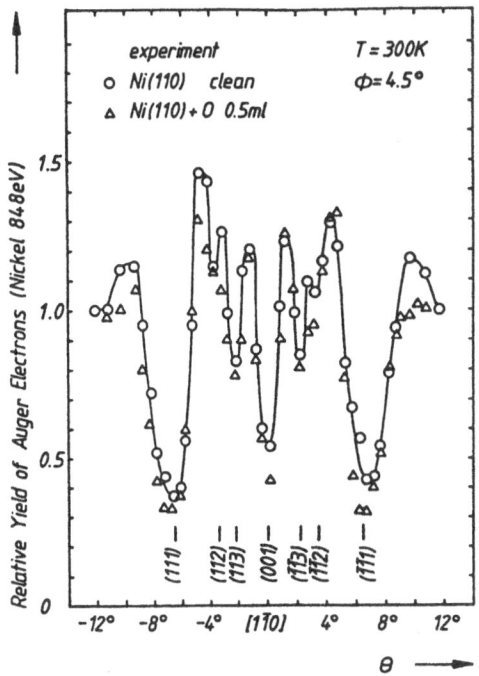

Relative Yield of Auger Electrons (Nickel 848eV)

experiment   T = 300K
o Ni(110) clean $\phi = 4.5°$
△ Ni(110)+O 0.5ml

1.5

1.0

0.5

0

(111) (112)(113) (001) (113)(112) (111)

-12°   -8°   -4°   [1Ī0]   4°   8°   12°

$\theta \longrightarrow$

Fig. 4.27. Yield profile of the nickel L_3VV (848 eV) Auger electrons induced by 150 keV protons versus the lateral angle of incidence θ with respect to the [1Ī0] direction for a clean (o) and an oxygen covered Ni(110) surface (△). The angle of incidence to the surface plane ϕ is 4.5° (planar channeling). The full line is a fit to the values measured with the clean surface [4.14]

tainties, the experimentally found channeling dips for clean and oxygen covered Ni(110) are in coincidence, although the surface is reconstructed. In axial surface channeling with $\phi = 1°$ along the [1Ī0] direction the angular-yield profile has changed drastically (Sect. 4.4.3). Therefore planar channeling (large angles ϕ) is less suited to study the surface structure than axial surface channeling (small angles ϕ). The condition for the selection of the most suitable angle of incidence ϕ to the surface for axial surface channeling is empirically found to be

$$\phi_c/2 < \phi \lesssim \phi_c \tag{4.41}$$

with ϕ_c given by (4.38).

4.6 Factors of Influence on the Angular Yield Profiles

In this section the influence of the radiation damage, of the thermal vibrations, of the mean-free path of the induced electrons, and the surface steps on the angular yield profiles is discussed.

4.6.1 Radiation Damage

The projectiles loose energy in their interaction with the target by inelastic (electronic) energy transfer to the core and valence electrons of the target (Sect. 4.3.2) and by elastic (nuclear) energy transfer to the target atoms. The nuclear energy loss gives rise to dislocations, sputtering of substrate atoms and desorption of adsorbates. The mean value of the inelastic energy loss of 150 eV protons in an amorphous nickel target or in random incidence is of about 0.4 eV/nm [4.28]. From energy and momentum considerations it is well known that the elastic energy transfer in a single projectile-target atom collision is related to the square of the reflection angle (for small angles). In axial channeling the energy loss due to such elastic collisions depends on the fourth power of ψ for a projectile with transverse energy $E_\perp = E\psi^2$,

$$-(dE/dz)_n \cong (M_1/M_2)Nd^2_{[hkl]}E\psi^4, \qquad (4.42)$$

where N is the atomic density, M_1 and M_2 the mass of projectile and target atom, and $d_{[hkl]}$ the interatomic distance along the axis of the channeling motion. The fourth power enters through the averaging over the various positions towards the row atoms a channeled projectile attains during its motion. Numerically, a 150 keV proton moving along the [1$\bar{1}$0] direction of the nickel lattice at the critical channeling angle (here, $\psi_c = 2.8°$), would loose only 0.08 eV/nm. The consequence of the low energy transfer to target atoms in nuclear encounters with channeled projectiles is the drastic reduction of radiation damage.

Computer simulations (for 150 keV protons, Ni(110) surface covered by 0.05 ML oxygen, $\phi = 1°$, and T = 300 K) give an estimate of the sputtering yield of 2.5×10^{-3} Ni atoms per proton for incidence parallel to the [1$\bar{1}$0] direction and 5×10^{-2} Ni atoms per proton for random incidence. The desorption yield of the adsorbed oxygen is 3.5×10^{-2} oxygen atoms per proton for incidence along the [1$\bar{1}$0] direction (the oxygen atoms are located between the [1$\bar{1}$0] nickel rows, Sect. 4.4.4) and 2×10^{-2} oxygen atoms per proton for random incidence. For a typical ion current density of 100 nA/mm^2 and an usual measuring time of 15 min for a complete angular θ-scan about 10^{15} protons/cm^2 are incident with $\phi = 1°$ on the crystal surface. Together with the low sputtering and desorption yields, this guarantees that the beam induced surface damage and the desorption of adsorbates is rather small. In the experiments presented above even after 1 h of proton bombardment neither a decrease in the Auger electron yield of the adsorbed oxygen nor an increase of the minimum of the angular field profile, which is a measure of the damage (Sect. 4.2.4), has been observed.

4.6.2 Thermal Lattice Vibrations

The influence of the target temperature is well-known from other channeling experiments and from theoretical investigations [4.1–3,7]. The angular-yield profile of the Auger electrons is influenced, with increasing target temperature, by two effects: (i) widely elongated target atoms disturb the continuum steering potential and cause a more homogeneously distributed ion flux density; (ii) the expanded distribution of the target atoms causes a broadening of the effective ionization probability function, (Sect. 4.3.1). Both the effects increase the minimum of the angular-yield profile. In addition, the attenua-

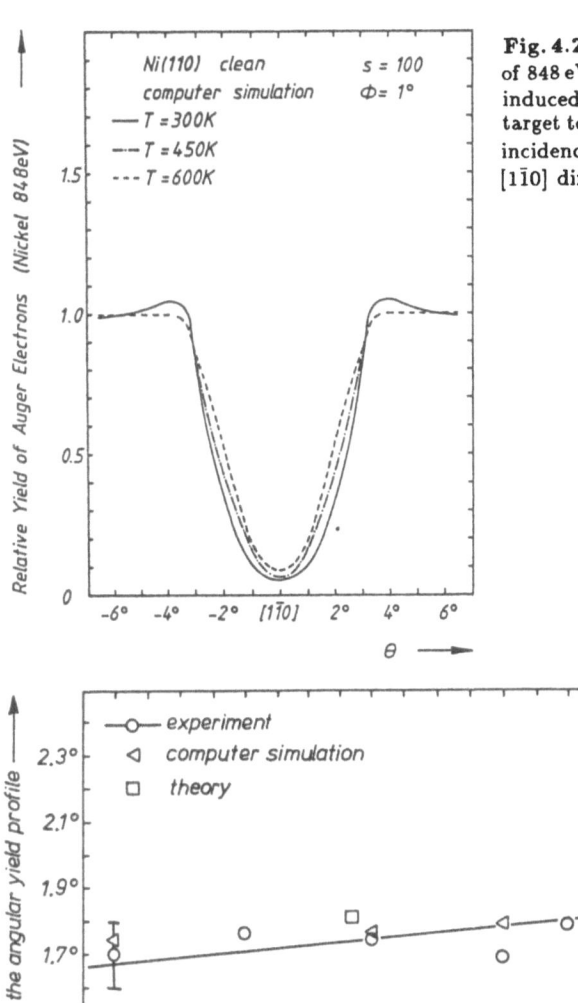

Fig. 4.28. Computed angular yield profiles of 848 eV Auger electrons (nickel L_3VV) induced by 150 keV protons for various target temperatures. The lateral angle of incidence θ is counted with respect to the [1$\bar{1}$0] direction in the Ni(110) surface [4.10]

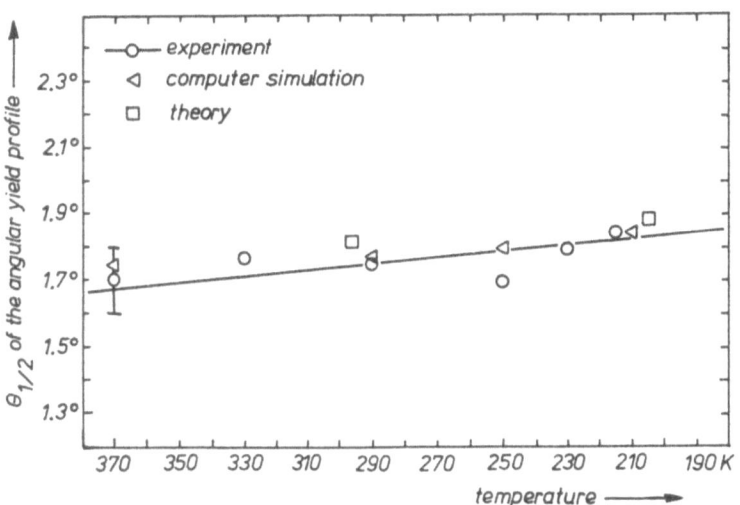

Fig. 4.29. Half width at half minimum of the angular yield profiles of 848 eV Auger electrons (nickel L_3VV) induced by 150 keV protons incident on a clean Ni(110) surface along the [001] direction versus the target temperature. The solid line is the best fit of the experimental results [4.39]. The values denoted by (□) are calculated according (4.8) and (4.16)

tion of the steering potential of the atomic rows decreases the HWHM $\theta_{1/2}$ of the angular yield profiles. But on the whole, the influence of the temperature is not very strong, as can be seen in Figs. 4.28 and 29. Because of the extended ionization probability function (Fig. 4.6) in channeling experiments using ion-induced Auger-electron emission the influence of the distribution of the thermally vibrating target atoms is less intense than in those with a nearly δ-shaped probability function, e.g. Rutherford backscattering. A similar result is obtained for inner shell excitation and x-ray measurements [4.40], which have identical excitation characteristics to those of the Auger-electron emission.

4.6.3 Mean-Free Path of Excited Electrons

As pointed out in Sect. 4.3, there exist several processes which prevent the Auger electrons produced in the crystal to leave the bulk with their characteristic energy. The escape probability depends on the depth y, in which the Auger electron is produced, and on the electron energy E_e, which determines the mean free path λ. The escape probability $P_E(y)$ is a weighting function which enters into the integral (4.23) of the Auger-electron yield over all the layers. Because of the rapid fall-off for increasing depth of the ion-flux density in surface channeling and of the similar shape of the angular production profile of the Auger electrons (Fig. 4.19) the integral angular-yield profile will exhibit no dependence on the mean-free path of the electrons. An example is given in Fig. 4.30. The angular-yield profile of nickel L_3VV Auger electrons, excited under the surface channeling condition does not change its appearance in a wide range of λ values.

The influence of the mean-free path on the angular-yield profile of Auger transitions with a large adiabatic distance, and for large angles of incidence ϕ to the surface is intense [4.14], because the intrinsic shape of the production profiles varies with the depth.

4.6.4 Surface Steps

Surface steps disturb the ion flux parallel to the surface. They level the flux density so that it becomes more homogeneous. Under surface channeling conditions, steps cause an increase of the minimum and a broadening of the width of the angular yield profiles. As an example, Fig. 4.31 shows the influence of the step density for proton channeling along the [001] direction. The distance of the steps (the terrace width) s is always given in units of the distance of the transverse planes to the low-index direction along the channeling motion. The height of one step corresponds to the periodicity length perpendicular to the surface plane.

A recent work [4.26] has shown that in surface channeling at surfaces with step distances s>30 units only the upward steps significantly influence the angular yield profiles. In any cases, it is sufficient to use a mean effective distance of upward steps as a constant fitting parameter in the computer simulation. To check, however, the influence of this mean effective step distance on surface channeling independent experiments are required. For example, ion scattering at grazing ion incidence is well appropriated for the determination of the mean effective step distance. In these experiments, the angular yield profile of the scattered ion at the crystal surface is measured. The scattering yield depends only on the inhomogeneous flux density distribution of the surface channeled

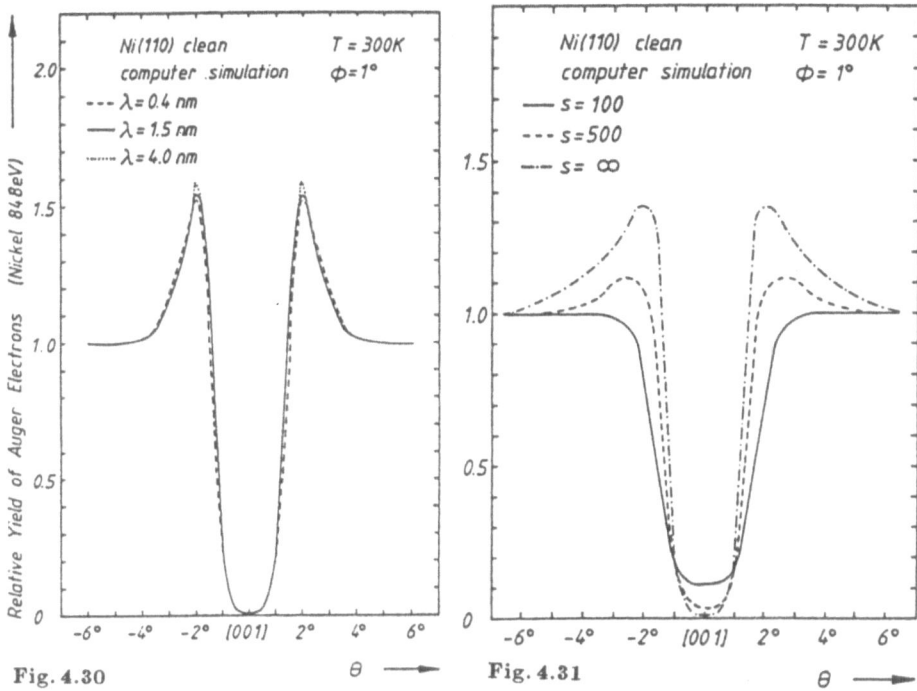

Fig. 4.30

Fig. 4.31

Fig. 4.30. Computed angular yield profiles of 848 eV Auger electrons (nickel L_3VV) induced by 150 keV protons incident along the [001] direction for various mean-free paths of the emitted Auger electrons [4.14]

Fig. 4.31. Computed angular yield profiles of 848 eV Auger electrons (nickel L_3VV) induced by 150 keV protons incident along the [001] direction for various step distances [4.14]. s is given in units of the distance of transverse planes to the [001] direction. $s = \infty$ represents a step free surface. The height of each step is taken 0.25 nm

ions, which varies with step density, and no more on the ionization and escape probability as the Auger-electron yield. The comparison of the experimentally observed scattering yield profiles with computer simulations, in which only the step distance is variable, provides the mean effective step distance, which actually influences the surface channeling dips [4.26].

4.7 Concluding Remarks

Since its first experimental realization only few years ago, the surface channeling has proved to be a very versatile method for surface analysis. It is sensitive to reconstructions of the substrate, to positions of adsorbates, and to the structure of thin epitaxial films. With this method it is possible to study all combinations of substrates and adsorbates and one is able to identify foreign atoms, to provide a straightforward estimate of the position of the adsorbates with an accuracy of about 10 pm, and simultaneously to obtain detailed information on the surface structure. To a large degree these informations are complementary to those provided by the various surface sensitive techniques.

Channeling concepts for surface analysis have given rise to two further scattering geometries. One geometry make use of ion backscattering of a beam aligned with or near a low-index direction into the bulk and the detector in a random direction; such an arrangement is often referred to as a single-alignment experiment. The fundamentals of this technique and its applications have extensively been described in a number of review articles [4.41,42]. In brief, the method is based on the fact that whenever a beam of energetic ions is incident on a perfect crystal surface along a low-index direction into the bulk, then only the first atom in the low-index row of atoms has a high probability of being hit by the ion. "Being hit" means, that the ion comes close enough to the atomic nucleus for it to participate in a close-encounter process such as, e.g., a nuclear reaction or wide-angle scattering. The probability of hitting an atom is equivalent to the yield of the close-encounter process. The deeper lying atoms are shadowed by the first atom, and the probability of hitting them is determined by the lattice vibrations, which cause them to spend part of their time outside the shadow. Displacements of the surface atoms away from the atomic rows cause an exposure of the second atom of the row to the beam and thereafter the hitting probability will be larger. This technique is sensitive to atomic translations in the plane of the surface, surface relaxation, and to the formation of a non-epitaxial monolayer.

Another geometry for surface analysis, developed by the FOM group [4.43, 44] make use of ion backscattering of a beam aligned with a low-index direction into the bulk, and the detector is along or near a second low-index direction (double-alignment experiment). In these experiments, also referred to as "channeling-blocking", the incident beam is aligned in a low-index direction into the bulk, such, that only the first two monolayers of the surface are "visible" to the beam, the deeper-lying atoms being shadowed by the surface atoms. The atomic structure of the outermost layer of a clean crystal can be obtained by measuring the directions in which projectiles, that are backscattered from the second layer, are blocked by the surface atoms. Thus, the second layer acts as an emitter of particles, the first layer blocks them in specific directions. From the resulting blocking pattern the atomic structure can be derived with the aid of simple geometry. Such measurements can reveal the structure of the surface, of an ordered array of adsorbates, and of an epitaxial thin film of less than 1 nm.

Acknowledgements. I am most grateful to my collaborator M. Schuster for the fruitful cooperation.

References

4.1   J. Lindhard: Kgl. Dan. Vid. Selsk. Mat. Fys. Medd. **34**, no. 14 (1965)

4.2   D.S. Gemmell: Rev. Modern Physics **46**, 1924 (1974)

4.3   D. Morgan (ed.): *Channeling-Theory, Observation and Application* (Wiley, New York 1973)

4.4   R. Sizmann, C. Varelas: In *Advances in Solid State Physics* **17**, 261 (Vieweg, Braunschweig 1977)

4.5   R. Sizmann, C. Varelas: Nucl. Instrum. Meth. **132**, 633 (1976)

4.6   P. Gombas: *Die statistische Theorie des Atoms* (Springer, Wien 1949)

4.7   C. Varelas, R. Sizmann: Rad. Eff. **16**, 211 (1972)

4.8   J. Lindhard, V. Nielsen, M. Scharff: Kgl. Dan. Vid. Selsk. Mat. Fys. Medd. **36**, no. 10 (1968)

4.9   G. Molière: Z. Naturforsch. **2a**, 133 (1947)

4.10   M. Schuster, C. Varelas: Surf. Sci. **134**, 195 (1983)

4.11   E. Bonderup, H. Esbensen, J.U. Andersen, H.E. Schiøtt: Rad. Eff. **12**, 261 (1972)

4.12   V.V.Beloshisky, M.A. Kumakov, V.A. Muralev: Rad. Eff. **13**, 9 (1972)

4.13   J.H. Barrett: Phys. Rev. B**3**, 1527 (1971)

4.14   M. Schuster, C. Varelas: Nucl. Instr. Meth. B**9**, 145 (1985)

4.15   D.H. Madison, E. Merzbacher: In *Atomic Inner-Shell Processes*, Vol. 1, ed. by B. Crasemann (Academic, New York 1975) p. 1

4.16   J.M. Hansteen, O.M. Johnsen, L. Kocbach: Atomic Data and Nuclear Data Tables **15**, 305 (1975)

4.17   W. Bambynek, B. Crasemann, R.W. Fink, H. Mark, C.D. Swift, R.E. Price, P.V. Rao: Rev. Mod. Phys. **44**, 716 (1972)

4.18   M.P. Seah: Surf. Sci. **32**, 703 (1972)

4.19   R.A. Baragiola: Rad. Eff. **61**, 47 (1982)

4.20   T. Koshikawa, R. Shimizu: J. Phys. D**7**, 1303 (1974)

4.21   S. Tongaard, P. Sigmund: Phys. Rev. B**25**, 4452 (1982)

4.22   V.M. Dwyer, J.A.D. Matthew: Surf. Sci. **143**, 57 (1984)

4.23   M. Hou, C. Varelas: Appl. Phys. A**33**, 121 (1984)

4.24   Chr. Lehmann, G. Leibfried: Z. Phys. **172**, 465 (1963)

4.25   C. Varelas, R. Sizmann: Rad. Eff. **25**, 163 (1975)

4.26   W. Graser, C. Varelas: Surf. Sci. **157**, 74 (1985)

4.27   O.S. Oen, M.T. Robinson: Nucl. Instr. Meth. **132**, 647 (1976)

4.28   J. Lindhard, M. Scharff, H.E. Schiøtt: Kgl. Dan. Vid. Selsk. Mat. Fys. Medd. **33**, no. 14 (1963)

4.29   C. Varelas, H.D. Carstanjen, R. Sizmann: Phys. Lett. **77**A, 469 (1980)

4.30   M. Schuster, C. Varelas: Nucl. Instr. Meth. B**2**, 299 (1984)

4.31   E. Sailer, C. Varelas: Nucl. Instr. Meth. B**2**, 326 (1984)

4.32   W. Graser, C. Varelas: Phys. Scripta T**6**, 153 (1983)

4.33   J.A. Van den Berg, L.K. Verheij, D.G. Armour: Surf. Sci. **91**, 218 (1980)

4.34   R.G. Smeenk, R.M. Tromp, F.W. Saris: Surf. Sci. **107**, 124 (1982)

4.35   H. Niehus, G. Comsa: Surf. Sci. **151**, L171 (1985)

4.36   U. Bill, R. Sizmann, C. Varelas, K.E. Rehm: Rad. Eff. **27**, 59 (1975)

4.37   C. Varelas, R. Sizmann: Surf. Sci. **71**, 51 (1978)

4.38   C. Varelas, G. Goltz, R. Sizmann: Surf. Sci. **80**, 524 (1979)

4.39   R. Pfandzelter: Diploma thesis, Universität München (1985)

4.40   J.U. Andersen, J.A. Davies: Nucl. Instr. Meth. **132**, 179 (1976)

4.41   L.C. Feldman: Nucl. Instr. Meth. **191**, 211 (1981)

4.42   L.C. Feldman, J.W. Mayer, S.T. Picraux: *Materials Analysis by Ion Channeling* (Academic, New York 1982)

4.43   W.C. Turkenburg, W. Sozska, F.W. Saris, H.H. Kersten, B.G. Colenbrander: Nucl. Instr. Meth. **132**, 587 (1976)

4.44   J.F. van der Veen, R.M. Tromp, R.G. Smeenk, F. Saris: Nucl. Instr. Meth. **171**, 143 (1980)

5. Dynamical Surface Properties in the Harmonic Approximation

J.E. Black
With 16 Figures

Recent advances in surface spectroscopy, in particular in electron energy loss spectroscopy (EELS) and helium beam scattering spectroscopy, have made it possible to measure the frequency of vibration of atoms at material surfaces. Before these advances the surface scientist was obliged to concentrate on averaged vibrational properties of surface atoms, such as mean-square displacement or to study waves of long wavelength. The theory needed for such studies is described in detail in a review article by *Wallis* [5.1]. The objective of the present chapter is to focus attention on the theory needed for the study of vibration frequencies, the theory of lattice dynamics of discrete systems. Where possible we also compare the theoretical results with those obtained in experiment.

Study of the dynamics of pure crystal surfaces will lead to an understanding of the interatomic forces at the surface, and show how they differ from the forces in the bulk. Moreover the details of the relaxation and reconstruction can be explored. While such studies are of interest to condensed-matter theorists there is a range of related problems that are of interest to chemists. These problems involve the dynamics of adatoms and admolecules on solid surfaces.

The outline of this chapter is as follows. In Sect. 5.1 we provide some background information. In Sect. 5.2 we deal with the theory of vibrations in the harmonic approximation. We treat cases with no periodicity parallel to the surface, cases with a bare but periodic surface, and finally cases in which a periodic layer of adatoms is deposited on a bare periodic substrate.

In Sect. 5.3 we treat experimental and theoretical results for bare metal surfaces. We first consider the methods used to calculate the vibration frequencies, we then examine the results of some rather general theoretical calculations, finally we examine theoretical results for specific metals, comparing them with experimental data where possible.

In Sect. 5.4 we treat metal surfaces with adatoms or admolecules present. We begin with a treatment of how mode frequencies are to be calculated. Here we treat both isolated adatoms and periodic overlayers of adatoms. We then consider theoretical results for various metal/periodic adatom systems. Where possible the results are compared with those obtained by experiment.

In Sect. 5.5 we present a few concluding remarks.

5.1 Introductory Remarks

The lattice dynamics of atoms at surfaces of various types of crystals has been studied. These materials include inert gas crystals, ionic crystals, semiconductor crystals and metal crystals. Each of these materials is interesting in its own right because of the rather different forces that bind the atoms together. The inert gas atoms can be thought of as coupled by Van der Waals attractive forces which are balanced by repulsive Pauli exclusive forces. The ionic crystals contain ions which are coupled by electrostatic forces. The semiconductor atoms are coupled to their nearest neighbour atoms by covalent bonds. Metal forces are the most complicated. The valence electrons leave the parent atoms when the metal atoms form a solid. These electrons move freely through the solid forming an electron gas, which holds the parent ions together. One can nevertheless think of these ions as interacting with one another, the interaction being screened by the intervening electron gas.

In this chapter we will concentrate on bare metal surfaces, metal surfaces with a variety of adatoms, and metal surfaces with layers of rare-gas atoms. The reader is referred to recent articles by *De Wette* [5.2], *Toennies* [5.3], and *Benedek* [5.4] for a review of the experimental and theoretical results for ionic crystals. For a review of semiconductor crystal surface vibration studies the reader is referred to *Kanellis* et al. [5.5], and *Allan* and *Mele* [5.6], and for atoms on graphite see *Ellis* et al. [5.7].

A small digression on the structure of surfaces is both appropriate at this point and will be useful later. In Fig. 5.1a a surface layer is shown. The vectors *a* and *b* are chosen so that

$$r = na + mb, \tag{5.1}$$

where n = 0, ±1, ±2, ±3... and m = 0, ±1, ±2, ±3..., is a vector to any atom on the surface. The vectors define a "primitive surface cell". There is one such cell for each surface atom. Figure 5.1a was drawn for arbitrary values of *a* and *b*. Such a lattice of points is known as an oblique lattice. There are four special lattices [5.8] which are invariant under rotation by $2\pi/3$, $2\pi/4$, $2\pi/6$ or under mirror reflection. They are also illustrated in Fig. 5.1. In Table 5.1 we illustrate the nature of the surface structure for face centered cubic (fcc) and body centered cubic (bcc) crystals in the absence of reconstruction. The Miller indices define a vector perpendicular to the surface.

In our models of crystals we will introduce no defects. However, the surface layers may have structure different from bulk layers. We may have "relaxation" in which the atoms in the surface layers are displaced by small amounts from the positions they would occupy in the bulk. The relaxation is believed to occur in the top layer of the nickel(100) surface, for example [5.9]. The top layer atoms are displaced downward by 3.2% of their bulk spacing above the layer below. We may also have the more dramatic "reconstruction" of upper layers. For example, the surface of iridium(110) is believed to reconstruct [5.10]. LEED studies indicate this reconstruction forms a (1×2) structure at the surface layer. The

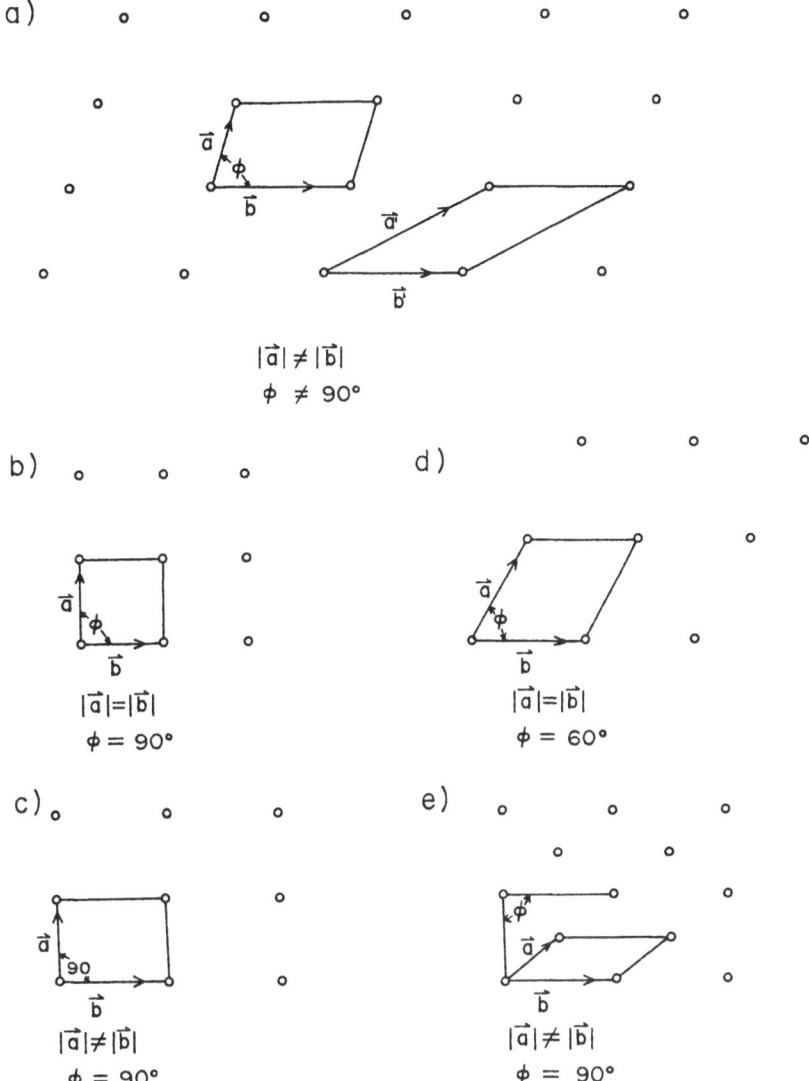

Fig. 5.1a–e. The 5 surface Bravais lattices (a) p-oblique, (b) p-square, (c) p-hexagonal, (d) p-rectangular, (e) c-rectangular. The vectors **a** and **b** are shown along with a primitive surface cell. An infinite number of choices of primitive cells can be made for each lattice type. The cell marked with **a'** and **b'** is one example

unreconstructed surface and one proposed model of the reconstructed surface are shown in Fig. 5.2. We see that the primitive cell needed to generate the entire (110) surface in the reconstructed case is two times longer in the y-direction than the cell in the unreconstructed case. Hence the designation (1×2) rather than (1×1) for this structure.

In this chapter we will explore the dynamics of adparticles on perfect crystal surfaces. While real processes take place on surfaces which are far from

Table 5.1. The structure of atoms at fcc and bcc surfaces for various Miller indices

Miller indices	fcc	bcc
100	p-square	p-square
110	p-rectangular	c-rectangular
111	p-hexagonal	p-hexagonal
210	c-rectangular	p-rectangular
211	p-rectangular	p-rectangular
221	p-rectangular	c-rectangular
310	p-rectangular	c-rectangular
311	c-rectangular	c-rectangular
320	c-rectangular	p-rectangular
321	p-oblique	p-oblique
322	p-rectangular	p-oblique
331	c-rectangular	c-rectangular
332	p-rectangular	p-rectangular

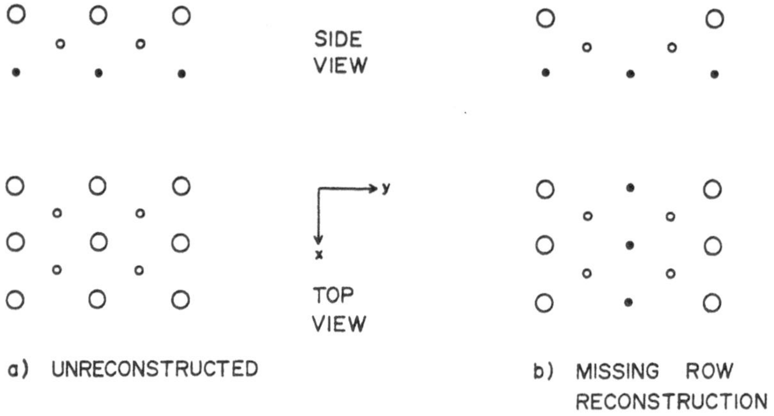

Fig. 5.2a,b. The first three layers of the face centered cubic (110) surface are shown: (a) no reconstruction, (b) missing row reconstruction. The large circles indicate atoms at the surface, smaller circles indicate atoms one layer down, and dots indicate atoms two layers down

those of perfect crystals it can be argued [5.11] that a study of the behaviour of adparticles on ideal faces of different Miller indices of a crystal can lead to an understanding of the adparticle dynamics on a real substance.

We deal with three cases when adparticles are introduced. In the first case the adparticles form a regular lattice on the host or substrate. In such cases the periodicity parallel to the surface is maintained. In Fig. 5.3 we show an example of oxygen on nickel(100). The interesting point here is that while the translation symmetry is preserved we now need to introduce a unit cell which is twice as long in two dimensions in order to describe the system. The adatoms form a $p(2\times2)$ or primitive (2×2) lattice. A second example of oxygen on nickel(100) is the $c(2\times2)$ or centered (2×2) structure. This is illustrated in Fig. 5.3a.

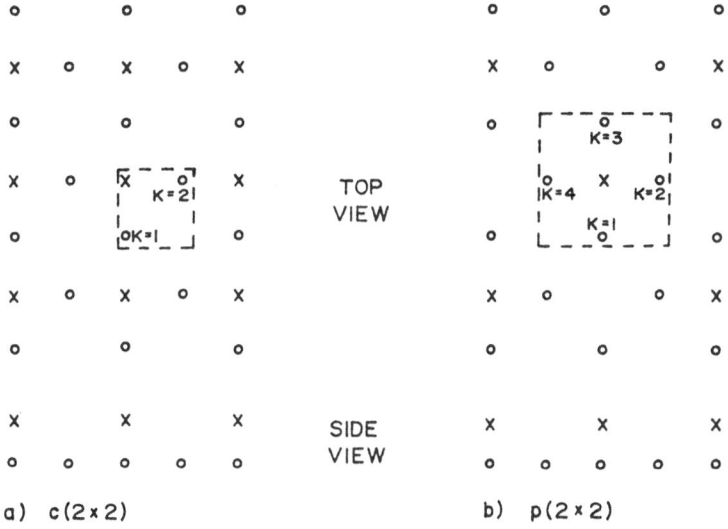

Fig. 5.3a,b. Open circles indicate the top layer of nickel atoms on a (100) surface. X's represent the overlying oxygen adatoms. (a) The c(2×2) structure with one oxygen for two nickel. (b) The p(2×2) structure with one oxygen for four nickel

In the second case the periodicity parallel to the surface is destroyed. Here we may think of a single adparticle on the surface, or a non-periodic arrangement of adparticles. In such cases we can no longer think of a periodic arrangement of adparticles, and therefore of a periodic arrangement of the entire crystal parallel to the surface, although the atoms in the layers below the adatoms are periodic.

The final case we consider is that in which several layers of rare-gas atoms are deposited on a metal substrate. Recent helium-beam scattering experiments have explored the dynamics of such systems. Periodicity of the system as a whole parallel to the crystal substrate surface is lost. This is because the rare gas atom spacing differs from the atom spacing in the metal substrate.

5.2 The Theory of Vibrations in the Harmonic Approximation

In this section we will set up the equations of motion for the various cases we wish to discuss. We begin with the case of no periodicity parallel to the surface. Then we treat the case of a bare periodic surface. Finally we examine the dynamics of metal with a periodic overlayer of adatoms.

5.2.1 The Case of No Periodicity

Our treatment of this situation is similar to that of *Wilson* et al. [5.12]. We begin with a finite crystal of N atoms (Fig. 5.4). At this point in our theory

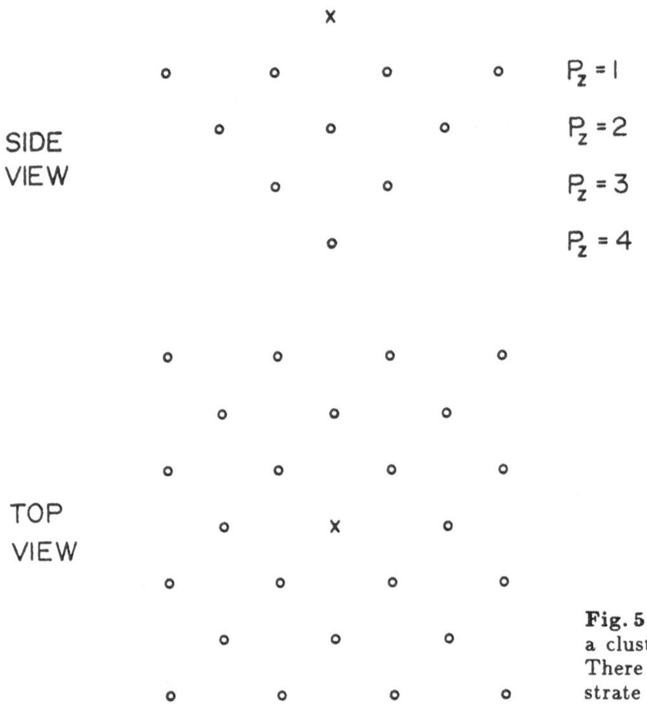

X

SIDE
VIEW

$P_z = 1$

$P_z = 2$

$P_z = 3$

$P_z = 4$

TOP
VIEW

X

Fig. 5.4. Isolated particle (x) on a cluster of substrate particles. There are 4 layers of the substrate shown

there is no need for any order in the atomic arrangement. There are, however, two important differences between this simple problem and the problem we wish eventually to address. First we wish eventually to examine the vibration modes of a crystal extending to infinity parallel to the surface and to infinity below the surface. Second we will eventually be interested in modes which our experimental tools are capable of detecting, namely modes which are localized in the vicinity of the surface. We must establish how they are to be selected from the infinite number of modes present.

Our first objective is to obtain the equations of motion of the atoms in terms of the potential energy of the system of atoms. We assume that the potential energy can be expressed as a function of the cartesian coordinates of the nucleii of the atoms. We write

$$\Phi = \Phi(\mathbf{R}(1), \mathbf{R}(2), \ldots, \mathbf{R}(N)), \tag{5.2}$$

where Φ is the total potential energy, $\mathbf{R}(n)$ is the location of the nth nucleus and may be written

$$\mathbf{R}(n) = \mathbf{R}^0(n) + \mathbf{u}(n), \tag{5.3}$$

where $\mathbf{R}^0(n)$ is the equilibrium position of the nucleus and $\mathbf{u}(n)$ is its displacement from equilibrium. We let $u_\alpha(n)$ be the αth cartesian component of the displacement.

For small values of displacement the energy may be expanded in a Taylor series about the equilibrium position. We then have

$$\Phi = \Phi(R^0(1), R^0(2), \ldots, R^0(N)) + \sum_{l,\alpha} \frac{\partial \Phi}{\partial u_\alpha(l)} \bigg|_{EQ.} u_\alpha(l)$$

$$+ \frac{1}{2} \sum_{l,\alpha} \sum_{l',\beta} \frac{\partial^2 \Phi}{\partial u_\alpha(l) \partial u_\beta(l')} \bigg|_{EQ.} u_\alpha(l) u_\beta(l') + \ldots, \tag{5.4}$$

where EQ. denotes evaluation at the equilibrium positions of the nucleii.

In the harmonic approximation the terms beyond those involving the second power of the displacement are neglected. The approximation is poorest at high temperatures, those close to the melting point of a solid, where the displacements from equilibrium are of the same order as the equilibrium interatomic spacing.

The force on the lth particle in the α direction may be written as the negative gradient of the potential energy

$$F_\alpha(l) = -\frac{\partial \Phi}{\partial u_\alpha(l)} \tag{5.5}$$

when all atoms are at their equilibrium sites they experience no force by definition. Thus

$$F_\alpha(l)|_{EQ.} = -\frac{\partial \Phi}{\partial u_\alpha(l)} \bigg|_{EQ.} = 0, \tag{5.6}$$

and (5.4) may be written, in the harmonic approximation, as

$$\Phi = \Phi(R(1), R(2), \ldots, R(N))|_{EQ.} + \frac{1}{2} \sum_{l,\alpha} \sum_{l',\beta} \Phi_{\alpha\beta}(l, l') u_\alpha(l) u_\beta(l'), \tag{5.7}$$

where $\Phi_{\alpha\beta}(l, l') \equiv \partial^2 \Phi / \partial u_\alpha(l) \partial u_\beta(l')|_{EQ.}$.

We can now use (5.5) to find the force by differentiation of (5.7), obtaining

$$F_\alpha(l) = -\sum_{l',\beta} \Phi_{\alpha\beta}(l, l') u_\beta(l'). \tag{5.8}$$

From Newton's law of motion we then have

$$m(l)\ddot{u}_\alpha(l) = -\sum_{l',\beta} \Phi_{\alpha\beta}(l, l') u_\beta(l'). \tag{5.9}$$

It is customary to introduce mass-weighted cartesian coordinates

$$W_\alpha(l) = m(l)^{1/2} u_\alpha(l) \tag{5.10}$$

and our equation of motion becomes

$$W_\alpha(l) = -\sum_{l',\beta} D_{\alpha\beta}(l, l') W_\beta(l'), \quad \text{where} \tag{5.11}$$

$$D_{\alpha\beta}(l, l') \equiv \Phi_{\alpha\beta}(l, l')[m(l)m(l')]^{-1/2} \tag{5.12}$$

159

is called the dynamical matrix. A solution of the system of equations (5.11) can be written

$$W_\alpha(l) \rightarrow W_\alpha(l) e^{i\omega t}. \tag{5.13}$$

Substitution into (5.11) yields

$$\omega^2 W_\alpha(l) = \sum_{l',\beta} D_{\alpha\beta}(l,l') W_\beta(l'), \tag{5.14}$$

where the time dependence is no longer implicitly present in $W_\alpha(l)$.

We have a system of linear homogeneous equations. These have nontrivial (non-zero) solutions only for those values of ω^2 for which the determinant of the coefficients vanishes. That is

$$|D_{\alpha\beta}(l,l') - \omega^2 \delta_{\alpha\beta} \delta_{ll'}| = 0. \tag{5.15}$$

We have here the characteristic equation of the matrix $\underset{\approx}{D}$. We can label the eigenvalues of the matrix by an index s that takes on the values s $= 1, 2, 3 \dots, 3N$, and rewrite (5.14) as

$$\omega_s^2 e_\alpha^{(s)}(l) = \sum_{l',\beta} D_{\alpha\beta}(l,l') e_\beta^{(s)}(l'). \tag{5.16}$$

Here $e_\alpha^{(s)}(l)$ is the eigenvector of the matrix $\underset{\approx}{D}$ corresponding to the eigenvalue ω_s^2. Because $\underset{\approx}{D}$ is a real symmetric matrix ω_s^2 is necessarily real. The eigenvectors $\{e_\alpha^{(s)}(l)\}$ can be chosen to be real with no loss of generality, and can be assumed to satisfy the orthonormality and completeness conditions

$$\sum_{l,\alpha} e_\alpha^{(s)}(l) e_\alpha^{(s')}(l) = \delta_{ss'}, \quad \sum_s e_\alpha^{(s)}(l) e_\beta^{(s)}(l') = \delta_{\alpha\beta} \delta_{ll'}. \tag{5.17}$$

Finally we note that the general solution of the problem can be written as a linear combination of the allowed solutions

$$u_\alpha(l) = \sum_s K_s \frac{e_\alpha^{(s)}(l)}{[m(l)]^{1/2}} e^{i\omega t}, \tag{5.18}$$

where K_s depend on the initial conditions.

In the examples presented in this chapter we will typically assume that the total potential energy of the crystal can be expressed as a sum over two body interactions $\phi_{ij}(R_{ij})$ so that

$$\Phi(\boldsymbol{R}(1), \boldsymbol{R}(2), \dots, \boldsymbol{R}(N)) = \frac{1}{2} \sum_i \sum_j \phi_{ij}(R_{ij}), \tag{5.19}$$

where R_{ij} is the separation between atoms i and j. If we now obtain the coefficients $\Phi_{\alpha\beta}(l,l')$, (5.7), we find

$$\Phi_{\alpha\beta}(l,l') = \delta_{l,l'} \sum_{m'}^{l} K_{\alpha\beta}(l,m') - (1 - \delta_{l,l'}) K_{\alpha\beta}(l,l'). \tag{5.20}$$

Here the prime denotes exclusion of the term $m' = l$ and

$$K_{\alpha\beta}(l,l') = \frac{\phi'_{ij}(R^0_{ij})}{R^0_{ij}}\delta_{\alpha\beta} + \left[\phi''_{ij}(R^0_{ij}) - \frac{\phi'_{ij}(R^0_{ij})}{R^0_{ij}}\right]\hat{n}_\alpha(l,l')\hat{n}_\beta(l,l'), \qquad (5.21)$$

where $\hat{n}_\alpha(l,l')$ is the αth cartesian component of a unit vector pointing from the equilibrium position of the atom labelled by l to that of the atom labelled by l'. The subscript ij appears on the first and second derivatives of $\phi_{ij}(R_{ij})$ because different two body interaction functions may be used at the surface than are used in the bulk, and because different functions may be used for nearest neighbour interaction than next-nearest neighbour interactions etc. The force-constant models used in practice will be discussed at appropriate places in this chapter.

5.2.2 The Case of a Bare Periodic Surface

Consider the situation in which periodicity is present parallel to the surface. Here we shall suppose these are N_L layers (that is a finite depth of the crystal) but now, unlike the situation in Fig. 5.4, we have the crystal extending to infinity in the two directions parallel to the surface (which we must do if we are to deal with periodicity in those two directions) and no adatom.

It is convenient to replace the notation of l, l' for the atomic positions with a notation which allows us to distinguish between directions parallel to the layers and perpendicular to the layers. Let $l = (l_\parallel, l_z)$ where $l_\parallel$ indicates the horizontal position of the atom in the layer and l_z is an integer representing the layer. We may then rewrite the equation of motion. Equation (5.14) becomes

$$\omega^2 W_\alpha(l) = \sum_{l',\beta} D_{\alpha\beta}(l,l') W_\beta(l'). \qquad (5.22)$$

At first sight this equation, with an infinite number of $W_\alpha(l)$, would seem to pose a difficult problem. We reduce the problem to one involving $3N_L$ coordinates as follows. We search for eigensolutions in the form of plane waves. Let $Q_\parallel$ represent a wave vector parallel to the surface. We introduce the solutions

$$W_\alpha(l) = W_\alpha(Q_\parallel; l_z) \exp\left[iQ_\parallel \cdot R^0(l)\right] \qquad (5.23)$$

into (5.22) and obtain

$$\omega^2 W_\alpha(Q_\parallel; l_z) = \sum_{l',\beta} D_{\alpha\beta}(l,l')$$
$$\times \exp\left\{(iQ_\parallel \cdot [R^0(l') - R^0(l)]\right\} W_\beta(Q_\parallel; l_z). \qquad (5.24)$$

Consider the sum

$$d_{\alpha\beta}(Q_\parallel; l_z, l'_z) \equiv \sum_{l'_\parallel} D_{\alpha\beta}(l,l') \exp\left\{iQ_\parallel \cdot [R^0(l) - R^0(l')]\right\}. \qquad (5.25)$$

Since the sum is over all $l_\parallel$, and since $D_{\alpha\beta}(l,l')$ has the periodicity of the surface, it is clear that the sum does not depend on $l_\parallel$, thus there is dependence of the matrix $\underset{\approx}{d}$ on $Q_{||}$, l_z and l_z' but not $l_{||}$. $\underset{\approx}{d}$ is called the partial Fourier transformed dynamical matrix (the full Fourier transformed dynamical matrix occurs in the lattice dynamics in the bulk of a crystal where there is periodicity in all three directions). Our equation now reads

$$\omega^2 W_\alpha(Q_{||}; l_z) = \sum_{l_z', \beta} d_{\alpha\beta}(Q_{||}; l_z, l_z') W_\beta(Q_{||}; l_z'). \tag{5.26}$$

We have a system of linear homogeneous equations in $W_\alpha(Q_{||}; l_z)$. These have a non-trivial solution (i.e., non-zero) for those values of ω^2 for which the determinant of the coefficients vanishes. That is

$$|d_{\alpha\beta}(Q_{||}; l_z, l_z') - \omega^2 \delta_{\alpha\beta} \delta_{l_z l_z'}| = 0. \tag{5.27}$$

We now have the characteristic equation of the matrix $\underset{\approx}{d}$. We label the eigenvalues of the matrix by an index s that takes on the values $s = 1, 2, \ldots, 3N_L$ and rewrite (5.26) as

$$\omega_s^2 e_\alpha^{(s)}(Q_{||}; l_z) = \sum_{l_z', \beta} d_{\alpha\beta}(Q_{||}; l_z, l_z') e_\beta^{(s)}(l_z'), \tag{5.28}$$

where $e_\alpha^{(s)}(Q_{||}; l_z)$ is the eigenvector of the matrix $\underset{\approx}{d}$ corresponding to the eigenvalue ω_s^2. Because $\underset{\approx}{d}$ is Hermitian, see (5.25), the eigenvalues may be chosen to satisfy the orthonormality and completeness conditions

$$\sum_{l, \alpha} e_\alpha^{(s)}(Q_{||}; l_z) e_\alpha^{(s')}(Q_{||}; l_z)^* = \delta_{ss'},$$

$$\sum_s e_\alpha^{(s)}(Q_{||}; l_z) e_\beta^{(s)}(Q_{||}; l_z')^* = \delta_{\alpha\beta} \delta_{l_z l_z'}. \tag{5.29}$$

Note that while the eigenvalues are still real, the eigenvectors may be complex.

If one compares (5.15), with its 3N solutions, and (5.27) with its $3N_L$ solutions it may seem at first glance that some reduction in the problem has been obtained. It should be remembered, however, that (5.27) must be solved for each value of $Q_{||}$. Owing to the discrete nature of the lattice not all wavevectors contribute to different solutions of (5.24). Only the vectors lying inside or at the boundaries of the "surface Brillouin zone" need to be considered.

Starting with the vectors of the lattice a and b of (5.1) we generate vectors A and B of a "reciprocal lattice" with the equations

$$A = 2\pi \frac{b \times \hat{z}}{|(a \times b) \cdot \hat{z}|}, \qquad B = 2\pi \frac{\hat{z} \times a}{|(a \times b) \cdot \hat{z}|}, \tag{5.30}$$

where $\hat{z}$ is a unit vector perpendicular to the surface. A little thought shows that substitution of a vector $Q_{||} + G$ into (5.23), where

$$G = MA + NB, \tag{5.31}$$

and M and N are $= 0, \pm 1, \pm 2 \ldots$, will yield the same result as if $Q_{||}$ only were

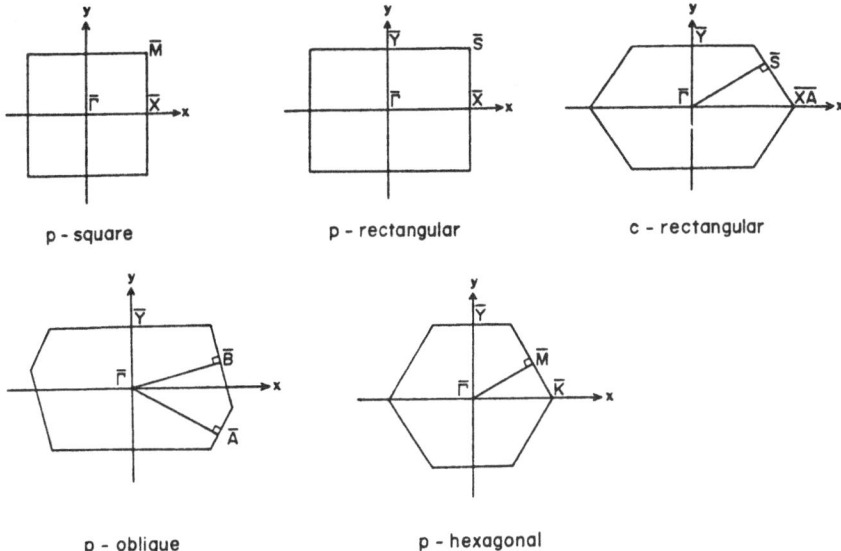

Fig. 5.5. The five surface Brillouin zone types corresponding to the five primitive surface cells in real space. Special points used to indicate the zone boundaries are marked $\overline{M}$ etc.

used. It is therefore possible to restrict the values of $Q_{\|}$ to a surface Brillouin zone. This zone is defined as the smallest area bounded by the right bisectors of the lines joining a point on the reciprocal lattice to adjacent reciprocal lattice points. There are five types of zones, corresponding to the five types of real space structures of Fig. 5.2. They are shown in Fig. 5.5. Note that special points at the zone center and zone boundary are labelled by capital letters, which will be referred to later.

If we restricted the crystal of Fig. 5.4 to N atoms with the provision that the wavelike solution repeated themselves parallel to the surface in adjacent blocks of N atoms (the so-called periodic boundary conditions) we would find that the number of allowed wavevectors $Q_{\|}$ inside the surface Brillouin zone times the number of layers was just equal to N so that in fact there were 3N solutions to the problem posed by (5.22). Thus we see that in a formal sense the introduction of the plane wave solutions has not reduced the number of eigenvalues and eigenvectors needed for a full description of the problem.

Note that $D_{\alpha\beta}(l,l')$ of (5.24) is the same as that described by (5.20) and (5.12) in the event that we wish to consider central forces.

5.2.3 The Case of a Periodic Overlayer

In situations in which an overlayer of adatoms is present on the crystal surface, the overlayer may not have a coverage of unity. In such cases the lattice dynamics must be modified. The arguments introduced in setting up the Fourier transformed dynamical matrix must be altered to allow for the fact that not all atoms in a given layer have the same environment and therefore the same displacements. We now break each layer into unit cells containing the inequivalent atoms, as illustrated in Fig.5.3 for oxygen on the Ni(100) surface. The inequivalent atoms in each layer are labelled by an index $K = 1, 2, \ldots, K_m$. The equations of motion now take the form

$$\omega^2 W_\alpha(Q_\parallel; l_z K, l'_z K') = \sum_{l'_z K', \beta} d_{\alpha\beta}(Q_\parallel; l_z K, l'_z K') W_\beta(Q_\parallel; l'_z K'), \qquad (5.32)$$

where

$$d_{\alpha\beta}(Q_\parallel; l_z K, l'_z K') \equiv \sum_{l'_\parallel} D_{\alpha\beta}(l, l') \exp\{iQ_\parallel \cdot [R^0(l'K') - R^0(lK)]\}. \qquad (5.33)$$

There are now $3N_L K_m$ eigensolutions. Closure and orthonormality relation similar to those of (5.29) apply. We have

$$\sum_{l_z, \alpha, k} e_\alpha^{(s)}(Q_\parallel; l_z K) e_\alpha^{(s')}(Q_\parallel; l_z K)^* = \delta_{ss'}$$

$$\sum_{s} e_\alpha^{(s)}(Q_\parallel; l_z K) e_\beta^{(s)}(Q_\parallel; l'_z K')^* = \delta_{\alpha\beta} \delta_{l_z l'_z} \delta_{KK'} \qquad (5.34)$$

From the point of view of keeping track of the equations the case dealt with here is no different from that of the one atom per unit cell. The various $l_z K$ pairs can simply be thought of as individual layers if one wishes to remove the more complicated notation.

5.3 Bare Metals

We begin with a discussion of the methods that have been used to determine the frequencies of surface modes and resonances. For purposes of simplifying this discussion we will assume the Fourier transformed dynamical matrix is known. We then proceed to discuss a number of rather general calculations of mode frequency, in each case also describing the method of determining the dynamical matrix. Finally we describe a number of experimental results for bare metal surfaces and along with these, where they exist, present the corresponding theoretical calculations.

5.3.1 Methods of Mode-Frequency Calculation

a) Continuum Methods

Methods for calculating dispersion relations when the material is approximated as being continuous have been in existence for many years. The work was begun by *Lord Rayleigh* [5.13]. In this subsection only a few brief comments about the results of the continuum method linking them with the discrete calculations, will be given. For more details the reader is referred to the review articles of *Farnell* [5.14] and *Wallis* [5.1].

The continuum approximation is valid at long wavelengths, that is in the vicinity of $\bar{\Gamma}$ at the center of the surface Brillouin zone. Lord Rayleigh discovered that waves which propagate parallel to the surface, but which decay exponentially into the bulk can propagate energy along the continuum surface. They are presently termed Rayleigh Waves. Lord Rayleigh took the material to be isotropic. If an anisotropic material is used than generalized Rayleigh waves can exist. These exhibit an exponentially damped sinusoidal decay into the bulk. Anisotropic materials also have solutions which decay exponentially parallel to the surface and slowly grow into the bulk. These solutions are, in effect, waves which propagate from the surface into the bulk at some angle to the normal to the surface. They are termed pseudo surface waves or leaky surface waves [5.14]. In discrete lattices the terms Rayleigh waves etc. are extended to large wavevectors as can be seen in the following discussion of the slab methods.

b) Slab Methods

This is by far the most easily applied method for calculation of surface modes. The semi-infinite metal is replaced by a metal slab of a finite number of layers. The modes of vibration of the slab are then determined by diagonalizing the matrix $d_{\alpha\beta}(Q_{\|}; l_z, l'_z)$ of (5.25). For a slab of N_L layers there are $3N_L$ modes for each wavevector $Q_{\|}$.

The results of a slab calculation based on only nearest neighbor interactions are shown for a 21 layer slab of nickel with (111) faces in Fig. 5.6. In their classical papers, *Allen* et al. [5.8,15] noted that there are modes closely spaced in frequency (indicated by the cross-hatched regions) which penetrate through the entire slab. These modes they termed "bulk modes" and the cross-hatched bands into which they divided were termed "bulk bands". Lying below or between the bulk bands are modes which penetrate only a few layers into the slab, and which are the "surface modes". These are labelled S_1, S_2, S_3 and S_4 in Fig. 5.6.

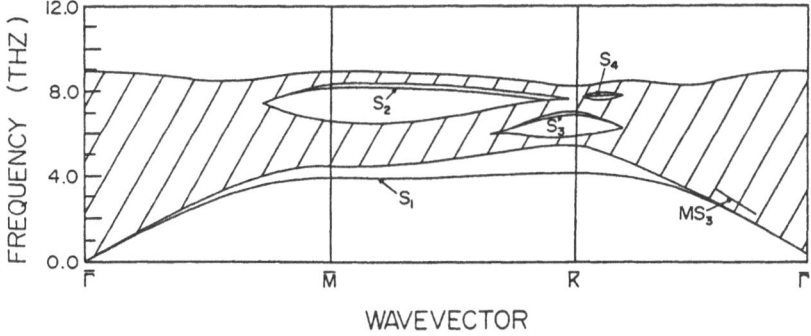

Fig. 5.6. Results of a nearest-neighbour slab calculation for 21 layers of nickel atoms: the (111) surface. Shaded regions correspond to the bulk bands. Surface modes are marked S_1 etc.

Along directions of high symmetry the bulk modes may divide or partition into sagittal modes (modes with longitudinal components and perpendicular to the surface components) and modes transverse to the direction of propagation and parallel to the surface. In such cases one may find surface modes of one symmetry type lying not outside or between the bands but inside a band of the other symmetry type.

Allen et al. noted that the frequencies which delineate the bulk bands and the frequencies of the surface modes converge rapidly as the number of layers of the slab increases. Slabs of as few as 10 layers provide useful information. Their paper was based on slabs of 21 layers (which other theorists have also found useful).

In a practical calculation one increases the number of layers until the mode or modes of interest have converged adequately at the required wave vectors. As one approaches the zone center modes such as S_1 (generalized Rayleigh wave) grow longer in wavelength and penetrate more deeply into the slab. What one finds along $\overline{\Gamma}\overline{M}$, for example, is illustrated in Table 5.2 for the 21 layer slab. The mode S_1 is present on both the top and bottom layers. These modes would be degenerate if the slab were infinitely thick. As it is the modes at the two surfaces couple to produce symmetric and antisymmetric modes with slightly different frequencies. Table 5.2 shows how the mode S_1 splits as the wavevector is reduced and also compares the slab results with exact values obtained by *Latkowski* and *Black* [5.16] using a method employing Gottlieb polynomials.

Table 5.2. A comparison of exact frequencies calculated with a 21 layer slab along the $\overline{\Gamma}\overline{M}$ direction. $Q_\parallel = 1.0$ corresponds to the zone boundary at $\overline{M}$

	Frequency of $S_1 \left[\mathrm{cm}^{-1}\right]$	
$Q_\parallel$	Exact	Slab
1.0	129.92	129.92
		129.92
0.5	91.84	91.82
		91.86
0.4	-	76.23
		76.48
0.3	-	58.60
		59.48
0.2	40.16	39.39
		41.72
0.1	20.35	19.04
		23.15

Allen et al. [5.15] describe the surface modes as peeling off from the bulk modes due to the reduction of the forces on the surface atoms when the atoms above them are removed to produce the surface. They note that there are some modes which do not shift out of the bulk bands. If they have the same symmetry as the bulk modes, they mix with the bulk bands producing what are termed "mixed modes" or "pseudo-surface modes". The mixed mode MS_3 is an extension of S_3 into the bulk bands. More recently the term "resonances" has been used to describe these cases. This seems appropriate, since the features have a finite width, unlike the surface modes, in the semi-infinite crystal.

In a slab calculation the resonances show up as small groups of modes of vibration localized near the slab surface. They are not always easy to identify as surface modes since they have the same symmetry as the bulk band in which they lie, and only a small number of slab modes have frequencies to couple with unless the slab is very thick.

We can summarize our comments about the slab method by noting that it is easy to use, provides excellent data on the frequency of the boundaries of the bulk modes and on the frequencies of surface modes. It must be used with care close to the zone center, where modes penetrate deeply into the crystal, or in cases when resonances exist.

c) Spectral Density Methods

In many problems it is necessary to know not only the surface mode frequencies and their eigenvectors, but also details of the resonances and the bulk bands in the form of a spectral density. The spectral density may be written as a simple frequency distribution function $f(\omega)$ such that $f(\omega)d\omega$ represents the number of modes lying between ω and $\omega + d\omega$. Of more use in studies of the scattering of electrons and helium atoms is the "effective frequency distribution function" of *Allen* et al. [5.15]

$$\varrho_{\alpha\beta}(l_z, l'_z; \mathbf{Q}_{||}, \omega) = \sum_s e_\alpha^{(s)}(\mathbf{Q}_{||}; l_z) e_\beta^{(s)}(\mathbf{Q}_{||}; l'_z)^* \delta(\omega - \omega_s). \tag{5.35}$$

A histogram of this density can be constructed using the modes of the slab [5.17]. However, the ability to detect the resonances and details of the bulk bands depends on how many layers are used. In general, many more layers are required to find the frequencies of the resonances than are needed to determine the surface modes frequencies.

To overcome this difficulty with the slab method a variety of techniques have been developed which allow one, starting with the dynamical matrix, to obtain exact solutions for the densities of (5.35) in the case of the semi-infinite crystal. In the long-wavelength or continuum limit a surface Green's function matching technique has been described for anisotropic and hexagonal systems by *Velasco* et al. [5.18,19]. More recently *Garcia-Molinar* et al. [5.20] have used a T-matrix method to study the Ni(111) surface using a nearest neighbour force constant model. They have also examined spectral densities for stacking faults.

Black et al. [5.21] presented the details of a Green's function calculation suitable for the calculation of spectral densities such as that shown in (5.35). The theory was set up for nearest and next-nearest neighbour interaction but could be extended to involve more neighbours. Results are shown in Fig. 5.7 for ϱ_{xx} and ϱ_{yy} at a wavevector 90% of the distance from $\overline{\Gamma}$ to $\overline{K}$ of Fig. 5.6. They represent a cross-section through the curves of Fig. 5.6. The modes S_3 and S_4 are clearly visible as are the bulk bands. Note the presence of a resonance

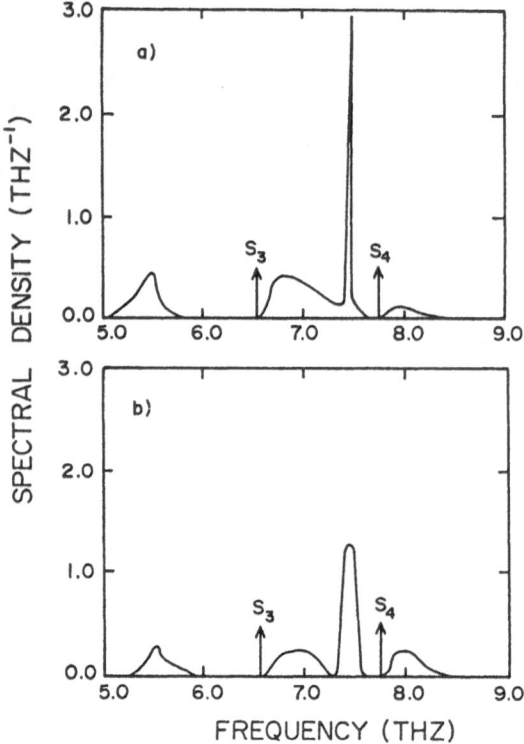

Fig. 5.7a,b. Results of spectral density calculations at a wavevector 90% of the distance from $\bar{\Gamma}$ to $\bar{K}$ (Fig. 5.6) (a) ϱ_{xx}, (b) ϱ_{yy}. Here the y-direction is longitudinal while the x-direction is transverse. The lower bulk band, the mode S_3 and the upper bulk band, with a gap containing S_4, are all evident. A resonance feature (not noticeable in Fig. 5.6) is also seen

feature of finite width in ϱ_{xx} in the bulk band. This represents only 0.5% of the total density of states in ϱ_{xx}, but 50% of the density in ϱ_{yy}. In a 21 layer slab calculation the mode is present as a single mode in the y-direction of vibration and is not detectable in the x-direction. (Incidentally along $\bar{\Gamma}\bar{K}$ the y-direction represents longitudinal vibration while the x-direction represents transverse vibration).

Croitoru [5.22] examined, using a Green's function method, the (110) surface of a simple cubic crystal. At the present time no general discussion of the face centered cubic (100) or (110) surfaces exists. Nor are there general discussion of body centered cubic surfaces.

Black et al. [5.23] described a continued fraction method which could be used to determine a spectral density with or without periodicity. The spectral density used is independent of wavevector. It has the form

$$\varrho_{\alpha\beta}(l, l'; \omega) = \sum_s e_\alpha^{(s)}(l) e_\beta^{(s)}(l')^* \delta(\omega - \omega_s),$$ (5.36)

where $e_\alpha^{(s)}(l)$ and $e_\beta^{(s)}(l')$ are as defined in (5.16). The density can be used to obtain the correlation function relating the motion of atom l in the α direction to atom l' in the β direction. We have

$$\langle u_\alpha(l)u_\beta(l')\rangle = \int_0^\infty d\omega \frac{\hbar}{2M\omega}[1 + 2\bar{n}\varrho_{\alpha\beta}(l,l';\omega)], \tag{5.37}$$

with M denoting the mass of the atom, $\bar{n}_\omega$ is the Bose-Einstein function, and $\hbar$ Planck's constant.

In the event that $\beta = \alpha$ and $l' = l$, (5.37) gives use the mean-square displacement of atom l in the α direction. We illustrate below how the continued fraction method can be used to find the density in this case. For $\alpha \neq \beta$ and $l \neq l'$ the reader is referred to *Black* et al. [5.23].

We define a Green's function

$$U_{\alpha\alpha}(l,l';z) \equiv \sum_s \frac{e_\alpha^{(s)}(l)e_\beta^{(s)}(l')}{\omega_s^2 - z^2}, \tag{5.38}$$

where z is a complex frequency. It is a straightforward matter to show that the spectral density can be expressed in terms of the Green's function by the relation

$$\varrho_{\alpha\alpha}(l,l';\omega) = \frac{\omega}{i\pi}[U_{\alpha\alpha}(l,l';\omega + i\varepsilon) - U_{\alpha\alpha}(l,l';\omega - i\varepsilon)], \tag{5.39}$$

where ε is a positive infinitesimal.

The equation of motion (5.14) can be expressed in terms of the Green's function and by comparing it with the tight binding results of *Haydock* et al. [5.24,25] one sees that the Green's function can be expressed in terms of a continued function

$$U_{\alpha\alpha}(l,l';z) = \cfrac{1}{z^2 - A_1 - \cfrac{|B_2|^2}{z^2 - A_2 - \cfrac{|B_3|^2}{z^2 - A_2 - \dots}}}. \tag{5.40}$$

The coefficients A_n, B_n are determined from an iterative procedure that can readily be programmed on a large computer. We introduce a 3N-dimensional column vector $|l\alpha\rangle = |1\rangle$ which has zeroes everywhere save for a single entry of unity at the atom and cartesian component whose spectral density is required, we set $B_1 = 0$, $|0\rangle = 0$ and obtain A_1 from the relation

$$A_1 \equiv \langle 1|\underset{\approx}{D}|1\rangle. \tag{5.41}$$

We then obtain the subsequent A_n, B_n from the equation

$$B_{N+1}|N+1\rangle = \underset{\approx}{D}|N\rangle - B_N|N-1\rangle - A_N|N\rangle, \tag{5.42}$$

where $\underset{\approx}{D}$ is the dynamical matrix. Note that we use

$$A_N = \langle N|\underset{\approx}{D}|N\rangle \tag{5.43}$$

and that $\langle N|N\rangle \approx 1.0$.

What makes the method work is that the A_n, B_n converge rather quickly (within 6 to 8 iterations) to some values we call A_∞, B_∞. It is then standard procedure to write the continued fraction as follows:

$$U_{\alpha\alpha}(l,l';z) = \cfrac{1}{z^2 - A_1 - \cfrac{|B_2|^2}{z^2 - A_2 - \ldots\ldots - |B_N|^2 T_\infty}} \qquad (5.44)$$

$$= \cfrac{1}{z^2 - A_1 - \cfrac{|B_2|^2}{z^2 - A_2 - \ldots\ldots - \cfrac{|B_N|^2}{z^2 - A_N - |B_{N+1}|^2 T_\infty}}} \qquad (5.45)$$

By comparing (5.44) and (5.45) we see that we can write

$$T_\infty = \frac{1}{z^2 - A_N - |B_{N+1}|^2 T_\infty} \qquad (5.46)$$

and that we can then solve for T_∞. Once T_∞ is known we can use (5.44) to solve for the spectral density. Incidentally the procedure described above can be used for a non-periodic cluster of atoms or for a periodic arrangement of atoms.

The spectral density obtained using the method described above can be used to determine mean square displacements of atoms at periodic bare metal surfaces. *Black* et al. [5.23] examined the W(100) surface. *Armand* [5.26] incidentally, used a Green's function method to explore mean square displacements of atoms at metal surfaces and correlations of these atoms. *Black* [5.27] applied the continued-fraction method to motion of atoms at the Ni(111) surface. *Mostoller* and *Landman* [5.28] used it at a stepped platinum surface, about which more will be said in Sect. 5.3.3e.

A similar procedure can be followed to set up spectral densities which are wavevector dependent, such as that shown in (5.35). The main difference is the replacement of the matrix $\underset{\approx}{D}$ of the non-periodic case, with the Fourier transformed dynamical matrix $\underset{\approx}{d}$ of (5.25) in the bare metal surface or $\underset{\approx}{d}$ of (5.33) in the periodic overlayer case, in (5.41,42 and 43). Unfortunately the A_n, B_n do not always converge in these cases. Difficulties occur when there are resonance modes in the bulk bands. At the present time a method of establishing whether or not the continued fraction will or will not work, short of actually trying it, is not known.

Incidentally it should be pointed out that the Green's function in the continued-fraction method is not used explicitly. It serves as a link to convert the continued fraction expression of (5.44) with the expression for the density of (5.39). Other methods that employ different techniques for obtaining the Green's function than that using the continued fraction are herein referred to as Green's function methods.

Details of the calculational methods of various authors will be described in what follows. The important point to keep in mind is that slab techniques may miss resonance features unless care is taken. Exact spectral density calculations, while more difficult than slab calculations, do not suffer from this limitation.

5.3.2 General Theoretical Calculations

The classic slab papers of *Allen* et al. [5.8,15] have already been cited. These authors presented figures similar to those of Fig. 5.6 for the (100) and (110) faces of face-centered cubic materials (Figs. 5.8 and 9). The atoms are taken to interact with a Lennard-Jones potential. Once this assumption is made the dynamical matrix can be constructed. Results are presented for two cases. In one case the surfaces are allowed to relax, while in the other case the interlayer spacing is fixed through the entire slab. The relaxation is found to "peel off" a few additional surface modes from the bulk bands. However, many of the features of the surface modes are not substantially different in the two cases. The papers are also useful in that they present a very comprehensive discussion of the theory underlying the slab technique along with results for three face centered surfaces.

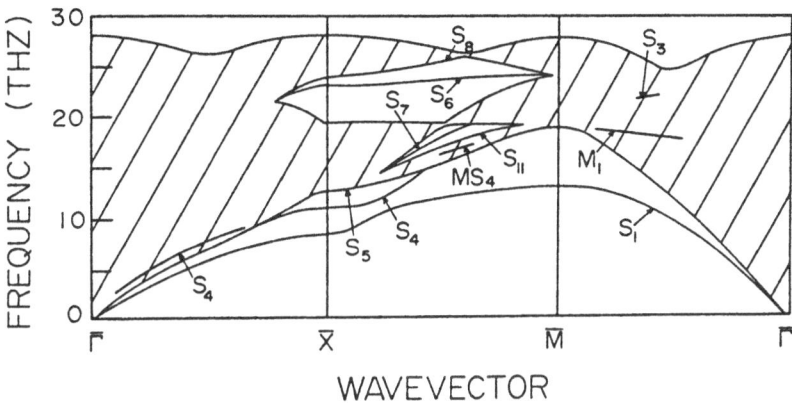

Fig. 5.8. Results of a nearest-neighbour slab calculation for 21 layers of nickel atoms: the (100) surface. Shaded regions correspond to bulk bands. Surface modes are marked S_1, etc.

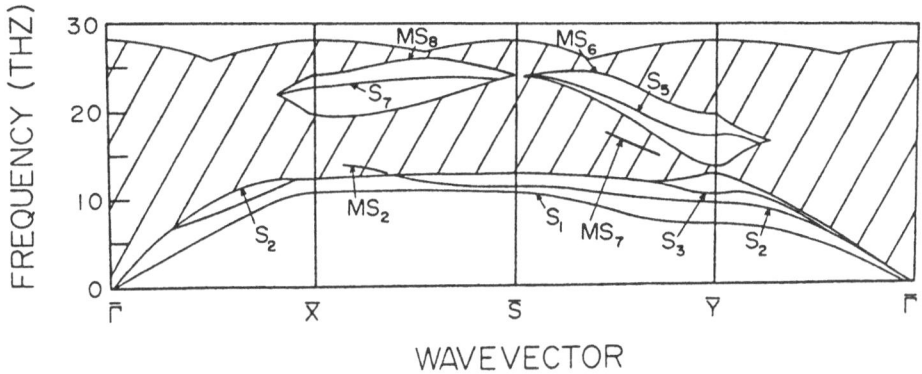

Fig. 5.9. Results of a nearest-neighbour slab calculation for 21 layers of nickel atoms: the (110) surface. Shaded regions correspond to bulk bands. Surface modes are marked S_1, etc.

Allen et al. found, "in addition to the generalized Rayleigh waves present at long wavelengths, other surface modes, which are characterized by the fact that they do not persist into the long-wavelength limit, and therefore cannot be obtained in the continuum approximation. The dependence of the amplitude of these modes upon distance from the surface is generally rather complex". The reader is referred to their papers for a discussion of the complexity. There are modes which are alternately longitudinal in one layer and perpendicular to the surface in the next, modes which are substantially larger in the second layer than in the first and so on. In other words, modes quite different from the simple exponentially decaying Rayleigh waves found at long wavelengths and described by *Farnell* [5.14]. *Allen* et al. [5.29] explored the surface modes of hcp crystals using techniques similar to those described above. They found little change from the results for the fcc crystals.

For the experimentalist interested in making contact between the theoretical work of Allen et al., and the experimental results from a specific material, the procedure is to fit one point on the theoretical curves to experiment and then scale other values of the theory using the same scale factor. (Strictly speaking, appropriate for nearest-neighbour model only). An alternative method of relating theory and experiment was introduced by *Black* et al. [5.30]. They presented theoretical data for the (100) surface of the face centered metals (Ag, Al, Au, Cu, Ir, Ni, Pd, Pt and Rh) and the body centered metals (Cr, Fe, K, Mo, Na, V and W). The data is presented for a variety of surface modes at the important points of the surface Brillouin zones. In this case the dynamical matrix was not obtained from a Lennard-Jones potential. Instead the potential was based on the assumption of up to third-neighbour central forces and angle-bending forces. A variety of techniques were used to obtain the force constants. To make contact with the actual metals the constants were obtained by fitting to known measured elastic constants, and to measured (neutron scattering) bulk phonon dispersion curves. The same forces were then used at the surface to obtain the surface-mode frequencies. No allowance was made for relaxation, reconstruction, or changes in force constants that might occur at the surface. In an earlier paper, *Castiel* et al. [5.31,32] using a two-neighbour central-force model and angle-bending forces fitted to bulk properties, examined surface modes at the (100) faces of Ag, Cu, Pd, Pd, Fe and W. Their data is presented in the form of dispersion curves similar to those of Fig. 5.6. In a subsequent paper *Black* et al. [5.33] presented theoretical data for the (111) surface of the face centered metals Ag, Al, Au, Cu, Ni, Pd and Pt.

Bortolani et al. [5.34,35] also examined the lattice dynamics of W(001) and Fe(001) for ideal and relaxed surfaces. They obtained, based on a force constant fit to the bulk phonons, a phonon spectrum. *Bortolani* et al. [5.36] then explored the relaxation of W(001) with a "microscopic force constant" approach.

Castiel et al. also examined the effects of varying force constants at the platinum surface. They found that "soft phonons" are possible and may account for reconstruction of the platinum surface. *Fasolino* et al. [5.37], on the other hand, explored the W(100) reconstruction problem using a method they termed "effective surface lattice dynamics". *Krakauer* et al. [5.38] also examined the question of the reconstruction of W(001). They suggested the transition might be driven electronically via the onset of a charge density wave accompanying the soft phonons. Finally *Pick* and *Tomasek* [5.39] addressed this problem for W(001), Mo(001) and Cr(001) surfaces. *Armand* and *Masri* [5.40] examined the surface mode and resonances for a (117) face of a face centered crystal. They used a nearest neighbour central force model. A general rule which enables one to predict qualitatively the shape of the dispersion relation for any vicinal surface of the type (11 m) is expressed. The calculations were based on Green's function technique. *Black* and *Bopp* [5.41] used the slab method to enumerate surface modes at the (100), (110), (111), (210), . . . , (332) face centered cubic surfaces and the (115), (117), (119) and (553) surfaces so that effects of increasing terrace size could be examined. A nearest neighbour central force model was used. As the Miller indices increased, so too did the number of layers needed to achieve a convergence of the surface mode frequencies. Roughly speaking it was not the 21 layers of Allen et al. that worked in these cases, but rather enough layers were needed to give 21 nearest neighbour spacings down into the crystal. Thus for the (322) surface about 100 layers were needed to achieve the convergence required.

More sophisticated models of interatomic forces have also been used. *Benedek* et al. [5.42] using a Green's function and computing electronic charge density deformations due to nuclear displacements method, calculated the surface phonon dispersion of Sn(001). *Calandra* et al. [5.43], using a pseudopotential perturbation theory, have calculated the surface phonons of Na(001) and K(001). Their results provide evidence of the importance of non-local screening effects in the determination of surface phonons. In the next few years we should see ab initio calculations of phonons for aluminum [5.44] and for tungsten [5.45].

5.3.3 Experimental Results and Theoretical Interpretation

We now present the results of experimental measurements of surface vibrations and their theoretical interpretation. Much of this work has been done in the last few years as a result of the development of two powerful techniques. One is electron energy loss spectroscopy (EELS). In this technique electrons of energies in the order of 100 eV are scattered from the surface of a metal and their energy loss to surface phonons is measured. The technique has been described in detail by *Ibach* and *Mills* [5.46]. Early experiments involved specular scattering, that is scattering from dipoles present at the surface and vibrating together. The scattering modes are at the $\bar{\Gamma}$ point of the surface Brillouin zone. More recent experiments involving off-specular scattering have allowed measurements away from the surface Brillouin zone center and out to the zone boundary.

After initial success with the alkali halides the technique of helium scattering has now been applied to metal surfaces. Here a beam of low-energy helium atoms is reflected from the metal surface and energy losses and gains to surface phonons determined by a time-of-flight technique. The technique was described by *Toennies* [5.47]. It is not limited to $\bar{\Gamma}$, but does tend to be more difficult to use out near the zone boundaries. In this section we now proceed to examine the various fcc surfaces that have been studied. As yet there is little experimental work on body-centered materials.

a) The (100) fcc Surfaces

Nickel. Lehwald et al. [5.48] measured the dispersion of the mode S_4 along the $\bar{\Gamma}\bar{X}$ direction. The measurement involved off-specular scattering of electrons. Their results at the zone boundary were higher than those predicted by the nearest-neighbour model of *Black* et al. [5.30], but lower than the results predicted by a two neighbour plus angle bending model. They suggested that the discrepancy with the nearest-neighbour model could be understood as due to the failure to include surface relaxation. By varying the force constant between atoms in the first and second layers of the nickel they found a fit to experiment if the force constant was increased by 20%.

The nearest-neighbour with relaxation fit to experiment was good. A calculation of the spectral density of modes perpendicular to the surface revealed a shoulder close to the bulk bands which was also seen in experiment.

As can be seen in Fig. 5.8, there are two other modes along the $\bar{\Gamma}\bar{X}$ direction. The mode S_1 is polarized in the surface and perpendicular to the direction of propagation. Thus it is not expected to scatter electrons. The mode S_6, on the other hand, while it is longitudinal, is expected to scatter electrons. Although estimates suggested its intensity would be 25 times lower than that of S_4, it should have been detected and was not positively identified in the experiment. Recently *Xu* et al. [5.49] re-examined the question of the S_6 mode. A rigid ion multiple scattering calculation based on slab eigenvectors for the modes S_4 and S_6 was used to predict actual intensities for the electron energy loss experiment. It was found that the ratio of the S_4 to S_6 intensity was very sensitive to the energy of the incident electrons for the given scattering angle of 65°. There are electron energies in the experimental range of 40 to 250 eV at which the S_6 intensity is comparable with that of S_4, and other energies at which S_6 is substantially depressed in intensity. It was at one of these energies that the original experiment was done. The experiment was repeated at an appropriate energy and, indeed, S_6 was found. Moreover the quantitative accord between theory and experiment was very good.

The intensity claculation was based on the nearest-neighbour model with a 20% stiffening of the force constant coupling the top layers of the slab. By varying the spacing between these layers it was possible to achieve even better accord between the experimental and theoretical intensities without altering the mode frequencies substantially. The best fit was obtained with a 2.6% contraction in the surface layer. It is of interest here to show the sorts of changes in frequency that occur with different models. These are presented in Table 5.3.

Table 5.3. Frequencies of the modes S_4 and S_6 from experiment and from several theoretical models of force constants

| Mode | Frequency $[\text{cm}^{-1}]$ | | | |
	Experiment	Nearest-neighbour and angle bending	Nearest-neighbour	Nearest-neighbour and stiffening
S_4	130 ± 2	138	125	132
S_6	255 ± 2	254	247	252

Confirmation of this choice of fitting parameter can be found in the work of *Frenken* et al. [5.9]. They measured the change in the spacing between the first and second layers of Ni(100) and found a contraction of 3.2%. A theoretical calculation, using methods described by *Allan* [5.50], gave a contraction of 3.2% and slab modes S_6 and S_4 in reasonable agreement with experiment, although S_4 was at $135 \,\text{cm}^{-1}$, higher than experimental results. The model then yielded a little too much stiffening of the force constant coupling the first two planes of atoms. The model is interesting in that, while some fitting is done to bulk properties of the nickel to establish some parameters, the relaxation and surface force constants emerge as a consequence of the existence of the surface. It is based on a treatment of the nickel d-band density of states described within a tight-binding approximation. This is used to set up a system of interatomic potentials which are then parameterized by fitting to bulk properties and allowing the surface to equilibrate.

Rocca et al. [5.51] have examined the $\overline{\Gamma}\overline{M}$ nickel(100) dispersion. Their results are in accord with the $\overline{\Gamma}\overline{X}$ results. They support the idea that the forces coupling the first and second nickel layers are increased by 20% over their bulk values.

Copper. Using the helium beam technique *Mason* and *Williams* [5.52] examined dispersion along the $\overline{\Gamma}\overline{M}$ direction from close to the zone center out to $0.9 \,\text{Å}^{-1}$. They found two modes, one which they identified with the Rayleigh wave S_1 below the bulk bands (Fig. 5.8) and the other which lay in the bulk bands. As yet there has been no detailed lattice dynamical study of their system.

A calculation using the *Black* et al. [5.30] force constant, gives agreement with the mode S_1 to within a $1 \,\text{cm}^{-1}$, but it is not obvious how the other mode is to be identified. *Sanz* and *Armand* [5.53], starting with the force constants of *Castiel* et al. [5.31] found that the frequency of S_1 along $\overline{\Gamma}\overline{M}$ could be lowered or raised by lowering or raising the force constants coupling the surface and one layer below it. In the case of copper no soft phonons were found. Such changes, in any case, are not needed to fit the Rayleigh mode of Mason and Williams.

Finally we consider the work of *Andersson* et al. [5.54]. They observed long-range scattering of slow electrons by phonon excitations at the Cu(100) surface (also Ni(100)). Their theoretical calculations of frequency, based on a nearest-neighbour (bulk fitted) model of the interactions, gave a good fit to experiment. Their measurements do not have the resolution of those of *Lehwald* et al. [5.48].

Platinum. As yet there are no measurements but *Sanz* and *Armand* [5.53] have done a theoretical study of the effects of relaxation on the surface modes. They show that lowering the interatomic surface force constants leads to soft phonons, and presumably to surface reconstruction.

b) The (111) fcc Surfaces

Silver. Doak et al. [5.55] scattered helium atoms in the $\overline{\Gamma}\,\overline{M}$ and $\overline{\Gamma}\,\overline{K}$ direction from the silver(111) surface. The results for the $\overline{\Gamma}\,\overline{M}$ direction are shown in Fig. 5.10 along with results for copper and gold to be discussed later [5.47]. They are qualitatively similar to those obtained by Mason and Williams for copper(100). That is, a Rayleigh wave is seen at low frequencies and additional modes are observed at higher frequencies lying in the bulk bands. *Doak* et al. [5.55] compared their early results with calculated values of *Armand* [5.26]. Armand based his slab calculations of the modes on a nearest-neighbour model of the silver fitted to the top frequency of the neutron dispersion data from the bulk. (Similar agreement is found with the nearest-neighbour model of *Black* et al. [5.33]).

While the model fits the Rayleigh wave it predicts a longitudinal wave in the bulk bands which is higher than the experimental values by about 35%. (The mode MS_3 in Fig. 5.6 is transverse and not expected to couple to the helium atoms). *Doak* et al. [5.55] suggested that their high frequency mode was anomalous.

Bortolani et al. [5.56] addressed the question of the anomaly along $\overline{\Gamma}\,\overline{M}$ in detail. They used a 69 layer slab and employed a force contact model of interatomic forces which contains parameters for a nearest-neighbour radial

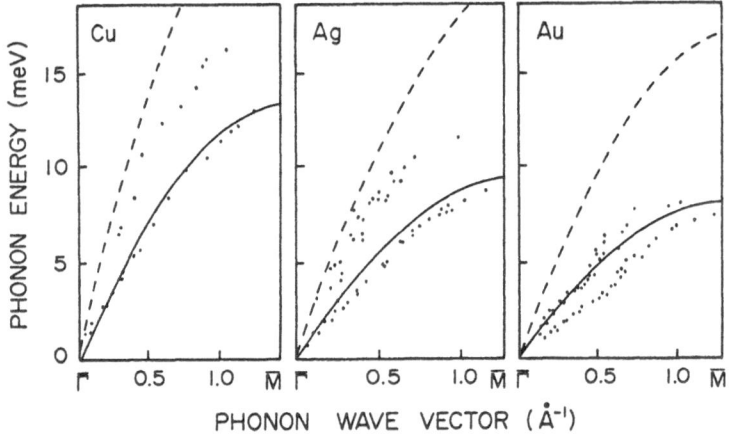

Fig. 5.10. Measured dispersion curves along the $\overline{\Gamma}\,\overline{M}$ direction for gold, silver and copper (111). The solid lines show the calculated Rayleigh waves, the dashed lines indicate the lower limit of the bulk longitudinal vibrations [5.47]

force constant, an angular force constant, and a tangential force constant. With these three constants fitted to bulk properties they obtain result comparable with Armand. If, however, they vary the constants to obtain agreement with experiment they find only certain values of the constants are permitted which leave the surface stable. A reduction of the nearest-neighbour force constant between atoms in the surface layer left the Rayleigh wave agreement intact and produced a new resonant mode below the bulk band at $\overline{M}$. Recently *Celli* et al. [5.57] calculated scattering amplitude in the framework of the distorted wave born approximation. They compare the results with experiment, and find the agreement to be very good.

Bortolani et al. [5.58] have also found the lowering of the first layer force constants to work for the $\overline{\Gamma}\,\overline{K}$ direction. Here there are three modes observed experimentally. One a Rayleigh wave and the other two in the bulk. One bulk mode they find to be a resonance while the other appears to be a pseudo Rayleigh wave.

Gold. Cates and *Miller* [5.59] measured the dispersion of the Rayleigh wave along the $\overline{\Gamma}\,\overline{M}$ direction. They compared their results with those of *Black* et al. [5.33] and *Nizzoli* [5.60]. The calculation of Nizzoli agrees with experiment. The nearest-neighbour frequencies of Black et al. and the more elaborate model with two neighbours and angle bending are both higher than experiment.

Recently *Bortolani* et al. [5.58] have examined gold(111) experimentally and theoretically. They found that they can fit the data using similar techniques to those used in the case of silver(111). For gold a larger lowering of the coupling between the atoms in the first layer is required than for silver. Arguments are presented for why this should be so.

Copper. The copper data of Fig. 5.10 show the best agreement of the experimental and calculated Rayleigh wave. As yet the resonance feature at higher frequencies has not been compared with results of a lattice dynamical calculations. The silver, gold and copper(111) surface phonons have been discussed in some detail in [5.61a].

Nickel. The only experimental data available is that of *Feuerbacher* and *Willis* [5.61]. These researchers examined the inelastic scattering of neon atoms from the nickel(111) surface. *Bortolani* et al. [5.62a] calculated the Rayleigh wave frequency along the $\overline{\Gamma}\,\overline{M}$ direction. They used a 41 layer slab and the same sort of force constants described for silver(111) fitted to bulk properties. The agreement between theory and experiment was not good but they argued it could be better if the experimental full width at half maximum of 14 meV were improved.

c) The (110) fcc Surfaces

Vibrational spectroscopy offers the possibility of distinguishing between various surface reconstructions. In Fig. 5.2 the unreconstructed and missing row

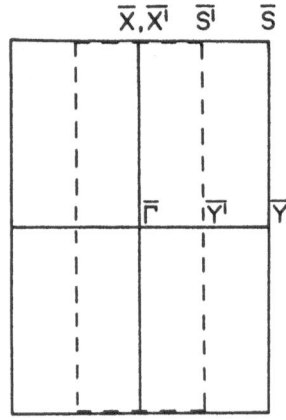

Fig. 5.11. Surface Brillouin zones for the (110) fcc case. Full lines correspond to the unreconstructed surface. Dashed lines indicate the missing row zone. Primes denote special points on the reconstructed zone

Table 5.4. Frequencies at $\overline{\Gamma}(\nu)$ and dominant polarization in the layers (*L*: at right angles to the channels, *SH*: parallel to the channels, *SV*: shear vertical, *O*: no motion)

No reconstruction			Missing row reconstruction			
$\nu\,[\text{cm}^{-1}]$	layer		$\nu\,[\text{cm}^{-1}]$	layer		
	1	2		1	2	3
41.40	L	SV				
48.36	SH	O				
53.41	SV	L				
68.39	O	L	59.24	O	SH	
68.39	O	SH	68.39	O	O	L
85.09	L	SV	68.39	O	O	SH

reconstruction of the (110) fcc metals surface are shown. The lattice dynamics of these two surfaces are quite different. The surface Brillouin zones are illustrated in Fig. 5.11. The zone of the missing row surface is reduced in size in the direction perpendicular to the channels because the real-space lattice is doubled in size in that direction. In Table 5.4 the surface modes at $\overline{\Gamma}$ are shown for the unreconstructed surface and the reconstructed surface. These are both obtained with the reconstructed unit cell. Thus the unreconstructed modes of $\overline{\Gamma}$ are the modes of $\overline{Y}$ folded back to the center of the zone. In a study of dispersion the unreconstructed surface would yield the frequencies shown at the wavevector $\overline{Y}$ while the reconstructed surface face would yield the modes shown at both $\overline{\Gamma}$ and $\overline{Y}$. Clearly the results are very different. There are no longitudinal or sheer vertical modes in the top layer in the reconstructed case. *Copper, Nickel. Mason* et al. [5.63] examined the surface modes along the $\overline{\Gamma}\,\overline{Y}$ direction from near the zone center to halfway to the $\overline{Y}$ point. Their data as yet has not been compared with any slab calculation. There seems to be two Rayleigh waves and a third mode lying in the bulk bands. Estimates using the nearest-neighbour model of *Black* et al. [5.33], in conjunction with unpublished Ni(110) nearest-neighbour calculations, suggest that the observed modes are

15% or more depressed below the calculated values. Thus there may be some softening at the Cu(110) surface.

A somewhat different approach to the study of the Cu(110) and Ni(110) surfaces has been described recently by *Strocio* et al. [5.64]. Their technique utilizes inelastic dipole scattering of electrons from surface vibration resonances of clean metal surfaces. Nearest-neighbour models fitted the bulk give good agreement with experiment but the resolutions of the experimental techniques is not as sharp as that of helium scattering or off specular EELS.

d) Other Surfaces

Pt(332). *Ibach* and *Bruchmann* [5.65] have used specular electron energy loss spectroscopy to study the Pt(332) surface, or stepped Pt(111) surface. This surface consists of (111) terraces of atoms linked by one atom steps. They found a broad loss feature which is peaked at $205\,\mathrm{cm}^{-1}$, about $10\,\mathrm{cm}^{-1}$ above the bulk bands of platinum. The feature is interpreted as due to a vibration of the atoms at the step. They argued that for the feature to split off from bulk bands requires a rather large change in the interatomic force constants at the step.

Molstoller and *Landman* [5.28] were able to fit the observed feature by introducing a 35% increase in the force constants of the step atoms over their bulk value. They used a continued fraction method to generate a spectral density. They examined ϱ_{zz} and found it peaked at $205\,\mathrm{cm}^{-1}$.

Allan [5.66], using the tight-binding approximation described earlier in Sect. 5.3.3a, was able to demonstrate that the atoms at the step should relax 0.05 interatomic distances. This relaxation gave rise to an increase in the interatomic force constants at the step large enough to give rise to a step localized phonon at a frequency above the bulk bands.

Pinas and *Maradudin* [5.67], using a different approximation for the atom interaction than Allan, found relaxation at the step but did not find evidence of a high-frequency surface phonon. They employed a continued-fraction method to generate the phonon density of states of the step atoms.

5.4 Metals with Adsorbates

An enormous amount of experimental work has been done, using the technique of electron energy loss spectroscopy, on modes of vibration of adsorbates at surfaces. We can distinguish between two rather different regions of the electron loss spectrum. On the one hand, there are the frequencies above the substrate bulk bands. These tell us primarily about the nature of the adparticles, their binding to the surface and their disposition on the surface. On the other hand, there are those frequencies at and below the top of the bulk bands. These too can tell us something about the adparticles but, in addition, may tell us about the geometry and interatomic forces in the first few layers of the substrate.

In this chapter emphasis will be placed on the frequency region below the top of the bulk bands. Thus a great many papers dealing with adsorbed species, in which low-frequency modes are not observed, are not discussed here. This omission allows us to concentrate on the problem of what can be learned in the low-frequency region, a non-trivial problem because we are getting information about both the adparticles and the substrate.

In Sect. 5.4.1 we consider the theoretical calculation of modes of vibration. It is shown that the calculations are rather straightforward in the high frequency region of the spectrum. In Sect. 5.4.2 we then concentrate on presenting examples of experimental results and the corresponding theory from the low-frequency regions. The systems considered all involve periodic overlayers of adsorbed atoms and molecules. In Sect. 5.4.3 we consider the example of rare gas adsorbed on metals when the gas and metal are out of registry.

5.4.1 The Calculation of Adsorbate Mode Frequencies

When an isolated adatom is present on a metal surface its modes of vibration (once some assumptions about interatomic forces have been made) can be determined by examining the modes of (5.15), that is the modes of a finite cluster of metal atoms to which the adatom is attached. Typically one is interested in a "surface mode", that is a mode in which the adatom vibration is very large. By increasing the size of the metal cluster the frequency of the mode of interest can be determined to some desired accuracy. In Table 5.5 we show the cluster size dependence of parallel and perpendicular vibrations of oxygen adsorbed on Ni(100). The oxygen is at 0.9 Å above the surface in a fourfold hollow site. The bulk bands end at $300\,\mathrm{cm}^{-1}$ so that the oxygen vibration frequencies are above the bulk bands.

Table 5.5. Oxygen binding mode frequency shifts for various sizes of a Ni(100) cluster. The unshifted frequency at the left is for the 4 nickel atoms infinitely massive

Mode	Cluster size (number of nickel atoms) 4[a]	4	5	13	21	25
perpendicular	371.0	64.5	65.1	68.6	73.1	73.1
parallel	524.6	44.4	44.5	45.8	47.6	47.6

[a]The four nickel atoms are taken to be infinitely massive.

The lowest frequency is that for oxygen adsorbed on four infinitely massive nickel atoms. The remaining frequencies (shown as the shift in frequency from the infinitely massive case) are for clusters in which the nickel atoms have the appropriate mass. The convergence to accuracies of the order of $1\,\mathrm{cm}^{-1}$ is fairly rapid. Examples of the convergence for a variety of adatoms are to be found in [5.68].

Black et al. [5.68] obtained the frequencies of the clusters of atoms using the Wilson's FG matrix method [5.12]. They employed a modified version of the computer program due to *Schachtschneider* and *Snyder* [5.69]. As a result their calculation could be used with the usual chemical valence forces such as stretch, stretch-stretch, stretch-bend, etc. Since the valence forces are readily available to chemists for a variety of molecules their use may be preferable to setting up the elements of the dynamical matrix of (5.12) and diagonalizing the matrix.

In the event that a reasonable size of cluster fails to produce the required convergence the continued fraction method may be used to obtain both the isolated atom mode frequencies and the spectral density. Here it is possible to achieve the equivalent of as many as 1500 atom clusters without diagonalizing a 4500×4500 matrix, as one would have to using the dynamical matrix of (5.15). Once the dynamical matrix has been set up, the continued fraction coefficients A_n, B_n can be obtained in a fairly simple manner as described by *Black* et al. [5.23].

Grimley [5.70] noted that if the adatom binding vibrational mode frequency was above the bulk phonon band then one could think of a local mode associated with the bond between the adatom and the substrate. If, on the other hand, the frequency lay below the top of the bulk phonon band then a nonlocalized mode would be added to the modes of the substrate. In our notation we would call such a mode a resonance. A fairly large cluster of substrate atoms might be needed to determine the frequency of such a resonance.

Now we consider a periodic array of adatoms present on a substrate. We may determine the modes, *at a given wavevector*, of a system consisting of the adlayer and several substrate layers. We use (5.32) and diagonalize the matrix $d_{\alpha\beta}(\boldsymbol{Q}_{\|}; l_z K, l'_z K')$. By increasing the number of substrate layers we can converge the modes of interest (those in which the adlayer vibration is large) to any desired accuracy. For modes above the bulk phonon band only a few layers of the substrate may be required. This method was discussed by *Black* [5.71] for layers of oxygen and CO on Ni(100). If the adatoms are well separated from one another then there will be little dispersion of the modes. On the other hand, if a dense coverage of adatoms is present, then the dispersion will be larger. In the closely packed c(2×2) oxygen on Ni(100), for example, the perpendicular vibration of the oxygen adatoms ranges from $400\,\text{cm}^{-1}$ at the zone center to $486\,\text{cm}^{-1}$ at the zone boundary (the $\overline{\text{X}}$ point). In the less tightly packed p(2×2) oxygen on Ni(100) the frequencies are 435 and $450\,\text{cm}^{-1}$, respectively. (This comparison is based on assuming the oxygen height and force constants are the same for both c(2×2) and p(2×2) overlayers.)

In effect, the method of calculating frequencies of periodic overlayers could be called a slab method. We are finding the modes of vibration of a periodically repeating small cluster consisting of an adatom and a few substrate atoms. An attempt was made by *Rahman* et al. [5.72] to use the continued fraction method to obtain $\varrho_{\alpha\alpha}(l, l'; \omega)$ for c(2×2) oxygen on Ni(100). By the time 1600 atoms had been introduced into the cluster the A_n, B_n had not converged. They were exhibiting oscillatory behaviour with a lengthy period, and probably only one oscillation had been completed. The calculation of the wavevector dependent spectral density

$$\varrho_{\alpha\beta}(l_zK, l_z'K'; \boldsymbol{Q}_{||}, \omega) \equiv \sum_s e_\alpha^{(s)}(\boldsymbol{Q}_{||}; l_zK)e_\beta^{(s)}(\boldsymbol{Q}_{||}; l_z'K')^*\delta(\omega - \omega_s)$$

was also explored, along with the wavevector dependent spectral density of (5.35) for the bare surface. The results are shown in Table 5.6. As can be seen the convergence of A_n, B_n at $\overline{\Gamma}$ for perpendicular motion of oxygen is almost immediate. The series for the perpendicular oxygen motion at $\overline{X}$ is periodic, but also converges rapidly. In this case an analytic termination, similar to that of (5.46), may be used. On the other hand, the periodic behaviour for parallel oxygen motion at $\overline{X}$ is not converging rapidly. Finally the behaviour of a nickel atom at $\overline{\Gamma}$ in the bare surface is seen to be oscillating with a long period and has not converged.

Table 5.6. Examples of the sequence (A_n, B_n) for various Green's functions for the $c(2\times2)$ adlayer of oxygen on Ni(100). For these calculations we have taken $R_\perp = 0.88$ Å

Atom and motion wave vector	Oxygen, $\perp$ motion $\overline{\Gamma}$ point		Oxygen, $\perp$ motion $\overline{X}$ point		Oxygen, $||$ motion $\overline{X}$ point		(Clean surface) Ni surf. atom, $\perp$ motion $\overline{\Gamma}$ point	
n	A_n	B_n	A_n	B_n	A_n	B_n	A_n	B_n
1	4878	0	4878	3611	9763	0	777	0
2	1445	1805	4617	550	3840	5109	1555	550
3	1555	777	1555	550	1244	869	1296	952
4	1555	777	2332	550	1796	733	1814	733
5			1555	550	1364	767	1555	673
6			2332		1708	794	1399	869
7					1433	743	1710	762
8					1654	812	1555	710
9					1477	737	1444	840
10					1616	817	1666	769

Lloyd and *Hemminger* [5.73] have considered the modes of periodic over-layers using a finite cluster of atoms. The cluster is bounded by square boundaries parallel to the surface and a finite number of substrate layers are used. Periodic boundary conditions are introduced by coupling atoms at the sides of the cluster to atoms at the opposite side of the cluster. The modes of the cluster can then be Fourier transformed so as to study dispersion. In their study of the $c(2\times2)$ oxygen on nickel system good agreement was found with the exact Green's function calculation. The finite cluster technique was also applied to calculations of spectral density, (5.36), at the bare nickel surface, and good agreement between their results and those obtained by *Black* [5.71] with a continued-fraction technique was found.

The cluster technique of Lloyd and Hemminger cannot yield better spectral densities or mode frequencies than the Green's function methods. It does, however, have a very practical strength. The technique is based on the Wilson's FG method [5.12]. In fact, the same computer programs that have been used by chemists to study the vibration modes of large molecules were employed by

Lloyd and Hemminger. As a result the usual chemical valence forces, such as stretch, stretch-stretch, bend-bend can be used directly. As noted previously, these forces are more readily available to chemists than the forces needed to construct the dynamical matrix.

In the periodic case, as for the isolated adatom, convergence is more rapid for high frequency modes and resonances can result if the adatom binding frequencies lie in the bulk bands. It is also true that the dispersion will be less for high-frequency modes provided the adparticles are not directly coupled together by high-frequency bonds or dipole coupling. *Andersson* and *Persson* [5.74] found dispersion in the C-O stretching vibration mode of CO on the Cu(100) surface. The C-O stretch frequency at $2100\,\mathrm{cm}^{-1}$ lies well above the top of the bulk phonon band of copper at $200\,\mathrm{cm}^{-1}$. They were able to show that the dispersion of about $30\,\mathrm{cm}^{-1}$ was due to dipole-dipole interactions between the adsorbed CO molecules (dispersion due to the substrate would be less than $0.1\,\mathrm{cm}^{-1}$ [5.71]).

In the discussion that follows we shall see that the slab method and Green's function methods have been used in the study of periodic overlayers. The Green's function methods are preferable and safer (since they accurately obtain resonance features) but the slab method is easy to apply. In fact, if the slab method has been set up with a dynamical matrix it is not very difficult to go from this to the evaluation of the continued fraction coefficients. This then may yield a good wavevector dependent spectral density if one is lucky, and is certainly a simpler procedure than that of setting up the Green's function.

5.4.2 Experimental and Theoretical Results – Commensurate Overlayers

a) The fcc(111) Substrate

There are very few general descriptions of the periodic overlayer problem. *Alldredge* et al. [5.75a] examined the effects of overlayers on the fcc(111) surfaces. Their monolayer had the same structure and potential as the substrate. They considered two cases. In one case the adsorbate atoms were heavier than the substrate atoms and, in the other case, the adsorbate atoms were lighter than the substrate atoms. Modes that were concentrated in the first layer were found to be most influenced by the choice of mass. Heavy adsorbed atoms enhanced the localization of the Rayleigh wave while light adsorbed atoms had the oppositer effect. For a summary of other preliminary work in this area, the reader is referred to [5.75b].

Ibach and *Bruchmann* [5.76] examined the spectra obtained with specularly scattered electrons from Ni(111) on which oxygen was adsorbed. The overlayers studied are designated p(2×2) and $(\sqrt{3}\times\sqrt{3})\mathrm{R}30°$, and correspond to coverages of 0.25 and 0.33, respectively. They are illustrated in Fig. 5.12. Electrons when scattered specularly can only interact with phonons at the $\overline{\Gamma}$ point of the adatom surface Brillouin zone (that is, all adatoms move in phase).

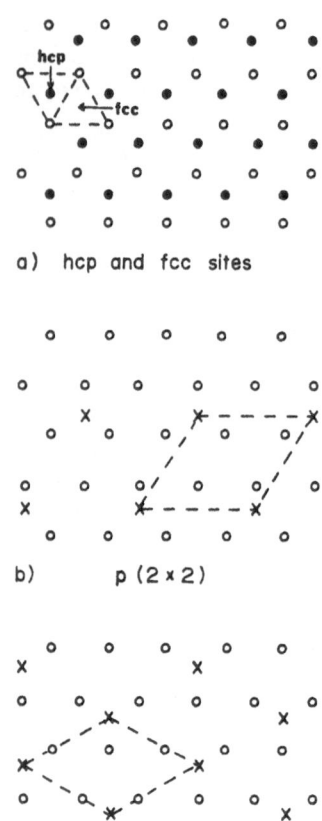

a) hcp and fcc sites

b)   p (2 x 2)

c)   (√3 x √3) R30°

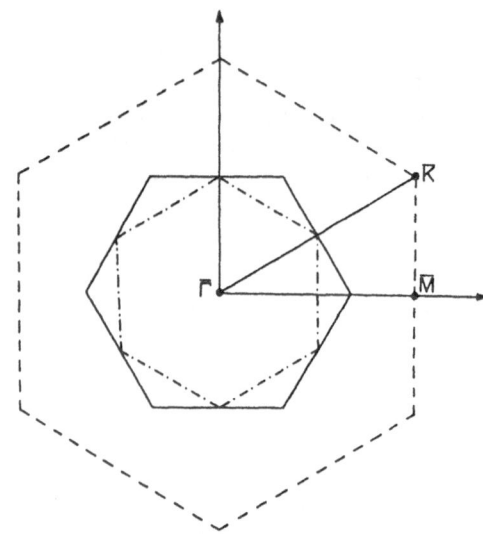

Fig. 5.13. The relation between the adsorbate Brillouin zones and the substrate Brillouin zones for the p(2×2) and (√3×√3)R30° overlayers. Dashed lines are for the substrate, solid lines are for the (√3×√3)R30° overlayer and dash-dot lines are for the p(2×2) overlayer

Fig. 5.12. Oxygen overlayers on Ni(111). (a) The open circles correspond to nickel atoms in the surface layer. Closed circles correspond to nickel atoms in the second nickel layer. hcp and fcc sites are indicated. (b) The p(2×2) oxygen overlayer is indicated with x-s. (c) The √3×√3)R30° oxygen overlayer is indicated with x-s. Dashed lines show the unit cells

The researchers noted that $\overline{\Gamma}$ for the overlayers in the p(2×2) case corresponded to the points $\overline{M}$ and $\overline{K}$ for the (111) substrate, while for the overlayer in the (√3×√3)R30° case corresponded to both $\overline{\Gamma}$ and $\overline{K}$ for the (111) substrate. This is illustrated in Fig. 5.13. They argued that oxygen motion perpendicular to the surface can drive the Ni(111) surface modes S_1 and S_2 at $\overline{M}$ in the p(2×2) case, and can drive S_2 at $\overline{K}$ in the (√3×√3)R30° case. This would account for two features at 135 cm^{-1} and 260 cm^{-1} in the p(2×2) case and one feature at 245 cm^{-1} in the (√3×√3)R30° case. They estimated (using the work of *Allen* et al. [5.15]) frequencies for these modes and found good agreement with experiment.

The arguments of Ibach and Bruchmann, while plausible, were not based on actual numerical calculations for oxygen on a nickel substrate. *Allan* and *Lopez* [5.77] examined the vibrations at $\overline{\Gamma}$ of oxygen perpendicular to the surface of six and twelve layers of nickel atoms anchored to an infinitely massive or frozen nickel substrate. They calculated the positions of the modes (actually they produced a weighted density) for the slab and found peaks consistent with the Ibach and Bruchmann data. However, they did find one extra peak in both the p(2×2) and ($\sqrt{3}\times\sqrt{3}$)R30° cases which they argued would not contribute to the specular reflection if a proper calculation of the EELS intensity were performed.

It is worth commenting on the procedure used by Allan and Lopez to obtain the force constants in their model. The substrate model is the same as that used in the work of Allan previously described and dealing with steps and surface relaxation in transition metals. Experimentally determined bulk properties of nickel are used to set up the interactions in the nickel substrate. The measured oxygen vibration frequency and height are used to set up the oxygen nickel interaction. There are a total of 6 parameters used in the final calculation. It is perhaps not surprising that some accord with experiment is found. The strength of this sort of procedure is that it can be carried out consistently, and that the agreement with experiment tells us that Ibach and Bruchmann offered a reasonable interpretation of the features as due to substrate surface phonons and not, for example, some adsorbed species of low binding frequency.

Bortolani et al. [5.78,79] also examined oxygen on Ni(111). They used a nearest-neighbour and next-nearest-neighbour central force model and also incorporated angle-bending forces in the substrate. These force constants were fitted to the known bulk phonons. They then introduced Ni-O stretch and Ni-O-Ni angle bending in order to treat the oxygen. These force constants were obtained by fitting the vibrations of a Ni_3O cluster to experimental data on the frequencies of low-coverage oxygen adatoms on Ni(111). They then tackled the dynamics of the ordered overlayers using the method of "symmetric coordinates". Their results were in general agreement with those obtained by Ibach and Bruchmann.

Bortolani et al. [5.78,79] were able to show that it was not the mode S_2 that was excited at $\overline{M}$ and $\overline{K}$ (as Ibach and Bruchmann had proposed). It was the mode A_2 of the Ni_3O cluster that was responsible. Moreover, since this mode lies in the bulk phonon band at $\overline{K}$ in the ($\sqrt{3}\times\sqrt{3}$)R30° case and in a gap between the bands in the p(2×2) case, this would explain why the experimental mode was narrower in the latter case than in the former.

The arguments cited above about the origin of the features in the EELS spectrum draw attention to an important question. Can one, by looking at the bare surface modes (obtained in a slab calculation, for example) reliably predict and/or explain the modes observed when an adlayer is present? One difficulty in doing this is that there may be more than one surface mode in the same frequency range with the correct symmetry. A second difficulty lies in the fact that high densities of bulk phonons may be present with which the adatoms binding frequency may resonate. These high densities can be seen only if spectral densities have been calculated. Thus there is great merit in performing a calculation such as that of *Bortolani* et al. [5.77] or *Allan* and *Lopez* [5.79] for the bare surface and overlayer with many layers. Such calculations give the vibration of the system as a whole. One need not try to decide if the adatoms are driving the substrate or the substrate is driving the adatoms. Moreover the dispersion of the high-frequency adatom modes and other features can also be obtained. We will return to this point later when we discuss off-specular EELS.

It is possible to determine something about the adsorption site from a study of low frequency modes. *Rahman* et al. [5.80], using a Green's function method, examined the spectral density for perpendicular motion for the $(\sqrt{3} \times \sqrt{3})R30°$ oxygen overlayer on Ni(111). They used a nearest-neighbour model of nickel-nickel interactions (based on a fit to the bulk phonons). For the oxygen they used a height of 0.9 Å and a nearest-neighbour oxygen-nickel interaction obtained from an ab initio cluster calculation of *Upton* and *Goddard* [5.81]. The results are in excellent agreement with experiment. They show a single high-frequency feature and a single mode below this in the bulk phonon band. The researchers noted that stiffening the nickel-nickel interaction by 20% between the first and second layers of the substrate could improve the agreement.

Rahman et al. examined what would happen if the oxygen were adsorbed in hcp sites instead of fcc sites on the Ni(111) surface. Calculations using an hcp site gave two modes in the bulk phonon band, in addition to the high frequency mode. The modes were sufficiently far apart that they could have been resolved in the EELS experiment. Thus they concluded that in fact oxygen adsorbed in the fcc sites (Fig. 5.12). The modes of bridge sites were also examined and found inconsistent with the experimental results. The above calculations, and those of Bortolani et al., do not yield an extra peak in the bulk phonon band as did the calculations of Allan and Lopez. One is led to suspect that their extra peak is an artifact of the slab calculation or due to the somewhat different force constant model used.

Strong et al. [5.82] have also adressed the question of using specular EELS data to determine overlayer structure. The system studied was oxygen on Al(111). Oxygen is known to adsorb both on top and below the first aluminum layer. The mode frequencies observed in this case lie at 320–400, 645 and 850 cm^{-1}. The top of the bulk bands is about 360 cm^{-1} so that the slab technique is appropriate here. They used a 13 layer aluminum slab (with a seven-neighbour model of the aluminum-aluminum interaction). They then explored various configurations of the surface and subsurface oxygen. By varying the oxygen-aluminum and oxygen-oxygen force constants they were able to obtain best fits to their data and thus to argue that only certain configurations were possible.

b) The fcc(100) Substrate

Andersson [5.83] studied the specular scattering of electrons from both sulfur and oxygen adsorbed on Ni(100). He observed high-frequency features in both the p(2×2) and c(2×2) overlayers. He also observed a shoulder at 260 cm^{-1} (the upper limit of the bulk nickel phonon bands is at 300 cm^{-1}) on the elastic peak in the p(2×2) case, which he attributed to cooperative nickel and oxygen motion.

Lehwald and *Ibach* [5.84] obtained specular EELS data for oxygen on Ni(100). They found one high-frequency line and two spectral features in the bulk phonon bands in the p(2×2) case. There was a single feature at the top of the bulk band in the c(2×2) case. *Allan* and *Lopez* [5.77], employing the approach described previously for the oxygen on Ni(111) case, used a slab method to explore the oxygen vibration perpendicular to the nickel surface. They obtained quantitative agreement with the results of experiment. As before they had extra features in their spectrum not observed in experiment. The researchers noted the problem that was to become central to the c(2×2) oxygen overlayer on Ni(100). The theory gave a frequency which was much higher than that observed in experiment.

Rahman et al. [5.72] adressed the oxygen on Ni(100) problem using a Green's function technique. A nearest-neighbour model was used for the substrate forces, and data for the oxygen on a (100) face of a twenty atom cluster from ab initio calculations of *Upton* and *Goddard* [5.81] was used for the oxygen-nickel nearest-neighbour force and the oxygen height. The p(2×2) agreement was excellent, but the c(2×2) peak was substantially too high in frequency. Upton and Goddard had found a second state of adsorbed oxygen at 0.55 Å above the nickel surface. This result failed to fit either the c(2×2) case or the p(2×2) case.

Upton and *Goddard* [5.81] had examined a nickel cluster, with an adsorbed oxygen atom, from which one electron had been removed. They suggested that this might mimic the effects of a closely packed oxygen layer such as the c(2×2) layer. They obtained an oxygen height of 0.26 Å. With this choice of the oxygen parameters excellent agreement with experiment was obtained. LEED experiments and other surface studies tended to support the height of 0.9 Å. Rahman et al. noted that if the oxygen were present at such a low height then there would be extremely high values of the frequency of vibration parallel to the surface. Thus EELS could provide a test of the height if parallel vibrations could be measured.

Bortolani et al. [5.79] also examined the case of oxygen on Ni(100) using techniques similar to those described for Ni(111). They were concerned about the low height used by Rahman et al. Using the slab method they were able to demonstrate that the experimental results could be fitted with oxygen at 0.9 Å provided the O-Ni stretch constant was 1.6 times as large in the p(2×2) case as in the c(2×2) case. They included angle bending forces in their model. *Armand* and *Lejay* [5.85] had noted that a twofold increase in force constant was needed if only nearest-neighbour forces were used.

The question of the low oxygen height was also adressed by *Andersson* et al. [5.86]. They examined the question of why oxygen and sulfur overlayers on Ni(100) exhibited such different spectra. A nearest-neighbour model of the interatomic forces was used. They fitted the model to the high frequency oxygen modes, the oxygen heights of 0.8 and 0.9 Å, and the bulk phonon spectrum. They were able to show that if the same oxygen height is used for both p(2×2) and c(2×2) overlayers, then some of the frequency shift between the two coverages can be understood as due to enhanced coupling to the substrate in the p(2×2) case, and not due entirely to different electronic structure. This point was developed earlier by *Ibach* and *Mills* [5.46]. They also found some details in the c(2×2) overlayer spectra that could be mimiced by introducing alternating changes in the force constants coupling the first and second nickel layers.

Szeftel et al. [5.87] described measured dispersion data along the $\overline{\Gamma}\,\overline{X}$ direction of the Brillouin zone for the c(2×2) oxygen overlayer (Fig. 5.3). This data was obtained with the off-specular scattering of the electrons. Four modes were observed. The original mode at $\overline{\Gamma}$ of 310 cm^{-1} was present. A new parallel mode was seen at about 450 cm^{-1}, well above the bulk band. This mode was consistent with a height for the oxygen adatom of 0.9 Å above the surface, not the height of 0.26 Å which yielded a much higher parallel frequency. Two other modes were also observed in the bulk band away from $\overline{\Gamma}$. It was possible to obtain a good fit to the data at an oxygen height of 0.9 Å and with a nearest neighbour model in which the oxygen-nickel force constant was fitted to the mode at 310 cm^{-1}, that is the perpendicular mode at $\overline{\Gamma}$. The results are shown in Fig. 5.14. The fit can be improved by introducing lateral interactions between

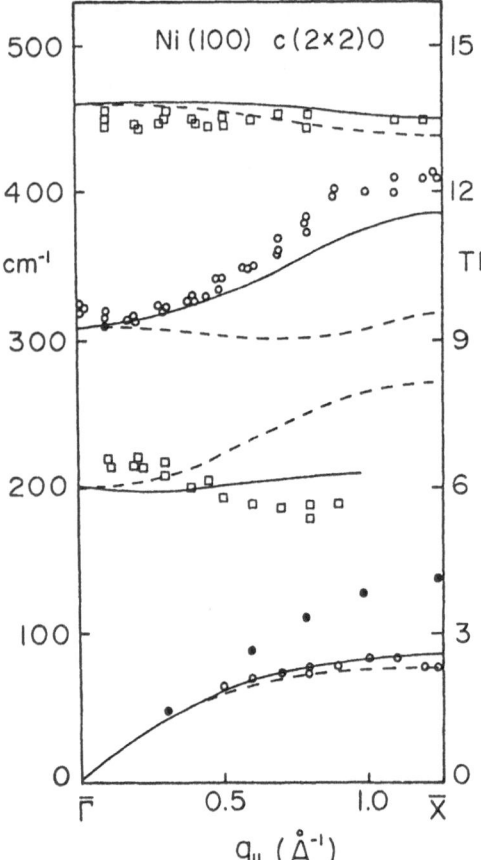

Fig. 5.14. Dispersion of adsorbate and substrate surface phonons for a Ni(100) surface with a c(2×2) oxygen overlayer. The solid-circle line is the S_4 phonon for a clean surface. Modes with predominantly perpendicular and predominantly longitudinal polarization are distinguished by the use of circles and squares for the data points, respectively. The solid lines are the theoretical calculations for oxygen 0.9 Å above the fourfold site. The dashed lines are for oxygen at 0.26 Å above the surface. In this case angle bending forces were included so that the high frequency perpendicular and longitudinal vibrations of oxygen could be fitted at $\overline{\Gamma}$ [5.87]

the oxygen adatoms and by reducing the interatomic forces between the first and second nickel layers to 30% of their bulk value. *Rahman* et al. [5.88] have discussed the oxygen c(2×2) overlayer in detail.

Frenken et al. [5.9] examined the data of *Szeftel* et al. [5.87]. They used the method of Allan described previously. The oxygen perpendicular vibration was fitted to the EELS specular frequency at $\overline{\Gamma}$. They also incorperated a 5.6% expansion of the outer nickel layer (which they had measured in shadowing and blocking experiments). The agreement between experiment and theory was not as good as that found in Szeftel et al. Although some reduction in the forces between the first and second layers of nickel was introduced by the change in separation it was not enough to lower the forces by the 30% needed to fit the data.

A final comment about the p(2×2) oxygen system is appropriate at this point. *Ibach* [5.89] has described data obtained by off-specular EELS at the wavevector $\overline{X}$. *Lloyd* and *Hemminger* [5.73] found that the model of *Rahman* et al. [5.72], while it fitted the data at $\overline{\Gamma}$, failed to fit the experimental data for the parallel mode at $\overline{X}$ (the experiment gave 640 cm^{-1} and the theory gave 572 cm^{-1}). They were able to fit the experimental data at $\overline{X}$ by introducing a stretch-stretch interaction coupling opposite nickel-oxygen pairs.

Recently *Lehwald* et al. [5.90] have completed an experimental and theoretical Green's function study of c(2×2) sulfur overlayers on Ni(100). In this case the low-frequency mode in the bulk band is fitted by theory when the force between the first two nickel layers is increased by 20% (rather than being decreased by 30% as in the c(2×2) oxygen case). This suggestion is then that oxygen weakens the nickel-nickel binding in the top layers, while the sulfur overlayers somehow strengthen it. This weakening in part also accounts for the softer Ni-O c(2×2) forces compared with the Ni-S forces.

Andersson and *Persson* [5.91,92] and more recently *Persson* [5.92] have examined CO, CH, O and S on Cu(100). Also they have examined the ordered oxygen overlayers on Ni(100) and Ni(111). They have obtained theoretical spectra using simple nearest-neighbour models. An efficient calculational scheme for the lattice dynamics of dipole active sulfur modes was presented by Persson. The thrust of these two papers has been to show that the results of the calculations can be interpreted in terms of symmetry restrictions for the dipole activity of surface phonons. They conclude that important knowledge about adsorption sites can be obtained from simple arguments concerning the symmetry properties of the substrate phonons and the symmetry of the adsorption sites.

Rahman and *Ibach* [5.93a] have recently examined the Ni(100) surface in the presence of carbon and nitrogen. They showed how the nickel surface can reconstruct when the nickel surface modes soften due to stresses between the nickel and the adsorbate.

The systematics of oxygen in the c(2×2) structure on Ni(100) have been explored by *Strong* and *Erskine* [5.93b]. They have explored the effects on frequency of altering the height and various force constants. This then indicates some of the systematics of fitting EELS spectra with variation of the parameters in the models of forces, and varying height of the adsorbate.

c) The fcc(110) Substrate

Baro and *Ollé* [5.94] performed EELS experiments with oxygen chemisorbed on Ni(110). They presented symmetry arguments which suggested that their results supported a missing-row model of the substrate with oxygen adatoms populating the low bridge sites of the reconstructed substrate. *Masuda* et al. [5.95] also performed EELS measurements on this system. *Allan* and *Lopez* [5.96] carried out lattice dynamical calculations similar to those described for oxygen on Ni(111). They examined several different oxygen geometries and found the best agreement with experiment when the long bridge site was used.

The hcp(001) Surface

Rahman et al. [5.97] presented experimental data obtained by specular EELS for ordered oxygen overlayers on Ru(001). The p(2×2) overlayer shows a single dipole active mode above the bulk phonon band. However, the p(1×2) overlayer shows an unusual feature. Two dipole active modes are present above the bulk phonon band. A Green's function type of calculation was performed. The ruthenium-ruthenium interaction was modelled with a nearest-neighbour force fitted to the bulk phonons. Oxygen-ruthenium bond lengths and force constants were estimated from chemical consideration and then adjusted to give a good fit to experiment. Calculations were performed for both fcc and hcp sites and little difference was found. The existence of a second mode above the bulk phonon band (this feature lies below the highest phonon frequency in bulk ruthenium, which is located at $310\,cm^{-1}$) could be understood as due to coupling of oxygen atom motion parallel to the surface and perpendicular to the close packed rows of the p(2×2) structure. This occurs due to the reduction of the symmetry from C_{3v} in the p(2×2) case to C_s in the p(1×2) case. The results above the bulk band would also be found if a slab calculation were performed. A lower-frequency feature at $240\,cm^{-1}$ predicted by theory was also seen in the experiment. (The top of the bulk band in Ru is at $310\,cm^{-1}$).

5.4.3 Experimental and Theoretical Results – Incommensurate Overlayers

Mason and *Williams* [5.98], using the He-beam scattering technique, examined the vibrations of monolayers of Ar, Kr, and Xe on Ag(110). They found in this case, and in the case of xenon on Cu(100) [5.99] that a single mode whose frequency does not depend on wavevector – that is a dispersionless mode – is present. They also see integral multiples of this mode.

Gibson et al. [5.100] examined monolayers of Ar, Kr and Xe on Ag(111). They too found dispersionless modes. When the study was extended to bilayers, trilayers and 25 layers of the rare gases then modes with dispersion were seen. The results for argon are illustrated in Fig. 5.15. One sees an evolution from a flat dispersionless feature to the Rayleigh wave S_1 of Fig. 5.6. Incidentally LEED studies show that the rare-gases overlayers, while incommensurate with the silver(111) surface, do form (111) layers whose orientation is the same as that of the substrate.

The results for rare gases can be understood qualitatively with a simple model in which nearest-neighbour forces are used to link the rare-gas atoms (fitted at $\overline{M}$ to the Rayleigh wave of the 25 layer case) and the first rare-gas layer is linked to a smooth, infinitely massive, substrate by perpendicular forces fitted to the monolayer mode frequency. The results of these calculations are the dashed lines of Fig. 5.15. The agreement with experiment is good. It suggests that, in fact, the He-beam scattering is detecting the modes of vibration of the various layers.

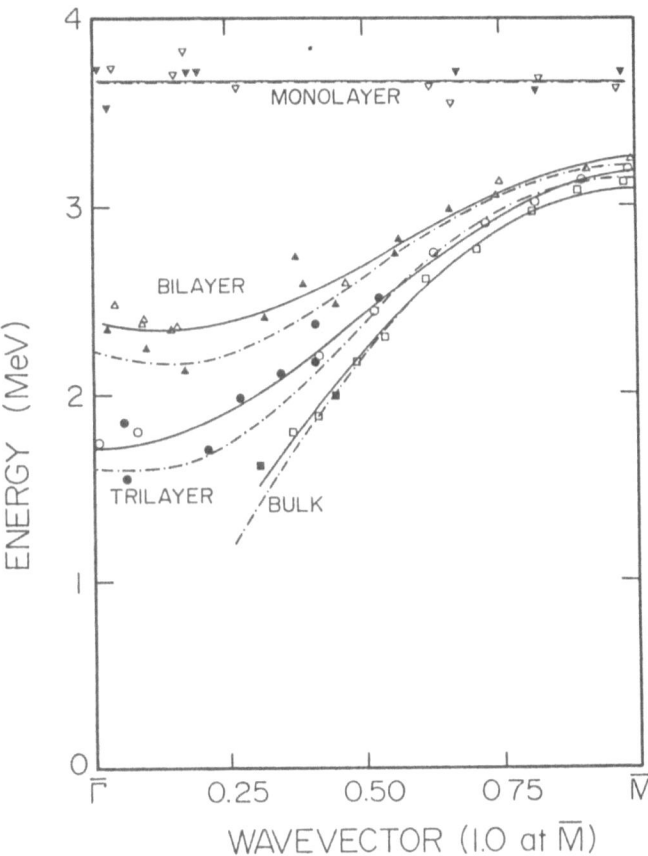

Fig. 5.15. Experimental results for argon monolayers, bilayers, trilayers and bulk (25 layers). The open squares correspond to phonon loss events (that is the incident helium atoms lose energy in the scattering process) while the closed squares correspond to phonon gain events. The full lines are a fit to the data, while the dash-dot lines are results of the simple nearest neighbour calculations [5.100]

The fit between theory and experiment can be improved by introducing the gas-phase potentials of *Barker* et al. [5.101] for xenon and krypton and *Barker* et al. [5.102] for argon. The rare-gas atoms are taken to interact out to five neighbours. In addition, the spacing within the layers is taken to be that obtained by experiment. Finally a model of *Vidali* et al. [5.103] is used to couple the silver to each layer of the rare gas and the layers are allowed to relax normal to the surface. Agreement between theory and experiment is good except for the argon trilayer in the vicinity of the zone center where theory is lower than experiment.

It has yet to be seen if the introduction of three body forces, such as Axilrod-Teller forces and substrate mediated forces of *McLachlan* [5.104] will improve the dispersion curve fits of theory and experiment. There are, however, some discrepancies between the theory and experiment of a different character. In Fig. 5.16 the results for a monolayer of argon are shown. Three modes are present. A longitudinal mode, a transverse mode and a mode whose vibration is perpendicular to the surface. (In systems of more than one layer the transverse modes are still pure transverse but the other modes are sagitally polarized in general). The theory of *Celli* et al. [5.57], while developed for bare metal(111) surfaces, does contain scattering from both the longitudinal and perpendicular modes at the surface. Preliminary calculations using their formulas suggest that the longitudinal mode should be detectable. More theoretical work is needed in this area to see why the longitudinal mode is not seen in experiment. Similar remarks apply in the case of bi- and trilayers.

One interesting problem that remains to be solved here is that of doing a better treatment of the coupling between the phonons of the substrate and

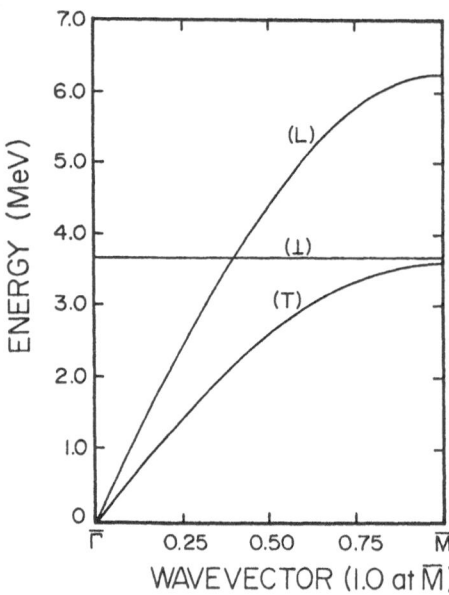

Fig. 5.16. Calculated dispersion of the longitudinal (L), transverse (T) and perpendicular modes ($\perp$) for a xenon monolayer

the phonons of the overlayer. This problem is not an easy one because of the incommensurateness of the substrate and overlayer. *Black* et al. [5.105] have made a start on the problem by considering the rare gases couple to a continuous substrate. They found that when the adsorbate modes lie inside the bulk bands of the substrate, then they are "leaky" modes and have a finite lifetime. Interesting effects occur when the overlayer modes cross the Rayleigh wave of the substrate.

In an alternate approach the modes of a xenon monolayer on silver were examined assuming the nearest-neighbour distances were in the ratio 1.5:1.0 (in the xenon-silver system the actual ratio differs by $\approx 3.0\%$ from this value). The results were consistent with those obtained with the continuous model of the substrate. It is clear that more work could usefully be done on the incommensurate problem. In particular, an experiment with increased resolution may detect the crossing, by the overlayer modes, of the Rayleigh waves.

5.5 Concluding Remarks

5.5.1 Bare Surfaces

We have seen that EELS and He-beam scattering techniques permit the accurate measurement of metal surface phonons. Moreover, theoretical methods of calculation of spectral intensities and time of flight spectra now exist for the techniques.

The ingredients that link theory and experiment are the inter-atomic forces. There is evidence that the forces between the surface atoms and atoms one layer below the surface are stiffer than in the bulk in the Ni(100) case. On the (111) surfaces of silver, gold and copper, on the other hand, there is evidence that the forces between atoms in the surface layer are reduced from their bulk values.

Simple theories using one or two neighbours and angle-bending forces have been used to fit experimental frequencies. More complicated models, involving, for example, some of the valence forces such as stretch-stretch coupling and other coordinates combinations, have not been needed. It is perhaps worth pointing out that when a few force constants are used in fitting the observed frequencies then the solution is not unique. Many sets of force constants may fit the data. On the other hand, as pointed out by *Leigh* et al. [5.106], detailed eigenvector measurement can resolve this uncertainty. Thus the work of *Tong* and collaborators [5.107] on EELS intensity (sensitive to eigenvector) may aid in choosing the correct force constant models.

There are still many metal surfaces on which even preliminary work does not exist. The body centered metals are an example. *Kesmodel* [5.108a] has begun work on Cu(100) and *Toennies* and collaborators [5.108b] have begun work on Rb(100).

It does seem that EELS of bare surfaces and He-beam scattering of bare surfaces, now that theoretical intensities can be calculated, have reached a point where ab initio calculation of inter-atomic forces can be fruitfully tested and contribute to our understanding of the behaviour of atoms and the electron gas at the metal surface. As yet no one has published experimental data on the bcc surfaces such as tungsten and sodium, nor on the simple fcc metal aluminum. These surfaces are of interest because, as mentioned previously, ab initio calculation of frequencies should be available in the near future [5.44,45]. In fact, at the 1985 March Meeting of the American Physical Society, *Bohnen* and *Ho* [5.109] presented ab initio dispersion data for Al(110) and *Reinecke* et al. [5.110] presented ab initio dispersion data for W(100).

5.5.2 Adsorbates

We have seen that the adsorbate modes, both above and below the top of the bulk phonon bands, can be used to distinguish between adsorbate sites and also to determine the height of adatoms above the metal surface. The low-frequency modes have also been used to identify changes in substrate interatomic forces produced by the overlayer.

Two rather different approaches to the interpretation of the modes exist. The formal, and computational, approach in which the spectral density is calculated, and the more intuitive approach in which the modes are interpreted by considering selection rules and symmetry arguments. The thrust of many papers has been to develop the latter approach for the (100) and (111) surfaces and their various overlayers by interpreting the results of the exact theory and experiment in terms of the symmetries of the substrate and adsorbate modes. There are difficulties in extending this intuitive sort of approach to new overlayers and surfaces of different Miller indices, and in such cases the formal lattice dynamical calculation may be warranted. Never-the-less the more intuitive approach is extremely useful to the chemist in identifying species and their geometry in the study of new materials.

It is clear that EELS has reached a point where it can assist in the identification of species adsorbed on metals, and that there are cases where the low-frequency modes can assist in this identification.

It is also clear that EELS can be used in the low-frequency region to tell us something about the effects of adsorbates on the forces between the layers of the substrate. Here, as in the case of bare metals, there is now the possibility of testing ab initio calculations of forces when periodic overlayers of atoms are adsorbed at metal surfaces. At the present time ab initio calculations have only been done for small clusters, and not for the periodic systems of infinite extent.

Another area that needs some development is that of adatoms on fcc surfaces of additional metals and on bcc surfaces. This would enable the experimentalist to build up the techniques for the interpretation of modes in the low-frequency region of the spectrum in which symmetry arguments were used, and not the full lattice dynamical calculations.

A final area that needs some theoretical study is the following. All results reported above for both bare and adsorbate covered surfaces, have dealt with the phonons along high-symmetry directions. It would be of interest to know if more could be learned about the systems under study if directions other than those of high symmetry were used.

Acknowledgements. A great deal of my knowledge of surface vibrations has come from conversations with H. Ibach, S.Y. Tong, R.F. Wallis and M. Wolfsberg. I am particularly indebted to D.L. Mills for introducing me to the subject of surface vibrations, and for his continued interest in, and encouragement of, my work in this area. Finally, I wish to acknowledge the financial support of the Natural Sciences and Engineering Research Council of Canada.

References

5.1   R.W. Wallis: Prog. Surf. Sci. **4**, 233 (1973)
5.2   F.W. de Wette: Comments Solid State Phys. (GB) **11**, 89 (1984)
5.3   J.P. Toennis: J. Vac. Sci. Technol. A**2**, 1055 (1984)
5.4   G. Benedek: Surf. Sci. **126**, 624 (1983)
5.5   G. Kanellis, M.F. Morehouse, M. Balkanski: Phys. Rev. B**21**, 1543 (1980)
5.6   D.C. Allan, E.J. Mele: Session EI5, American Physical Society March Meeting, Baltimore, Maryland, USA (1985)
5.7   T.H. Ellis, G. Scoles, U. Valbusa: Chem. Phys. Lett. **94**, 247 (1983)
5.8   R.E. Allen, G.P. Alldredge, F.W. de Wette: Phys. Rev. B**4**, 1648 (1971-I)
5.9   J.W. Frenken, J.F. Van der Veen, G. Allan: Phys. Rev. Lett. **51**, 1876 (1983)
5.10   M.A. Van Hove: Surf. Sci. **103**, 189 (1981)
5.11   G. Somorjai: Surf. Sci. **89**, 496 (1979)
5.12   E.B. Wilson, J.C. Decius, P.C. Cross: *Molecular Vibrations* (McGraw-Hill, New York 1955)
5.13   Lord Rayleigh: Math. Soc. Proc. (London) **17**, 4 (1887);
        E.A. Ash, E.G.S. Paige (eds.): *Rayleigh-Wave Theory and Applications*, Springer Ser. Wave Phen., Vol. 2 (Springer, Berlin, Heidelberg 1985)
5.14   G.W. Farnell: In *Physical Accoustics* 6, 109 (Academic, New York 1970)
5.15   R.E. Allen, G.P. Alldredge, F.W. de Wette: Phys. Rev. B**4**, 1661 (1971-II)
5.16   J. Latkowski, J.E. Black: Unpublished
5.17   V. Roundy, D.L. Mills: Phys. Rev. B**5**, 1347 (1972)
5.18   V.R. Velasco, F. Yndurain: Surf. Sci. **85**, 107 (1977)
5.19   V.R. Velasco, F. Garcia-Molinar: Surf. Sci. **83**, 376 (1979)
5.20   F. Garcia-Molinar, G. Platero, V.R. Velasco: Surf. Sci. **136**, 601 (1984)
5.21   J.E. Black, T.S. Rahman, D.L. Mills: Phys. Rev. B**27**, 4072 (1983)
5.22   M. Croitoru: Rev. Romm. Phys. (Romania) **20**, 1067 (1975)
5.23   J.E. Black, B. Laks, D.L. Mills: Phys. Rev. B**22**, 1818 (1980)
5.24   R. Haydock, V. Heine, M.J. Kelly: J. Phys. C**5**, 2845 (1972)
5.25   R. Haydock, V. Heine, M.J. Kelly: J. Phys. C**8**, 2591 (1975)
5.26   G. Armand: Solid State Commun. **48**, 261 (1983)
5.27   J.E. Black: Surf. Sci. **105**, 359 (1981)
5.28   M. Mostoller, U. Landman: Phys. Rev. B**20**, 1755 (1979)
5.29   R.E. Allen, G.P. Alldredge, F.W. de Wette: Phys. Rev. B**6**, 632 (1972)
5.30   J.E. Black, D.A. Campbell, R.F. Wallis: Surf. Sci. **115**, 161 (1982)
5.31   D. Castiel, L. Dobrzynski, D. Spanjaard: Surf. Sci. **59**, 252 (1976)
5.32   D. Castiel, L. Dobrzynski, D. Spanjaard: Surf. Sci. **63**, 21 (1977)
5.33   J.E. Black, F.C. Shanes, R.F. Wallis: Surf. Sci. **133**, 199 (1983)
5.34   V. Bortolani, F. Nizzoli, G. Santoro: Transition Metals, Toronto, Canada, 1977 (Inst. Physics, London 1978) p. 236
5.35   V. Bortolani, F. Nizzoli, G. Santoro: Proc. Intern. Conf. on Lattice Dynamics 1977 (Flammarion, Paris 1978) p. 302
5.36   V. Bortolani, F. Nizzoli, G. Santoro, E. Tosatti: Solid State Commun. **26**, 507 (1978)

5.37   A. Fasolino, G. Santoro, E. Tosatti: Surf. Sci. **125**, 317 (1983)
5.38   N. Krakauer, M. Posternak, A.J. Freeman: Ordering in Two Dimensions (North Holland, Amsterdam 1981) p. 47
5.39   S. Pick, M. Tomasek: Surf. Sci. **130**, L307 (1983)
5.40   G. Armand, P. Masri: Surf. Sci. **130**, 89 (1983)
5.41   J.E. Black, P. Bopp: Surf. Sci. **140**, 275 (1984)
5.42   G. Benedek, M. Miura, W. Kress, H. Bilz: Phys. Rev. Lett. **52**, 1907 (1984)
5.43   C. Calandra, A. Catellani, C. Beatrice: Surf. Sci. **148**, 90 (1984)
5.44   A. Eguilez, A. Maradudin, R.F. Wallis: Private communication
5.45   E. Wimmer, A.J. Freeman, M. Weinert, H. Krakauer, J.R. Hiskes, A.M. Karo: Phys. Rev. Lett. **48**, 1128 (1982)
5.46   H. Ibach, D.L. Mills: *Electron Energy Loss Spectroscopy and Surface Vibrations* (Academic, New York 1982)
5.47   J.P. Toennies: Proc. 6th General Conf. of the European Physical Society "Trends in Physics", Prague (1984)
5.48   S. Lehwald, J.M. Szeftel, H. Ibach, T.S. Rahman, D.L. Mills: Phys. Rev. Lett. **50**, 518 (1983)
5.49   Mu-Liang Xu, B.M. Hall, S.Y. Tong, M. Rocca, H. Ibach, S. Lehwald, J.E. Black: Phys. Rev. Lett. **54**, 1171 (1985)
5.50   G. Allan: Surf. Sci. **85**, 37 (1979)
5.51   M. Rocca, S. Lehwald, H. Ibach, T.S. Rahman: Surf. Sci. Lett. **138**, L123 (1984)
5.52   B.F. Mason, B.R. Williams: Phys. Rev. Lett. **46**, 1138 (1981)
5.53   J.G. Sanz, G. Armand: Surf. Sci. **85**, 197 (1979)
5.54   S. Andersson, B.N.J. Persson, M. Persson, N.D. Lang: Phys. Rev. Lett. **52**, 2073 (1984)
5.55   R.B. Doak, U. Harten, J.P. Toennies: Phys. Rev. Lett. **51**, 578 (1983)
5.56   V. Bortolani, A. Franchini, F. Nizzoli, G. Santoro: Phys. Rev. Lett. **52**, 429 (1984)
5.57   V. Celli, G. Benedak, U. Harten, J.P. Toennies, R.B. Doak, V. Bortolani: Surf. Sci. **143**, L376 (1984)
5.58   V. Bortolani, G. Santoro, U. Harten, J.P. Toennies: Surf. Sci. **148**, 82 (1984)
5.59   M. Cates, D.R. Miller: Phys. Rev. B**28**, 3615 (1983)
5.60   F. Nizzoli: Private communication to D.R. Miller
5.61   U. Harten, J.P. Toennies, Ch. Wöll: Faraday Disc.Chem. Soc. **80** (1985);
        B. Feuerbacher, R.F. Willis: Phys. Rev. Lett. **47**, 526 (1981)
5.62   V. Bortolani, A. Franchini, F. Nizzoli, G. Santoro, G. Benedek, V. Celli: Surf. Sci. **128**, 249 (1983);
        U. Harten, J.P. Toennies, Ch. Wöll, G. Zhang: Phys. Rev. Lett. **55**, 2308 (1985)
        G. Comsa: Private Communication;
        D. Neuhaus, F. Joo, B. Feuerbacher: To be published
5.63   B.F. Mason, K. McGreer, B.R. Williams: Surf. Sci. **130**, 282 (1983)
5.64   J.A. Strocio, S.R. Bare, M. Persson, W. Ho: To be published
5.65   H. Ibach, D. Bruchmann: Phys. Rev. Lett. **41**, 958 (1978)
5.66   G. Allan: Surf. Sci. **89**, 142 (1979)
5.67   G.J. Pinas, A.A. Maradudin: J. Electron. Spectrosc. and Relat. Phenon. (Neth) **30**, 131 (1983)
5.68   J.E. Black, P. Bopp, K. Lutzendirchen, M. Wolfsberg: J. Chem. Phys. **76**, 6431 (1982)
5.69   J.H. Schachtschneider, R.G. Snyder: Spectrochim. Acta **19**, 117 (1963)
5.70   T.B. Grimley: Proc. Phys. Soc. **79**, 1203 (1962)
5.71   J.E. Black: Surf. Sci. **116**, 240 (1982)
5.72   T.S. Rahman, J.E. Black, D.L. Mills: Phys. Rev. B**25**, 883 (1982)
5.73   K.G. Lloyd, J.C. Hemminger: Surf. Sci. **143**, 509 (1984)
5.74   S. Andersson, B.N.J. Persson: Phys. Rev. Lett. **45**, 1421 (1980)
5.75   G.P. Alldredge, R.E. Allen, F.W. de Wette: Phys. Rev. B**4**, 1682 (1971);
        R.F. Willis (ed.): *Vibrational Spectroscopy of Adsorbates*, Springer Ser. Chem. Phys., Vol. 15 (Springer, Berlin, Heidelberg 1980) Chap. 4
5.76   H. Ibach, D. Bruchmann: Phys. Rev. Lett. **44**, 36 (1980)
5.77   G. Allan, J. Lopez: Surf. Sci. **95**, 214 (1980)
5.78   V. Bortolani, A. Franchini, F. Ninzzoli, G. Santoro: Sol. State. Commun. **41**, 369 (1982)

5.79   V. Bortolani, A. Franchini, F. Ninzzoli, G. Santoro: J. Elect. Spectr. and Related Phenomena **29**, 219 (1983)

5.80   T.S. Rahman, D.L. Mills, J.E. Black: Phys. Rev. B**27**, 4059 (1983)

5.81   T. Upton, W.A. Goddard: Phys. Rev. Lett. **46**, 1635 (1981)

5.82   R.L. Strong, B. Firey, F.W. de Wette, J.L. Erskine: Phys. Rev. B**26**, 3483 (1982) [see also Erratum: Phys. Rev. B**27**, 3893 (1982) for correction]

5.83   S. Andersson: Surf. Sci. **79**, 385 (1979)

5.84   S. Lehwald, H. Ibach: In *Vibrations at Surfaces*, ed. by R. Caudano, J.M. Gilles, A.A. Lucas (Plenum, New York 1982)

5.85   G. Armand, Y. Lejay: Solid State Commun. **24**, 321 (1977)

5.86   S. Andersson, P.-A. Karlsson, M. Persson: Phys. Rev. Lett. **51**, 2378 (1983)

5.87   J.M. Szeftel, S. Lehwald, H. Ibach, T.S. Rahman, J.E. Black, D.L. Mills: Phys. Rev. Lett. **51**, 268 (1983)

5.88   T.S. Rahman, D.L. Mills, J.E. Black, J.M. Szeftel, S. Lehwald, H. Ibach: Phys. Rev. B**30**, 589 (1984)

5.89   H. Ibach: Proc. 9th Intern. Vacuum Cong. and 5th Intern. Conf. on Solid Surfaces, Madrid (1983)

5.90   S. Lehwald, M. Rocca, H. Ibach, T.S. Rahman: Phys. Rev. B**31**, 3477 (1985)

5.91   S. Andersson, M. Persson: Phys. Rev. B**24**, 3659 (1981);
        S. Andersson, M. Persson: Surf. Sci. **117**, 352 (1982)

5.92   M. Persson: Phys. Scripta **29**, 181 (1984)

5.93   T.S. Rahman, H. Ibach: Private Communication;
        R.L. Strong, J.L. Erskine: Phys. Rev. B**31**, 6305 (1985)

5.94   A.M. Baro, L. Olle: Surf. Sci. **126**, 170 (1983)

5.95   S. Musuda, M. Nishijima, Y. Sakisaka, M. Onchi: Phys. Rev. B**25**, 863 (1982)

5.96   G. Allen, J. Lopez: Unpublished

5.97   T.S. Rahman, A.B. Anton, N.R. Avery, W.H. Weinberg: Phys. Rev. Lett. **51**, 1979 (1983)

5.98   B.F. Mason, B.R. Williams: Surf. Sci. **139**, 173 (1984)

5.99   B.F. Mason, B.R. Williams: Phys. Rev. Lett. **46**, 1138 (1981)

5.100   K. Gibson, S.J. Sibener, B. Hall, D.L. Mills, J.E. Black: J. Chem. Phys. **83**, 4256 (1985)

5.101   J.A. Barker, M.L. Klein, M.V. Bobatic: IBM J. Res. Devel. **20**, 222 (1976)

5.102   J.A. Barker, R.A. Fisher, R.O. Watts: Molecular Phys. **21**, 657 (1971)

5.103   C. Vidali, M.W. Cole, J.R. Klein: Phys. Rev. B**28**, 3064 (1983)

5.104   A.D. McLachlan: Mol. Phys. **7**, 381 (1964)

5.105   B. Hall, D.L. Mills, J.E. Black: Phys. Rev. B**32**, 4932 (1985)

5.106   R.S. Leigh, B. Szigeti, V.K. Tewary: Proc. Roy. Soc. (London) A**1320**, 505 (1971)

5.107   S.Y. Tong: Phys. Today **37**, 50 (August 1984)

5.108   L. Kesmodel: Private communication (1984);
        U. Harten: Private communication

5.109   K.P. Bohnen, K.M. Ho: Session HI 11 American Physical Society March Meeting, Baltimore, MA USA (1985)

5.110   T.L. Reinecke, S.C. Ying, S. Tiersten: Session E17 American Physical Society March Meeting, Baltimore, MA USA (1985)

Additional References with Titles

Fu, C.L., A.J. Freeman, E. Wimmer, M. Weinert: Frozen-phonon total-energy determination of structural Surface Phase Transitions: W(001), Phys. Rev. Lett. **54**, 2261 (1985)

Hall, B.M., D.L. Mills: Electron energy loss studies of surface vibration on Ni(100): Surface and Bulk Phonon Contribution, to be published

Harten, U., J. P. Toennies, C. Wöll, G. Zhang: Observation of a Kohn anomaly in the surface phonon dispersion curves of Pt(111), Phys. Rev. Lett. **55**, 2308 (1985)

Ho, K.M., K.P. Bohnen: First-principles calculation of surface phonons on the Al(110) surface, Phys. Rev. Lett. **56**, 934 (1986)

Kern, K., R. David, R.L. Palmer, G. Comsa, J. He, T.S. Rahman: Adsorbate-induced Rayleigh-phonon gap of p(2×2)0/Pt(111), Phys. Rev. Lett. **56**, 2064 (1986)

Kern, K., R. David, R.L. Palmer, G. Comsa, T.S. Rahman: Surface phonon dispersion of platinum(111), Phys. Rev. B33, 4334 (1986)

Kesmodel, L.L., M.L. Xu, S.Y. Tong: Cross-section analysis of surface and bulk phonons by scattering from Cu(100), to be published

Neuhaus, D., F. Joo, B. Feuerbacher: Anomalies in the surface phonon dispersion Pt(111), Surf. Sci. Lett. 165, L90 (1986)

Rocca, M., S. Lehwald, H. Ibach, T.S. Rahman: EELS study of the dynamics of clean Ni(100): Surface Phonons and Surface Resonances, to be published

6. Molecular Dynamics and the Study of Anharmonic Surface Effects

W. Schommers

With 35 Figures

The role of molecular dynamics (MD) in many-particle physics is discussed. In particular, it is pointed out that "simple models" are not available for solving various problems in the physics of condensed matter (including surface physics). In such cases MD calculations are important. For example, *disordered* systems with strong *anharmonicities* can be treated by MD without approximations provided a realistic interaction potential is given.

The MD method is discussed in connection with statistical mechanics. It is pointed out that the *thermodynamic limit* is in most cases observed for models with some hundred particles.

Geometrical models for the bulk, and models with surfaces and interfaces are introduced. All these models are capable of simulating *infinite* systems with a *finite* (and sufficiently small!) number of particles. The evolution with time of such systems is investigated; in particular, the velocity distribution and the temperature fluctuations are studied.

The MD method is applied to problems in many-particle physics with particular emphasis placed on surface physics. The following problems are discussed:

- Phase transitions in two-dimensional systems.
- Diatomic molecules adsorbed on a surface.
- Melting transition of near monolayer xenon films on graphite.
- Microscopic behavior of krypton atoms at the surface.

6.1 Introductory Remarks

In systems with surfaces there is only a *two-dimensional* translational invariance of the force constants and this implies that the normal-mode solutions of the equation of motion have the form of two-dimensional Bloch functions. A central problem in the study of phonons (Chap. 5) is the determination of the force constants which can be evaluated in the following ways:

1. The force constants of particles near the surface are assumed to be the same as those for particles in the bulk.
2. The force constants are evaluated at the static-equilibrium positions; these are obtained by minimizing the *static energy*. In this case the relaxation of the surface particles, and the resulting changes in the force constants are taken into account.

3. The force constants are evaluated with respect to the positions corresponding to *uniform* thermal expansion throughout the crystal.
4. In the determination of the force constants the additional (additional to the uniform thermal expansion in the bulk) thermal expansion near the surface is taken into account. In this case, the mean positions of the particles are obtained by minimizing the *free energy* of the system with a surface.

The mean-square amplitudes of the particles are significantly larger at the surface of the crystal than in the bulk and, therefore, the harmonic approximation is limited to low temperatures; for example, in the case of noble-gas crystals below one sixth of the melting temperature T_m (as compared to about $1/3 \, T_m$ for the bulk). Thus, in a more general treatment of surface phenomena *anharmonic* effects have to be taken into consideration. MD calculations are important in studying classical many-particle systems with strong anharmonicities, since anharmonicity is treated without approximation.

The study of surface phenomena with the help of MD calculations is just in its initial phase. Thus, the principles of the method will be explained in this chapter mainly by the example of bulk systems.

6.2 Analysis of Many-Particle Systems

6.2.1 Pair Potential Approximation

The description of the *macroscopic properties* of many-particle systems from the *interaction potentials* is a major objective of the theory of condensed matter. For example, the elastic constants and compressibilities of metals may be expressed in a relatively simple manner in terms of the interaction potential. Furthermore, pair correlation functions in statistical mechanics of disordered systems, which can be measured in scattering experiments, are connected to the interaction potential.

For the description of interaction not only the pair potential but also many-body forces may be of importance; it is shown in [6.1–4] that the influence of three-body forces on *dynamical* correlation functions cannot be neglected. So, the Hamiltonian for N particles of a classical system has to be expanded in terms of pair and many-body contributions:

$$H = \sum_{i=1}^{N} \frac{p_i^2}{2m_i} + U \quad \text{with} \tag{6.1}$$

$$U = \frac{1}{2} \sum_{\substack{i,j=1 \\ i \neq j}}^{N} u_{ij} + \frac{1}{6} \sum_{\substack{i,j,k=1 \\ i \neq j \neq k}}^{N} u_{ijk} + \cdots \tag{6.2}$$

Clearly, the most important term in (6.2) is the pair-potential contribution. In order to consider approximately the many-body forces in (6.2) one often uses "effective" pair potentials [6.5–7], which simulate these contributions. In terms of the effective pair potential v_{ij} the potential energy U is given by

$$U = \frac{1}{2} \cdot \sum_{\substack{i,j=1 \\ i \neq j}} v_{ij}. \tag{6.3}$$

Most of the pair-potential functions are derived in a phenomenological way and are expressed in terms of parameters which must be determined using experimental data; these potential functions should also be considered as effective pair potentials. In this chapter we will not discuss many-body forces, and all pair potentials used here will be designated by $v(r) \equiv v(r_{ij})$, where r_{ij} is the distance between the particles i and j. Examples for $v(r)$ are given in Fig. 6.1.

Figure 6.1a shows the Lennard-Jones (6.12) potential

$$v(r) = 4\varepsilon \left[\left(\frac{\sigma}{r} \right)^{12} - \left(\frac{\sigma}{r} \right)^{6} \right], \tag{6.4}$$

which is often used in the description of *noble-gas* many-particle systems. Figure 6.1b shows the Coulomb potential which describes the interaction in pure

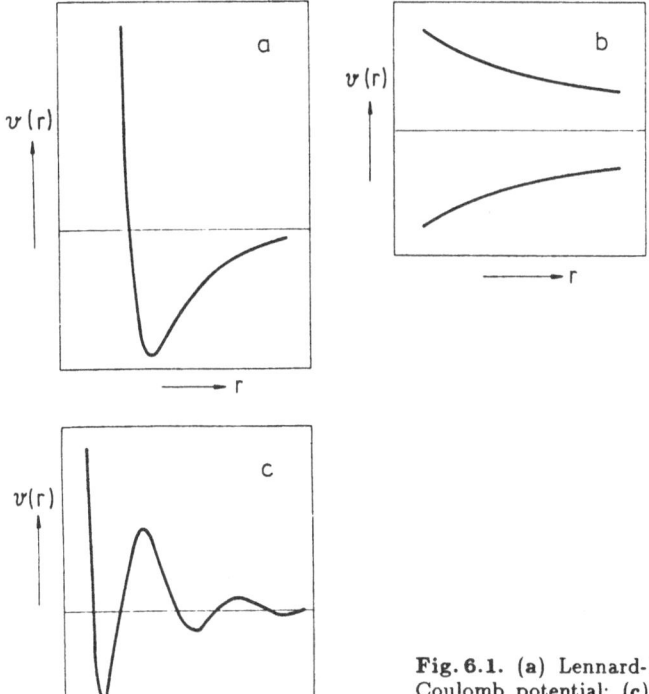

Fig. 6.1. (a) Lennard-Jones (6,12) potential; (b) Coulomb potential; (c) ion-ion interaction in the bulk of a metal

ionic systems. In Fig. 6.1c the ion-ion potential for a *metal* is shown; this pair potential has been calculated from the pseudopotential theory [6.8]. The oscillations of its long-range part are given by

$$v(r) \sim \frac{\cos(2k_F r)}{r^3},$$ (6.5)

where k_F is the Fermi wave-number.

In noble gases and ionic systems the properties at the *surface* can be determined in a very good approximation from the pair potentials of the bulk. However, this is not possible in the case of *metals*; the pseudopotentials (an example is given in Fig. 6.1c) have two parts: the *core contribution* which is essentially the same in the bulk and at the surface, and the *electronic contribution*, which depends on the electronic arrangement. The electronic arrangement at the surface is, however, different from that in the bulk. Thus, in the case of metals the pair potential at the surface is in general different from that in the bulk (Fig. 6.1.c). More details on pair potentials will be given in Sect. 6.3.4.

6.2.2 "Simple Models"

How can we determine with the Hamiltonian the properties of a many-particle system consisting of $\sim 10^{23}$ particles? In the "conventional" treatment of this problem we need "*simple models*". For example, the "simple model" in solid-state physics is the *crystalline* solid in the *harmonic* approximation. On the basis of this model one is able to determine the macroscopic properties of a lot of materials. However, there are also a lot of cases where this "simple model" is not applicable. For example, the silver subsystem of the solid electrolyte α-AgI is highly *disordered* and shows strongly *anharmonic* behavior [6.5-7]; a "simple model" for α-AgI and similar materials is not available. Also in the case of liquids and gases "simple models" have not been found. It should be mentioned that even in the case of gases with *low densities* we cannot restrict ourselves to the first terms in the virial expansion for the pressure P. This virial series can be written as

$$\frac{P}{k_B T} = \varrho + \varrho^2 B(T) + \ldots,$$ (6.6)

where k_B is the Boltzmann constant, T is the temperature, and ϱ is the density. The virial coefficient $B(T)$ is given by

$$B(T) = -\frac{1}{2} \int f(r) dr \quad \text{with}$$ (6.7)

$$f(r) = \exp\left(-\frac{v(r)}{k_B T}\right) - 1.$$ (6.8)

Also the other virial coefficients which belong to the terms in (6.6) with ϱ^3, ϱ^4, $\ldots, \varrho^n, \ldots$ can be expressed in terms of the pair potential $v(r)$ but the expressions are getting complicated with increasing n, and in practical calculations

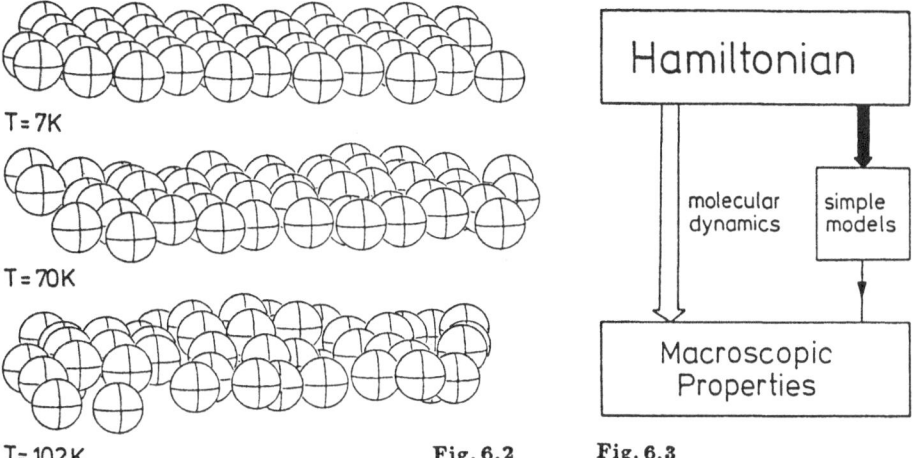

T = 7K

T = 70K

T = 102 K

Hamiltonian

molecular dynamics | simple models

Macroscopic Properties

Fig. 6.2   Fig. 6.3

Fig. 6.2. The structure of the outermost layer of a krypton crystal for the temperatures of T = 7 K, 70 K and 102 K. The melting temperature is 116 K

Fig. 6.3. If a "simple model" is available, macroscopic properties can be determined on the basis of such a model; in most cases additional assumptions are necessary. The advantage of molecular dynamics is that this description can be done without "simple models" and other additional assumptions

only the first few terms are accessible. It turned out that the expansion (6.6) converges slowly and even in the case of low density one has to consider more than two terms in (6.6) [6.3].

Also the quantitative treatment of *surface phenomena* on the basis of a "simple model" is not possible in general. One reason is that at the surface *anharmonic* effects have to be taken into consideration, because the mean-square amplitudes of the particles are significantly larger at the surface of crystals than in the bulk, and this effect can be observed even at relatively low temperatures. For example, a realistic calculation (details are given in Sect. 6.4.4) for a krypton crystal with a surface shows that due to the strong anharmonicities the disturbance of the periodical structure parallel to the surface it getting large with increasing temperature. This is demonstrated in Fig. 6.2 for the outermost layer of the krypton crystal. In the calculation for T = 102 K a periodical structure parallel to the surface can hardly be recognized. Although the temperature is still 14 K below the melting point, the outermost layer is extensively disordered, and it is in a *liquid-like* state (effect of surface pre-melting).

In summary, it can be said that only in the case of the crystalline solid in the harmonic approximation a "simple model" is available: this model is applicable to surfaces for sufficiently low temperatures only.

6.2.3 Molecular Dynamics

The general description of macroscopic properties of classical many-particle systems *without* the use of "simple models" and other simplifying assumptions can be done by means of MD calculations (Fig. 6.3): Hamilton's equations

$$\dot{p}_i = -\frac{\partial H}{\partial q_i},$$

$$\dot{q}_i = \frac{\partial H}{\partial p_i}; \quad i = 1, 2, \ldots, 3N \tag{6.9}$$

are solved by iteration with the help of a high-speed computer, and we obtain the following information

$$q(t_1), p(t_1)$$
$$q(t_2), p(t_2)$$
$$\vdots \tag{6.10}$$
$$q(t_i), p(t_i)$$
$$\vdots, \quad \text{where}$$

$$q(t_i) = (q_1, \ldots, q_{3N}, t_i),$$
$$p(t_i) = (p_1, \ldots, p_{3N}, t_i). \tag{6.11}$$

The *time step* in the iteration process is $\Delta t = t_{i+1} - t_i$. With k iteration steps the information about a system consisting of N particles is given in the time interval $\tau = k \cdot \Delta t$. Equation (6.10) contains the *total information* of the many-particle system. On the basis of this information macroscopic properties and other experimental data can be determined (at least in principle).

In the analysis of many-particle systems *correlation functions*

$$\langle a(t)b(t') \rangle \tag{6.12}$$

of two quantities a(t) and b(t') are of interest, where

$$a(t) = a(q(t), p(t)), \quad b(t') = b(q(t'), p(t')). \tag{6.13}$$

The angular brackets in (6.12) denote a thermodynamic average (Sect. 6.3). The time evolution of the system is given in statistical mechanics by the operator $\hat{S}_N(t)$:

$$a(q(t), p(t)) = \hat{S}_N(t - t')a(q(t'), p(t')), \tag{6.14}$$

where $\hat{S}_N(t)$ can be written in terms of the Liouville operator $\hat{L}_N$:

$$\hat{S}_N(t) = \exp(i\hat{L}_N t) \quad \text{with} \tag{6.15}$$

$$\hat{L}_N = i \sum_{i=1}^{3N} \left(\frac{\partial H}{\partial q_i} \frac{\partial}{\partial p_i} - \frac{\partial H}{\partial p_i} \frac{\partial}{\partial q_i} \right). \tag{6.16}$$

The information (6.10) can be used for the determination of correlation functions, see (6.12), and their time evolution.

6.3 Average Values

6.3.1 Density in Phase Space

Imagine a space of 6N dimensions whose points are determined by the 3N co-ordinates $q = (q_1, \ldots, q_{3N})$ and the 3N momenta $p = (p_1, \ldots, p_{3N})$. This space is the so-called *phase space*, and each point at time t corresponds to a *mechanical state* of the system. The *evolution with time* of the system is completely determined by Hamilton's equations (6.9), and is represented by a trajectory in phase space (Fig. 6.4a). The trajectory passes through the element dq dp at point (q, p) of the phase space, and the points in Fig. 6.4b indicate *how often* the elements of the phase space have passed through by the trajectory given in Fig. 6.4a. In other words, instead of the trajectory (Fig. 6.4a) we have now a "cloud" of phase points. The "cloud" is a great number of systems of the same nature, but differing in the configurations and momenta which they have at a given instant. In summary, instead of considering a single dynamic system (Fig. 6.4a), we consider a collection of systems (Fig. 6.4b), all corresponding to the same Hamiltonian. This collection of systems is the so-called *statistical ensemble* (see also, for example, [6.9–16]). The introduction of the statistical ensemble is very useful as regards the relationship between *dynamics* and *thermodynamics*.

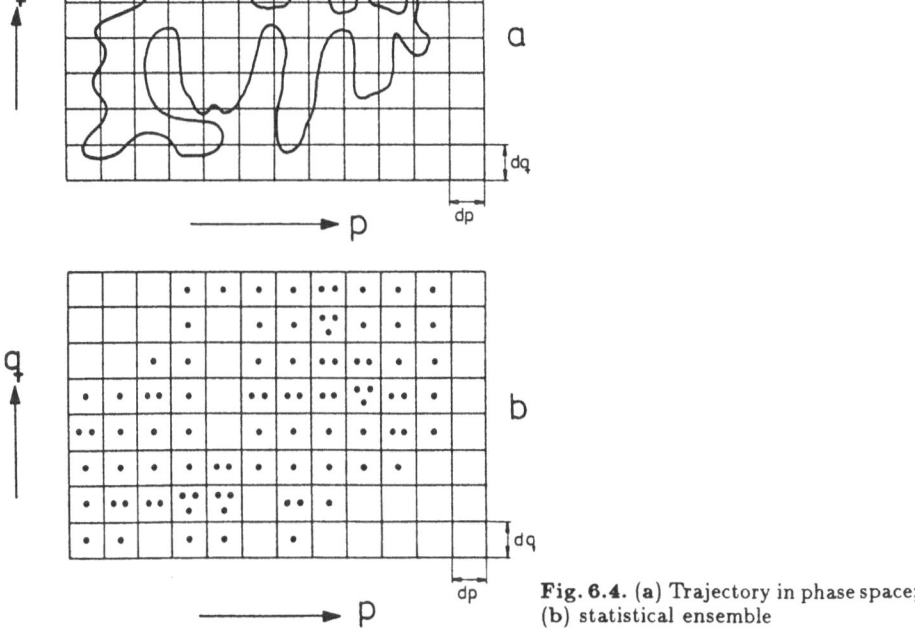

Fig. 6.4. (a) Trajectory in phase space; (b) statistical ensemble

The statistical ensemble can be described by a density

$$\varrho(q_1, \ldots, q_{3N}, p_1, \ldots, p_{3N}, t) \tag{6.17}$$

in phase space. The number of points in the statistical ensemble is arbitrary and, therefore, $\varrho(q, p, t)$ will be normalized as

$$\int \varrho(q, p, t) dq \, dp = 1. \tag{6.18}$$

The quantity

$$\varrho(q, p, t) dq \, dp \tag{6.19}$$

can be considered as the *probability* of finding at time t a system of the ensemble in the element dq dp at the point (q, p) of the phase space. The introduction of the density (6.17) is meaningful only, if the density takes an asymptotic value, for long times t

$$\lim_{t \to \infty} \varrho(q, p, t) = \varrho(q, p), \tag{6.20}$$

where $\varrho(q, p)$ is the density function of the statistical ensemble. When the trajectory $q(t)$, $p(t)$ is able to produce, over a long period of time, the function $\varrho(q, p)$ (ergodic hypothesis) then we have *equivalence* between the average over the time and the average over the statistical ensemble of a function $f(q(t), p(t))$

$$\langle f \rangle = \frac{\int f(q, p) \varrho(q, p) dq \, dp}{\int \varrho(q, p) dq \, dp}, \tag{6.21}$$

$$\langle f \rangle = \lim_{\tau \to \infty} \frac{1}{\tau} \int_0^\tau f(t) dt. \tag{6.22}$$

6.3.2 Statistical-Mechanical Ensembles

Different expressions exist for the density function $\varrho(q, p)$. These expressions depend on the *thermodynamic environment*. Without going into detail we shall list here the results for the three most important situations (see, for example, [6.15]):

1. *Microcanonical ensemble* (an isolated system; N, V and E given, where E is the energy)

    $$\varrho(q, p) = \begin{cases} \varrho_0 = \text{const} & \text{for} \quad E < H(q, p) < E + \Delta \\ 0 & \text{otherwise} \end{cases} \tag{6.23}$$

    where

    $$\Delta \ll E. \tag{6.24}$$

    With (6.21) we obtain

    $$\langle f \rangle = \frac{\displaystyle\int_{E < H < E + \Delta} dq \, dp f(q, p)}{\displaystyle\int_{E < H < E + \Delta} dq \, dp}. \tag{6.25}$$

2. *Canonical ensemble* (a closed, isothermal system; N, V and T given)

$$\varrho(q, p) = \frac{\exp\left(-\frac{H(q,p)}{k_B T}\right)}{\int \exp\left(-\frac{H(q,p)}{k_B T}\right) dq\, dp}. \tag{6.26}$$

3. *Grand canonical ensemble* (V, T and μ given, where μ is the chemical potential)

$$\varrho(q, p) = \frac{\exp\left\{-\frac{1}{k_B T}[H(q,p) - \mu N]\right\}}{\int \exp\left\{-\frac{1}{k_B T}[H(q,p) - \mu N]\right\} dq\, dp}. \tag{6.27}$$

Within the *thermodynamic limit*

$$N \to \infty,$$
$$V \to \infty,$$
$$\frac{N}{V} = \text{const} \tag{6.28}$$

the microcanonical, the canonical, and the grand canonical ensemble are equivalent, i.e. within the thermodynamic limit we obtain in all three cases the same value for the statistical average $\langle f \rangle$ of $f(q, p)$.

In connection with MD calculations two points are important:

i) Since in MD calculations we are restricted to *finite* systems (details are given in Sect. 6.4) the following question arises: How many particles N must a system with the density N/V have in order to fulfill condition (6.28) in a good approximation?

ii) What does "a long period of time $[\tau \to \infty$ in (6.22)] mean in connection with MD calculations?

Dicussion of (i). We have to investigate by means of a suitable quantity $g(q, p)$ whether the statistical average $\langle g \rangle$ is dependent on the particle number N. In other words, we have to perform MD calculations with various particle numbers and if there is

$$\langle g \rangle_{N'} = \langle g \rangle_{N''} = \langle g \rangle_{N'''} \quad \text{with} \tag{6.29}$$

$$N' < N'' < N''' \tag{6.30}$$

we may perform the calculations with a particle number of $N \geq N'$, and the results can be considered as independent of N.

In an MD calculation N, V and E are given and, therefore, the MD model represents a *microcanonical ensemble*. In the case of $N \geq N'$ we may compare the results of an MD calculation with those obtained in an analytical approach based on a canonical or grand canonical ensemble.

Various MD calculations showed that for solving most of the problems it is sufficient to work with models of *some hundred particles*.

Discussion of (ii). Let us consider a single particle of the system and let be v(t) its velocity at time t. Using (6.21 and 26), we obtain for the mean-square velocity

$$\langle v^2 \rangle = 4\pi \left(\frac{m}{2\pi k_B T} \right)^{3/2} \int_0^\infty v^4 \exp \left(-\frac{mv^2}{2k_B T} \right) dv$$

$$= 3\frac{k_B T}{m}. \tag{6.31}$$

If we define a function $v^2(\tau)$ by

$$v^2(\tau) = \frac{1}{\tau} \int_0^\tau v^2(t) dt \tag{6.32}$$

we expect [in according to (6.21,22)] that

$$\lim_{\tau \to \infty} v^2(\tau) = \langle v^2 \rangle. \tag{6.33}$$

In other words, we expect that after a sufficiently large time τ the velocity v(t) of a single particle of the system has passed through all the states given by the Maxwell distribution.

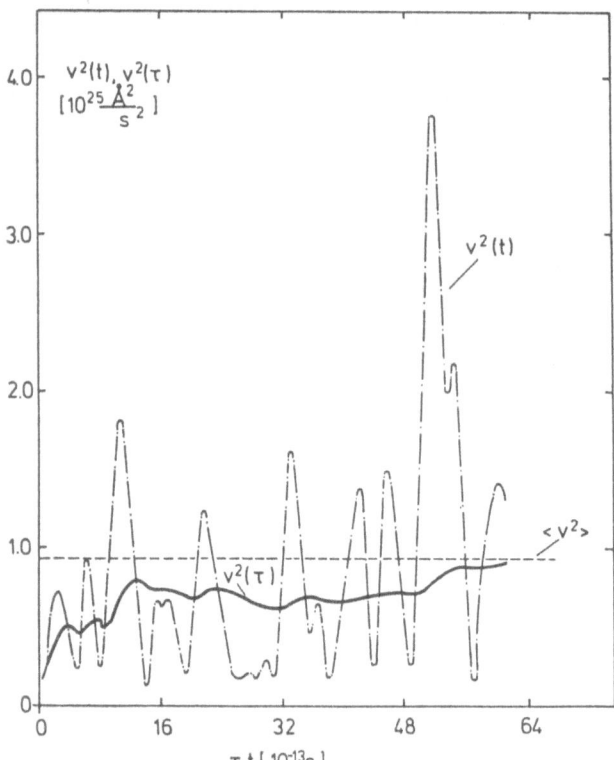

Fig. 6.5. $v^2(t)$ is the velocity squared at time t of a single particle (arbitrarily chosen) of a rubidium model consisting of 686 particles. $\langle v^2 \rangle$ and $v^2(\tau)$ are defined by (6.31,32)

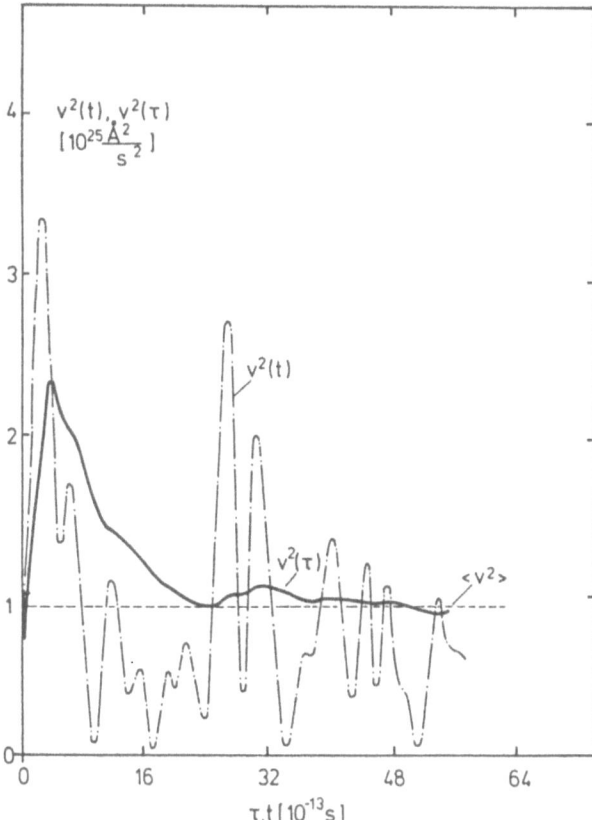

Fig. 6.6. $v^2(t)$ is the velocity squared of another particle (different from that in Fig. 6.5) of the rubidium system. $\langle v^2 \rangle$ and $v^2(\tau)$ are defined by (6.31,32)

We did such calculations on the basis of a realistic rubidium system consisting of 686 particles [6.17]. $v^2(t)$, $v^2(\tau)$ and $\langle v^2 \rangle$ are shown in Fig. 6.5 for an arbitrarily chosen particle. In Fig. 6.6 the same quantities are plotted for another particle. It can be seen from Figs. 6.5 and 6.6 that the condition $\tau \rightarrow \infty$, see (6.33), is already fulfilled in a good approximation at $\tau = 6 \times 10^{-12}$s.

6.3.3 Measureable Quantities

The microscopic information (6.10) is not directly accessible to measurements by only *averaged* quantities. Thus, also the theoretical expressions should be *averages* and are normally formulated within one of the statistical ensembles (microcanonical ensemble, etc.) listed above. In most cases these expressions are very complicated and have to be simplified on the basis of more or less *uncontrolled* approximations. In the case of MD calculations, however, we are able to express measureable quantities directly and without approximations using information (6.10). As an example, let us briefly discuss the pair correlation function.

Let $r = |\mathbf{r}_1 - \mathbf{r}_2|$ be the relative distance between a particle with the position vector $\mathbf{r}_1 = (q_1, q_2, q_3)$ and another particle with $\mathbf{r}_2 = (q_4, q_5, q_6)$. In a *canonical ensemble* in which the particles interact by a pair potential $v(r)$, the pair correlation function $g(r)$ for a monoatomic system is given by (see, for example, [6.18])

$$g(r) = \frac{V^2}{Z} \int d\mathbf{r}_3 \ldots d\mathbf{r}_N \exp\left[-\frac{1}{k_B T}\frac{1}{2}\sum_{\substack{i,j=1\\i\neq j}}^{N} v(r_{ij})\right], \tag{6.34}$$

where

$$Z = \int d\mathbf{r}_1 \ldots d\mathbf{r}_N \exp\left[-\frac{1}{k_B T}\frac{1}{2}\sum_{\substack{i,j=1\\i\neq j}}^{N} v(r_{ij})\right] \tag{6.35}$$

with

$$(\mathbf{r}_3, \ldots, \mathbf{r}_N) = (q_7, \ldots, q_{3N}). \tag{6.36}$$

The determination of $g(r)$ from the pair potential $v(r)$ on the basis of (6.34) is only possible with the help of *simplifying assumptions*, which are more or less uncontrolled approximations. According to the assumption we obtain various statistical theories which have often been used in the determination of $g(r)$ from $v(r)$ and vice versa [6.19–25]. Because the approximations are more or less uncontrolled all these statistical theories do not work reliably.

The determination of $g(r)$ by means of MD can be done without any uncertainty: on the basis of $v(r)$ we can calculate $g(r)$ in controlled steps. $g(r)$ is

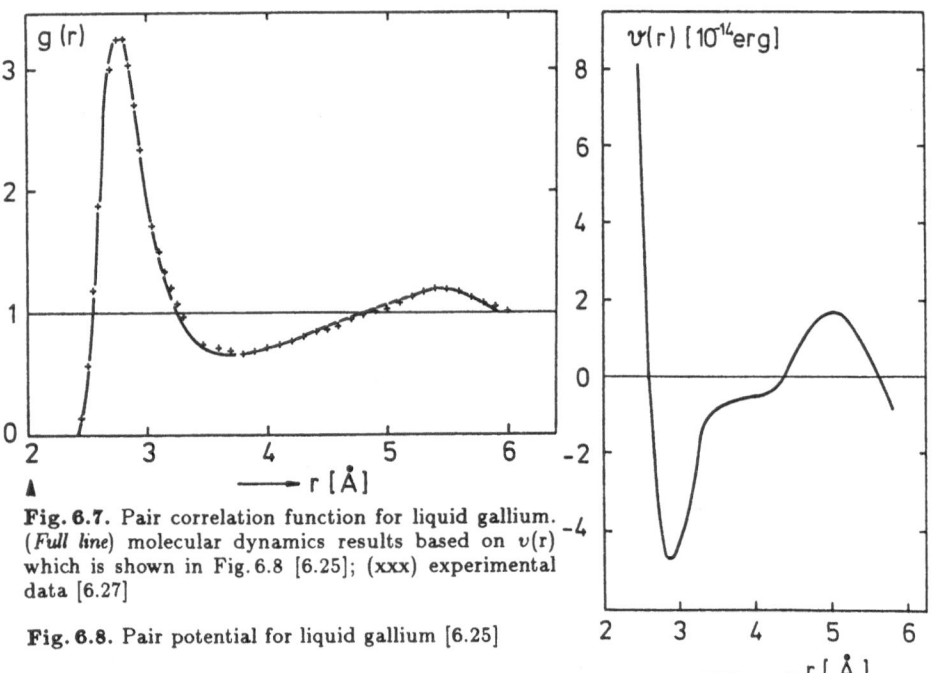

Fig. 6.7. Pair correlation function for liquid gallium. (*Full line*) molecular dynamics results based on $v(r)$ which is shown in Fig. 6.8 [6.25]; (xxx) experimental data [6.27]

Fig. 6.8. Pair potential for liquid gallium [6.25]

a measure of the probability that two particles of the system have the distance r, and we can simply compute g(r) on the basis of the information (6.10) by [6.26]

$$g(r) = \frac{1}{4\pi r^2 \Delta r} \frac{n(r, \Delta r)}{\varrho}, \tag{6.37}$$

where ϱ is again the macroscopic density, and $n(r, \Delta r)$ is the density in the spherical shell around a particle having the radii r and $r + \Delta r$. An example (liquid gallium in the bulk) is given in Fig. 6.7; the pair potential is shown in Fig. 6.8 [6.25].

The pair correlation function is not only of importance for the description of the liquid in the bulk and liquid surfaces. Also for temperatures *below* the melting point the surface structure can be strongly disturbed and has to be described by the pair correlation function; in connection with Fig. 6.2 we have pointed out that the *outermost* layer of a realistic krypton system is in a liquid-like state 14 K *below* the melting temperature. In principle, the Fourier transform of the pair correlation function can be measured at the surface in atomic beam experiments (Chap. 2).

6.4 Molecular Dynamics Systems

6.4.1 Models for the Bulk

MD calculations are normally based on the assumption that a *two-body* interaction will give a reasonable description of the properties of the system. This assumption is not a necessary restriction [6.1–4] but is generally employed because it implies a great simplification. Another restriction is that we can only use a relatively small number of particles, and this is due to the limitation imposed by the computers. In connection with MD calculations it is important to construct models which simulate *infinite* systems with a finite (and sufficiently small!) number of particles. Let us discuss this point in more detail.

Due to the finite number of particles the bulk properties are superimposed by "surface effects" which must be eliminated if we want to study pure *bulk properties*. For particles with $r_L < r_c$ (r_L being the distance of the particle from the boundary of the array, and r_c the range of interaction) the interaction with their surroundings is *asymmetric*, and, on the other hand, it is symmetric for particles with $r_L > r_c$ (Fig. 6.9). Clearly, such "surface effects" are getting small with increasing particle number N. However, in MD calculations N is a relatively small number and the effect of the cell boundaries can be eliminated by imposing the so-called *periodic boundary conditions* (PBC): Let us consider N particles which are arranged, for example, in a cubic box of length L. By adding or subtracting L from each of the 3N coordinates we obtain the particle coordinates of 26 "images" of the basic cubic array; a two-dimensional system with PBC (having 8 "images") is shown in Fig. 6.10. Due to the PBC the system has the following features:

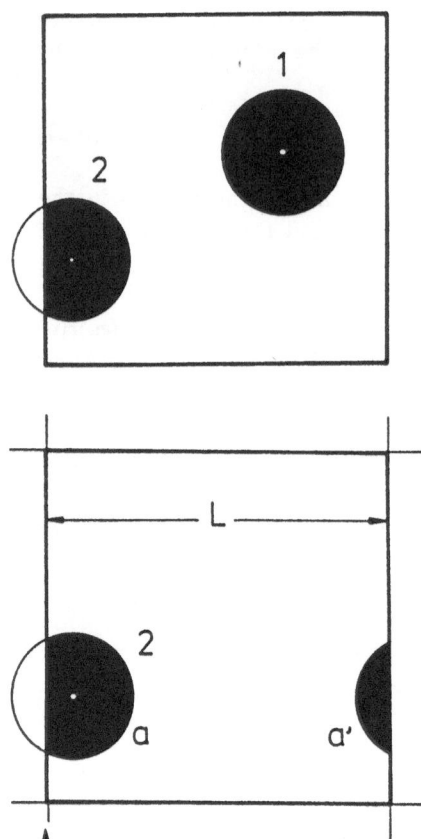

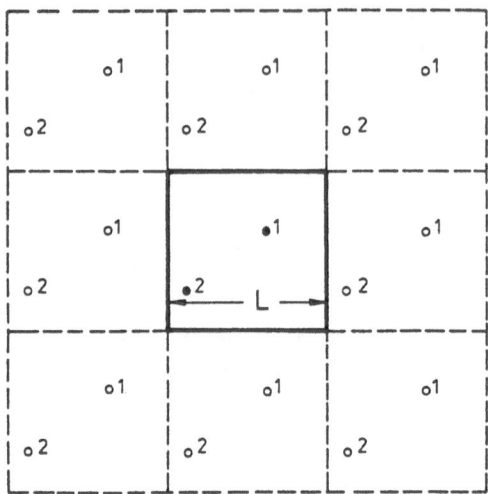

◀ Fig. 6.9. Two of the N particles (called 1 and 2) of the system. The spheres around each particle are the interaction ranges having a radius of r_c. The interaction is effective within the black areas of the spheres. Near the cell boundaries the interaction is asymmetric (particle 2), and it is symmetric in the inner of the cell (particle 1)

Fig. 6.10. A two-dimensional box of length L with periodic boundary conditions. 1 and 2 are two of the N particles

Fig. 6.11. Due to the periodic boundary conditions the interaction of particle 2 with its surrounding is now given by the black regions a and a′ (without PBC only by a) leading to a symmetric situation as in the case of an inner particle (for example, particle 1 in Fig. 6.9). Boundary effects are avoided in this way

i)   A particle which leaves the box from one face enters the box from the opposite face and, therefore, the density and energy of the system are conserved.

ii) Boundary effects are avoided. This is discussed in Fig. 6.11 with a schematic two-dimensional box. From Fig. 6.11 follows that the dimensions of the array must be such that there is no overlap between the regions a and a′. This means that the length L of the box must be considerably greater than the cut-off radius r_c of the interaction potential. In particular, the following inequality must be fulfilled

$$L > 2r_c. \tag{6.38}$$

In [6.28] *stochastic boundary conditions* have been suggested. This means: A particle which leaves the box wall is lost from the system, and new particles are introduced across the walls

212

– at *random* intervals
– in *random* positions, and
– at *random* velocities.

Such statistic boundary conditions cause the particle number of the system to vary and the system to constitute a grand canonical ensemble. In the following paragraphs we shall not discuss MD system with stochastic boundary conditions but exclusively MD models with PBC because it is more convenient to work with such an arrangement for solving most of the problems.

In the study of crystals the PBC should generate a lattice appropriate to the system under investigation. Thus, the shape of the model and the particle number N have to be chosen in such a manner that generation of the correct lattice is possible. For example, rubidium crystallizes in a bcc lattice, and for this material we should use a cubic array with a particle number of $N = 2n^3$, where n is an integer, i.e. $N = 2, 16, 54, 128, \ldots$.

6.4.2 Models with Surfaces and Interfaces

On the basis of the MD model for the bulk (Fig. 6.10) it is easy to establish a MD model for a system with free surfaces: it is simply a three-dimensional box with periodic boundary conditions (PBC) in two dimensions only (Fig. 6.12). In this way, we obtain a *slab* with two free surfaces. Slap-shaped crystals have often been used to calculate various dynamical surface properties [6.28–40]. In Sect. 6.5.4 we shall discuss in detail the structure and the dynamics of a slab-shaped krypton crystal.

The static and dynamic properties of *adsorbed layers* are often discussed on the basis of the following arrangement: one or several layers are placed on a *rigid* semi-infinite lattice which plays the role of the substrate. Examples will be discussed in Sect. 6.5.

The MD method should also be a powerful tool in studying *interfaces*. A possible model for the study of crystal-vapor (or crystal-liquid) interfaces is shown in Fig. 6.13. In principle, such an arrangement could be used for the study of layer growth.

6.4.3 Interaction Potentials

a) Long-range Effects

As discussed below, MD calculations are important for systems with *strong* anharmonicities. For example, a lot of MD calculations have been done for liquids and liquid-like systems. In the case of a liquid the particles perform a diffusive motion and no lattice sites are defined; we have already mentioned in Sect. 6.2.2 that the surface (in Fig. 6.2 the outermost layer) of crystals can be in a liquid-like state *below* the melting temperature.

The liquid in the bulk has often been described on the basis of *hard* spheres, i.e. the interaction between the particles is given by (see also Fig. 6.14a)

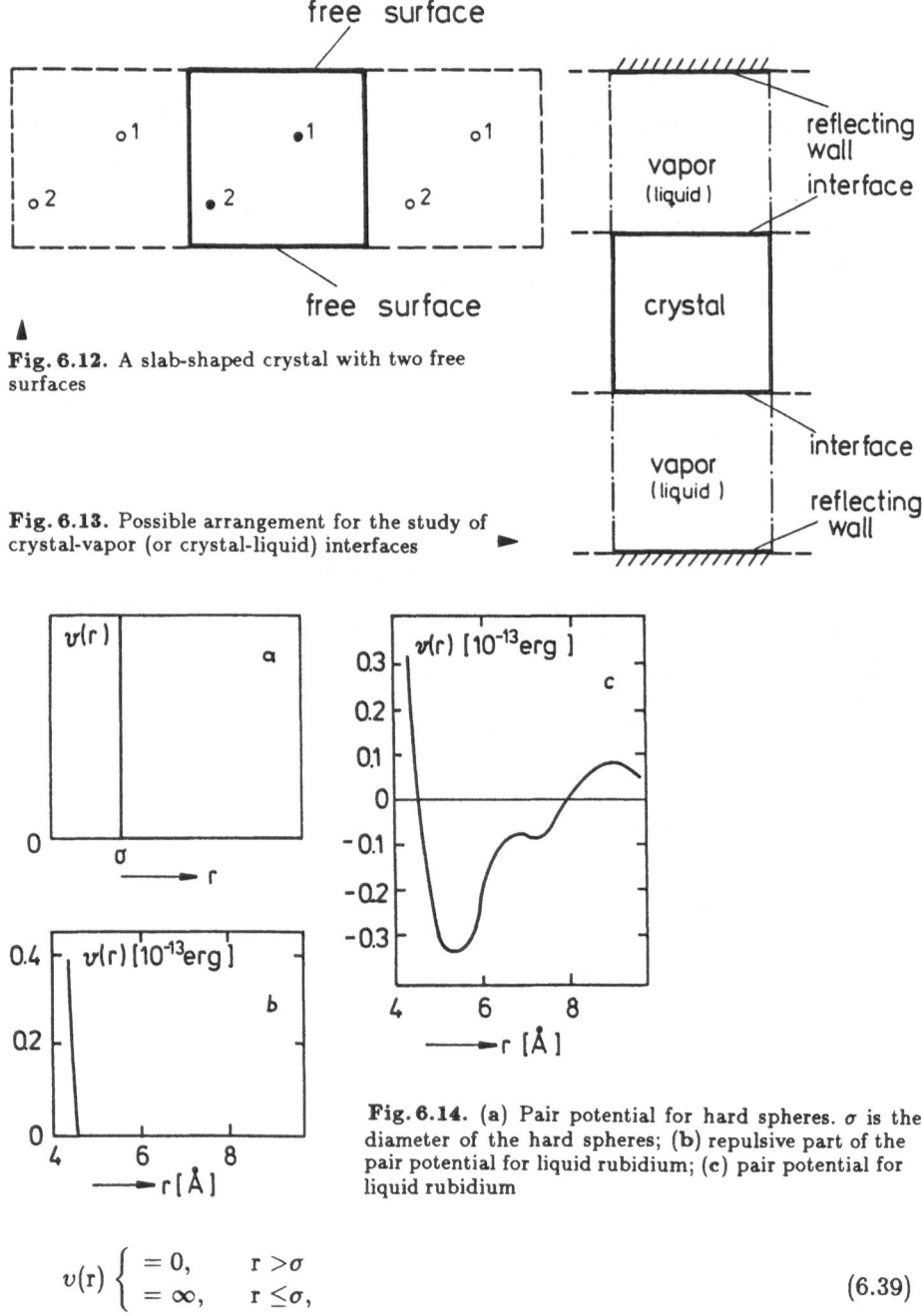

free surface

Fig. 6.12. A slab-shaped crystal with two free surfaces

Fig. 6.13. Possible arrangement for the study of crystal-vapor (or crystal-liquid) interfaces ►

Fig. 6.14. (a) Pair potential for hard spheres. σ is the diameter of the hard spheres; (b) repulsive part of the pair potential for liquid rubidium; (c) pair potential for liquid rubidium

$$v(r) \begin{cases} = 0, & r > \sigma \\ = \infty, & r \leq \sigma, \end{cases} \tag{6.39}$$

where σ is the hard-sphere diameter. The hard-sphere model is able to describe *qualitatively* a lot of liquid properties (e.g., the pair correlation function). However, for a *quantitative* description the long-range part of the pair potential must be considered. Let us briefly discuss this point by a *realistic* example.

In [6.41] MD calculations were performed for the following model: $N = 432$ rubidium atoms are arranged in a cubical box with the density $\varrho = 0.0107\,\text{Å}^{-3}$. The pair interaction (Fig. 6.14c) between the rubidium atoms has been calculated without parameters from the pair correlation function [6.42 and 43]. This model is able to describe very well experimental data [6.41,43]. Let us discuss here the influence of the *long-range* part of the potential given in Fig. 6.14c on the velocity autocorrelation function.

The velocity autocorrelation function $\psi(t)$ is defined by

$$\psi(t) = \frac{\langle v(0) \cdot v(t) \rangle}{\langle v(0)^2 \rangle}, \qquad (6.40)$$

where $v(t)$ is the velocity at time t for one atom of the ensemble. Within the canonical ensemble the brackets $\langle \ldots \rangle$ mean, see (6.26),

$$\langle v(0) \cdot v(t) \rangle = \frac{\int dq_1 \ldots dp_{3N}\, v(0) \cdot v(t) \, \exp\left(-H/k_B T\right)}{\int dq_1 \ldots dp_{3N} \, \exp\left(-H/k_B T\right)}. \qquad (6.41)$$

The Fourier transform of $\psi(t)$ yields a frequency spectrum $f(\omega)$ which in the case of the *harmonic* solid is the frequency spectrum of the normal modes, i.e., the phonons. The frequency spectrum is given by [6.44]

$$f(\omega) = \frac{2}{\pi} \int_0^\infty \psi(t) \, \cos \omega t \, dt, \qquad (6.42)$$

where $f(\omega)$ is normalized to unity

$$\int_0^\infty f(\omega) d\omega = 1. \qquad (6.43)$$

With the velocities $v_i(t)$, $i = 1, 2, \ldots, N$, calculated by MD, $\psi(t)$ can be simply determined by

$$\psi(t) = \frac{1}{N_\tau} \frac{1}{N} \sum_{j=1}^{N_\tau} \sum_{i=1}^{N} \frac{v_i(\tau_j) \cdot v_i(t + \tau_j)}{v_i(\tau_j)^2}, \qquad (6.44)$$

$\psi(t)$ being independent of the initial time τ_j (Liouville theorem). The number of initial times, N_τ, was chosen to be 25. According to *Zwanzig* and *Ailawadi* [6.45], the statistical error of $\psi(t)$ is then smaller than 1%.

In order to study the influence of the long-range part of the pair potential, $\psi(t)$ and $f(\omega)$ have been calculated *with* (cut-off radius r_c: 9.8 Å, Fig. 6.14c) and *without* (cut-off radius: 4.5 Å, Fig. 6.14b) the long-range part of the pair potential; the results are shown in Fig. 6.15. In the case without long-range part, $\psi(t)$ shows a less pronounced first minimum. In the related frequency spectrum (Fig. 6.15) the maximum moves from $\omega = 4 \times 10^{12} \text{s}^{-1}$ for $r_c = 9.8\,\text{Å}$ to $\omega = 2.3 \times 10^{12} \text{s}^{-1}$ for $r_c = 4.5\,\text{Å}$. The diffusion constant D expressed by [6.44]

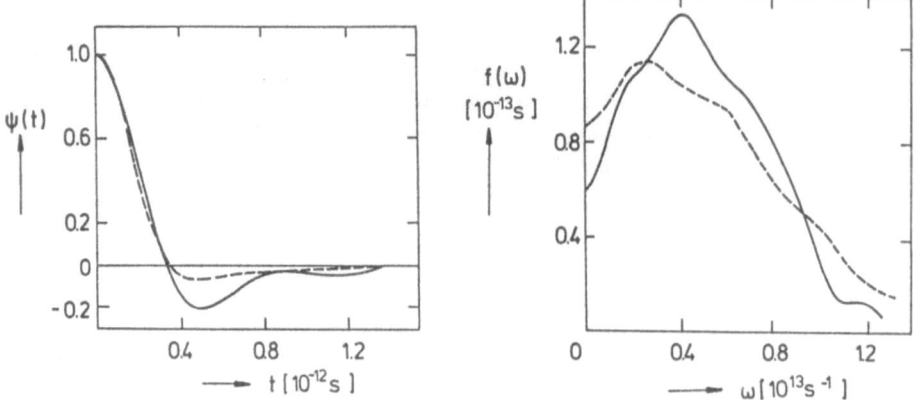

Fig. 6.15. (a) Velocity autocorrelation function for liquid rubidium, (*full line*) $r_c = 9.8$ Å, (*dashed line*) $r_c = 4.5$ Å; **(b)** frequency spectrum for liquid rubidium, (*full line*) $r_c = 9.8$ Å, (*dashed line*) $r_c = 4.5$ Å

$$D = \frac{\pi}{2} \frac{k_B T}{m} f(\omega = 0) \qquad (6.45)$$

increases from $D = 2.89 \times 10^{-5} \text{cm}^2/\text{s}$ ($r_c = 9.8$ Å) to $D = 4.19 \times 10^{-5} \text{cm}^2/\text{s}$ ($r_c = 4.5$ Å). The experimental value [6.46] for D is $2.82 \times 10^{-5} \text{cm}^2/\text{s}$.

In conclusion, it can be stated that already the repulsive part of the pair potential (Fig. 6.14b) is able to describe *qualitatively* a dense liquid in the bulk. However, in the *quantitative* description of such systems the long-range part of the potential cannot be neglected.

Interaction at Surfaces. The situation at the surface is characterized by two facts:

i) If the attractive (long-range) part of the potential is not considered, a system with free surface would fly apart. Thus, in the study of surface properties the long-range part of the potential is necessary.

ii) The surface properties are obviously much more sensitive to small variations in the pair potential than bulk properties; this point will be discussed in more detail in Sect. 6.5.4.

Thus, in the reliable description of surface properties the precise knowledge of the interaction (both its *repulsive* and its *attractive* parts) is required. However, obtaining an *accurate* pair potential for materials of interest (e.g., metals and materials with covalent bonding) is rather difficult even for the bulk, and the surface problem is much harder because of the change in electronic states and other properties near the surface. This difficulty is the reason why most of the MD studies on surfaces have been performed on the basis of *simple model systems,* i.e. systems for which the interaction is more or less well known (e.g., noble gases). Examples will be discussed in Sect. 6.5.

As pointed out in [6.47], the properties of *internal surfaces* seem to be quite well described by potentials used in the description of defect properties in the bulk. Several types of comparisons with experimental data are possible, such as internal surface energy as a function of misorientation and impurity content. The determination of potentials for the description of *external surface* properties became rather difficult by the fact that pertinent experimental data are obtained for several different surface configurations [6.47]: for atoms in smooth, low-index surfaces; for atoms at bumps or steps on surfaces; for adatoms on an otherwise smooth surface. The properties of a free surface are determined by the potential for the particles within a free surface, and only a few reliable potential functions exist.

b) Metal Surfaces

The *electron density* in metals must drop from a characteristic interior average value within the lattice to zero outside the surface (Fig. 6.16). The ion-ion potential in metals is dependent on the electron density, and the *charge redistribution* at the surface has the effect that the potential at the surface is different from that in the bulk of metals. How can we determine the pair potential at metal surfaces?

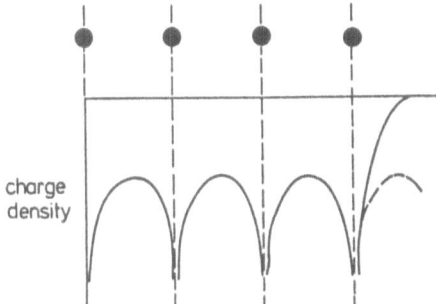

charge density

Fig. 6.16. Schematic charge distribution at a crystal surface

First-principle Calculations. For sufficiently low temperatures the pair interactions *outside* and *inside* the metal surface have been calculated microscopically in [6.48] on the basis of the *static electronic response function*: An ensemble of ions having the positions R_i, $i = 1, \dots, N$, are surrounded by a neutralizing electron charge of density $n(r)$. An infinitesimal displacement of ion k with the position R_k introduces a perturbating potential $\delta V_k(r - R_k)$ and an energy change given by

$$\Delta E = -\frac{1}{2} \sum_{R_i, R_j} \int dr \int dr' \delta V_i(r - R_i) \chi(r, r') \delta V_j(r - R_j), \qquad (6.46)$$

where $\chi(r, r')$ is the static electronic response function appropriate to the electron density $n(r)$. $n(r)$ can be determined by the *density-functional formalism*.

By this approach, instead of the complicated wave-function $\psi(\mathbf{r}_1,\ldots,\mathbf{r}_N)$ for the N particles the electron density

$$n(\mathbf{r}) = \frac{\int d\mathbf{r}_1 \ldots d\mathbf{r}_N \sum_{i=1}^{N} \delta(\mathbf{r} - \mathbf{r}_i)\psi^*(\mathbf{r}_1,\ldots,\mathbf{r}_N)\psi(\mathbf{r}_1,\ldots,\mathbf{r}_N)}{\psi^*(\mathbf{r}_1,\ldots,\mathbf{r}_N)\psi(\mathbf{r}_1,\ldots,\mathbf{r}_N)} \qquad (6.47)$$

is introduced as the variable. According to *Kohn* and *Sham* [6.49] $n(\mathbf{r})$ is given in the ground state by

$$n(\mathbf{r}) = \sum_{i=1}^{N} \phi_i^*(\mathbf{r})\phi_i(\mathbf{r}) \quad \text{with} \qquad (6.48)$$

$$\int n(\mathbf{r})d\mathbf{r} = N. \qquad (6.49)$$

The wave functions $\phi_i(\mathbf{r})$ satisfy

$$\left[-\frac{\hbar^2}{2m}\nabla^2 + V_{\mathrm{eff}}(\mathbf{r}) \right]\phi_i(\mathbf{r}) = \varepsilon_i\phi_i(\mathbf{r}), \quad \text{where} \qquad (6.50)$$

$$V_{\mathrm{eff}}(\mathbf{r}) \equiv V(\mathbf{r}) + V_{\mathrm{H}}(n(\mathbf{r})) + V_{\mathrm{xc}}(n(\mathbf{r})), \qquad (6.51)$$

$V(\mathbf{r})$ being an arbitrary external potential. $V_{\mathrm{H}}(n(\mathbf{r}))$ and $V_{\mathrm{xc}}(n(\mathbf{r}))$ are the usual Hartree and exchange-correlation potentials which are functionals of $n(\mathbf{r})$. The static susceptibility $\chi(\mathbf{r},\mathbf{r}')$ in (6.46) is a ground-state response and, therefore, it is a functional of $n(\mathbf{r})$ and can be described by $\phi_i(\mathbf{r})$. On the basis of (6.46) the ion-ion interaction *outside* and *inside* a *metal jellium surface* has been calculated [6.48]. In order to check the reliability of such potentials it would be most helpful to use them in MD calculations and compare the MD results with experimental data; in this way, a valuable insight into the validity of such potentials can be achieved.

Phenomenological Potentials. The potential calculations from first principles [6.50–54] can be performed without introducing any surface-dependent parameter. The parameters of *phenomenological* potential functions for the description of surface properties have generally to be fitted to certain surface properties, and it is hoped that the potential, which is successful in one surface situation, will be applicable to another surface problem as well. In the case of *noble gases* the parameters of the potentials can be fitted by bulk properties, and the resulting potential functions can be applied without doubt to the determination of surface-properties; an example will be discussed in Sect. 6.5.4.

As an example of a phenomenological potential let us discuss here the potential of *Morse* [6.55] which has been extensively used in the study of lattice dynamics [6.56], the defect structure in metals [6.57–65], the inert gases in metals [6.66,67], the equation of state [6.68,69], elastic properties of metals [6.69,70], and the interaction between gas atoms and crystal surfaces [6.71], etc.

In order to calculate the energy levels for *diatomic molecules, Morse* required the following conditions for the interatomic potential of the atoms of the molecule [6.72]:

1. $\lim_{r \to \infty} v(r) = 0$,
2. $v(r)$ has a minimum if $r = r_0$, where r_0 is the intermolecular separation,
3. $\lim_{r \to 0} v(r) = \infty$,
4. $v(r)$ should have the same energy levels as those given by equation

$$W(n) = -D + \hbar\omega_0 \left[\left(n + \frac{1}{2} \right) - x \left(n + \frac{1}{2} \right)^2 \right], \qquad (6.52)$$

which describes the spectroscopic data of molecules. Morse chose the following potential function

$$v(r) = \alpha_0 \exp\left[-2\alpha(r - r_0)\right] - 2\alpha_0 \exp\left[-\alpha(r - r_0)\right]. \qquad (6.53)$$

A solution of the radial part of Schrödinger's equation using this potential yielded the energy level representation of the type given by (6.52). However, as we have seen above [6.56–71], the Morse potential is not restricted to molecular energy level applications but has also been used extensively in the study of various metal properties.

On account of the redistribution of charge at metal surfaces it is doubtful whether the Morse potential, as well as other phenomenological potentials can describe properly *surface* properties if the parameters α_0, α and r_0 are fitted to the *bulk* properties of the metal. In the case of metal surfaces the parameters of potentials should be fitted by "certain" surface properties (e.g., the structure), and one should then use them in the determination of other surface properties (e.g., the dynamics).

6.4.4 Time Evolution of Molecular Dynamics Systems

In Sects. 6.4.1,2 we have discussed models which have often been used in MD calculations; in most cases the particle number is of the order of 10^2–10^3 simulating a very large system with $N \cong 10^{23}$ particles. In Sect. 6.4.3 we have pointed out that a *realistic* description of reality provides a careful construction of the interaction between the particles. Let us assume that the shape of the system, its particle number N and the interaction potential, are fixed; let us now investigate the *evolution* with time of MD systems.

a) Initial Values for the Coordinates and the Momenta

From the solution of Hamilton's equations (6.9) we obtain the coordinates and the momenta (velocities) of all N particles as a function of time, see (6.10); this is the total *microscopic* information about the system under investigation. However, the solution of the Hamilton equations provides that *initial values* for the *coordinates* and the *velocities* of the N particles are available.

Initial Values for the Coordinates. In the case of liquids and gases the particles can be distributed randomly with the appropriate density. In the case of crystals (with and without surfaces) the particles will be situated within the array so that the perfect lattice structure appropriate to the system under investigation is generated.

Initial Values of the Velocities. When there are no external forces acting on the system the directions of the velocities

$$\frac{v_i}{|v_i|}; \quad i = 1, \ldots, N \tag{6.54}$$

at the initial time t_0 should be distributed randomly so that

$$\sum_{i=1}^{N} \frac{v_i}{|v_i|} = 0, \tag{6.55}$$

(6.55) must be fulfilled for all times t (conservation of momentum).

In thermal equilibrium the *magnitudes* of the particle velocities are distributed according to the Maxwell distribution. It is, however, more convenient to choose for all particles the same magnitude of velocities. If the initial velocities are expressed by the mean-square velocities we have at time t_0

$$|v_i| = (\langle v^2 \rangle)^{1/2} = \left(\frac{3k_BT}{m}\right)^{1/2}; \quad i = 1, \ldots, N. \tag{6.56}$$

In this case the distribution of the velocities is not a Maxwellian but is given by

$$\frac{dn}{n} = g(v)dv, \quad \text{where} \tag{6.57}$$

$$g(v) = \left(\frac{m}{3k_BT}\right)^{3/2} v^2 \delta(1 - v^2 m/3k_BT) \quad \text{with} \tag{6.58}$$

$$\int_0^\infty \frac{dn}{n} = 1 \quad \text{and} \tag{6.59}$$

$$\int_0^\infty v^2 \frac{dn}{n} = \langle v^2 \rangle = \frac{3k_BT}{m}. \tag{6.60}$$

The fraction dn/n of the particles whose velocities are between v and $v + dv$ is initially given by a delta-function (Fig. 6.17), and this means that the system is initially *not in thermal equilibrium*. With the help of the function

$$\alpha(t) = \frac{\frac{1}{N} \sum_{i=1}^{N} [v_i(t)^2]^2}{\left[\frac{1}{N} \sum_{i=1}^{N} v_i(t)^2\right]^2}, \tag{6.61}$$

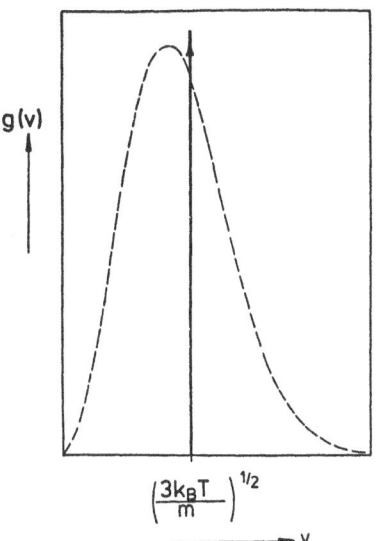

$g(v)$

$\left(\dfrac{3k_BT}{m}\right)^{1/2}$

v

Fig. 6.17. The solid line shows the initial distribution for the velocities [delta-function, see (6.58)]; the system is initially not in equilibrium. The distribution for the velocities of particles which are in thermal equilibrium is given by the Maxwell distribution (*dashed line*)

where $v_i(t)$, $i = 1, \ldots, N$, are again the velocities obtained from the MD calculations, we can study at which point in time the Maxwell distribution is reached. In the case of the Maxwell distribution, $g(v)$ in (6.57) is given by

$$g(v) = \frac{4}{\sqrt{\pi}} v^2 \left(\frac{m}{2k_BT} \right)^{3/2} \exp \left(-\frac{mv^2}{2k_BT} \right) \tag{6.62}$$

and the function $\alpha(t)$ in (6.61) takes the value of

$$\alpha(t) = \frac{5}{3} \tag{6.63}$$

for all times t. With our initial distribution function, see (6.58), we get at the time t_0

$$\alpha(t_0) = 1. \tag{6.64}$$

In Fig. 6.18 the function $\alpha(t)$ is plotted within the time region

$$t_0 = 0 \leq t \leq 5 \times 10^{-12} s$$

for an MD model consisting of 686 rubidium atoms [6.17]. It can be seen from Fig. 6.18 that *thermal equilibrium* $(\alpha(t) = 5/3)$ is reached after $\sim 10^{-12}$ s. Due to the finite number of particles $\alpha(t)$ fluctuates around its equilibrium value. Clearly, this fluctuations are getting small with increasing N. However, a system with *finite* N should have the following property

$$\lim_{\theta \to \infty} \frac{1}{\theta - \theta_1} \int_{\theta_1}^{\theta} \alpha(t) dt = \frac{5}{3}. \tag{6.65}$$

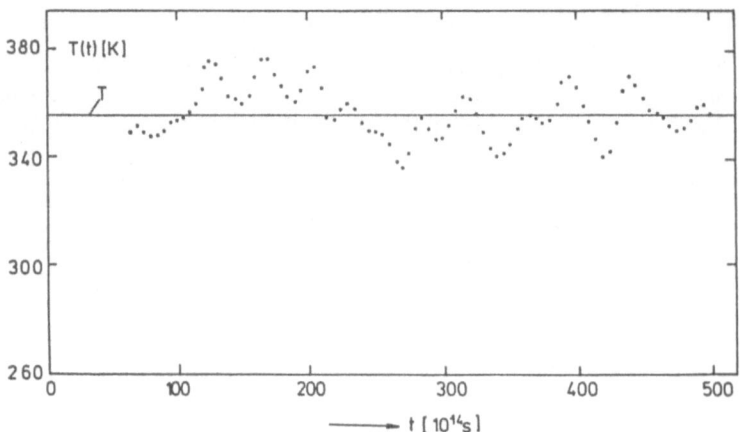

Fig. 6.18. The system (consisting of 686 particles) is initially ($t_0 = 0$) not in thermal equilibrium ($\alpha \neq 5/3$). After $\sim 10^{-12}$s Maxwell's distribution is reached. Due to the finite number of particles the system fluctuates around $\alpha = 5/3$

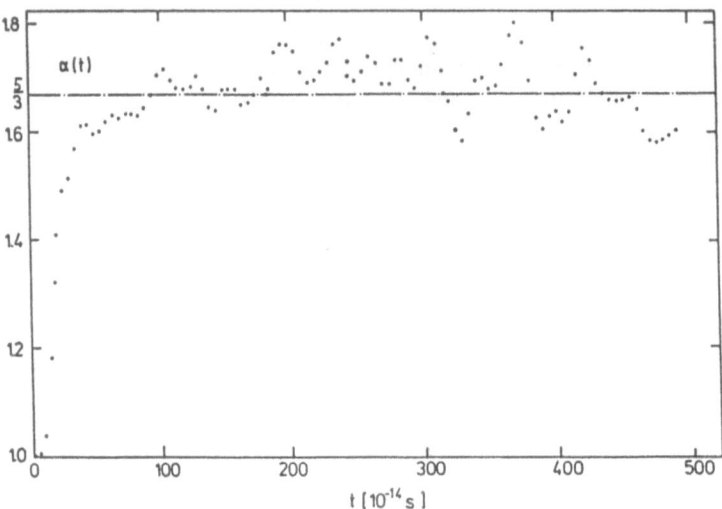

Fig. 6.19. The temperature as a function of time for a molecular dynamics model consisting of 686 particles. The mean temperature is $T = 355$ K

In our example θ_1 takes a value of $\sim 10^{-12}$s. Setting $\theta = \infty$ we are implicitely assuming that

$$\theta \ll \text{Poincare' period.}$$

The recurrence of initial states after an enormously long period ($\sim 10^{150}$ years for a macroscopic system) is called the Poincare' period.

Temperature of Molecular Dynamics Systems. Let us define the temperature of the system by

$$T(t) = \frac{m}{3Nk_B} \sum_{i=1}^{N} v_i(t)^2. \tag{6.66}$$

In most cases the computed temperature is different from the desired temperature; so, the velocities of all N particles have to be multiplied by a constant factor chosen to bring the system closer to the desired temperature. As in the case of $\alpha(t)$ also the temperature $T(t)$ in finite systems fluctuates around the mean temperature

$$T = \lim_{\theta \to \infty} \frac{1}{\theta - \theta_1} \int_{\theta_1}^{\theta} T(t)\, dt. \tag{6.67}$$

In Fig. 6.19 the function $T(t)$ is plotted within the time region of

$$\theta_1 \cong 10^{-12}s < t < 5 \times 10^{-12}s$$

for the rubidium model with 686 particles [6.17]; the mean temperature expressed by (6.67) was 355 K.

From the *fluctuations* of the temperature $T(t)$ we can extract the *specific heat* at constant volume (per particle)

$$c_v = \frac{1}{N} \left(\frac{\partial E}{\partial T} \right)_v. \tag{6.68}$$

For a microcanonical ensemble we obtain [6.73]

$$c_v = \frac{3}{2} k_B \left[1 - \frac{3N}{2T^2} (\overline{T^2} - T^2) \right]^{-1}, \quad \text{where} \tag{6.69}$$

$$\overline{T^2} = \lim_{\theta \to \infty} \frac{1}{\theta - \theta_1} \int_{\theta_1}^{\theta} T^2(t)\, dt. \tag{6.70}$$

In other words, the *temperature fluctuations* (see, e.g., Fig. 6.19) is a direct measure of the specific heat. Since the energy is a constant for a microcanonical ensemble, the *potential energy fluctuations* δU must be related to the temperature fluctuations δT by

$$\delta U = -\frac{3}{2} k_B \delta T. \tag{6.71}$$

Lebowitz et al. [6.73] used (6.69) to compute c_v from MD results for T and $\overline{T^2}$, and found good agreement with experimental data. Also the corresponding results for the liquid rubidium model discussed here agrees well with the experimental value for the specific heat. It should be emphasized that the specific heat for liquids and liquid-like systems is not so easy to compute even in the case where the pair potential $v(r)$ is not dependent on temperature. In this case we obtain the well-known formula [6.44]

$$c_v = \frac{3}{2}k_B + \frac{3}{2} \int \left(\frac{\partial g(r)}{\partial T}\right)_T v(r)dr. \qquad (6.72)$$

The problem consists in determining precisely the variation of the pair correlation function $g(r)$ with T before (6.72) is useful. In the determination of $(\partial g(r)/\partial T)_v$ the triplet correlation function [6.18]

$$g_3(r_1, r_2, r_3) = V^2 \frac{\int dr_4...dr_N \exp\left[-\frac{1}{k_B T}\frac{1}{2}\sum_{\substack{i,j=1\\i\neq j}}^{N} v(r_{ij})\right]}{\int dr_1...dr_N \exp\left[-\frac{1}{k_B T}\frac{1}{2}\sum_{\substack{i,j=1\\i\neq j}}^{N} v(r_{ij})\right]} \qquad (6.73)$$

is involved, and there are no reliable models available for $g_3(r_1, r_2, r_3)$ [6.74–87].

6.5 Applications of the Molecular Dynamics Method

In the preceding sections we have studied in detail the principles of MD. In particular, the following topics have been discussed:

i)   The role of MD in statistical physics.
ii) Construction of MD models.
iii) Evolution with time of MD systems.

Let us now apply the MD method to problems in many-particle physics with particular emphasis on surface physics.

6.5.1 Phase Transitions in Two-Dimensional Systems

The crystallization of a *two-dimensional* system (*without substrate*) was studied using MD [6.88]. The system consisted of 400 particles which were interacting through a Lennard-Jones potential. The crystallization could be produced in an entirely free system: there were neither walls enclosing the 400 particles nor periodic boundary conditions. Initially the particles were placed at *random* positions (Fig. 6.20a). The atoms move to the positions of lower potential energy and, because the energy of the system is conserved, the kinetic energy must increase. This increase of kinetic energy was removed by a damping term in the equations of motion, performing the function of a heat bath held at constant temperature. After an initial crystallization ($\sim 2.45 \times 10^{-12}$s) the damping force was turned off, so that after that time the particles moved freely under the influence of their interaction, and it turned out that several defects disappeared. After $\sim 1.2 \times 10^{-11}$s a relatively well-ordered crystal was obtained (Fig. 6.20b).

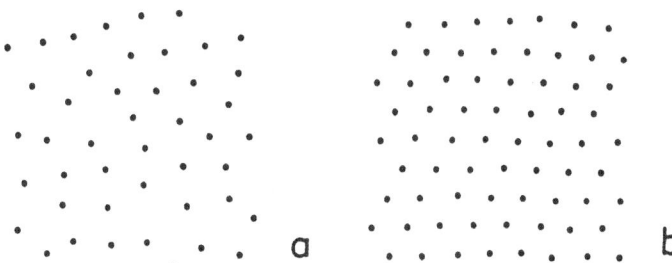

Fig. 6.20a,b. Crystallization of a two-dimensional system of particles interacting through a Lennard-Jones potential. (**a**) Initial configuration (small section of [Ref. 6.88, Fig. 1a]); (**b**) configuration after $\sim 1.2 \times 10^{-11}$s (small section of [Ref. 6.88, Fig. 1d])

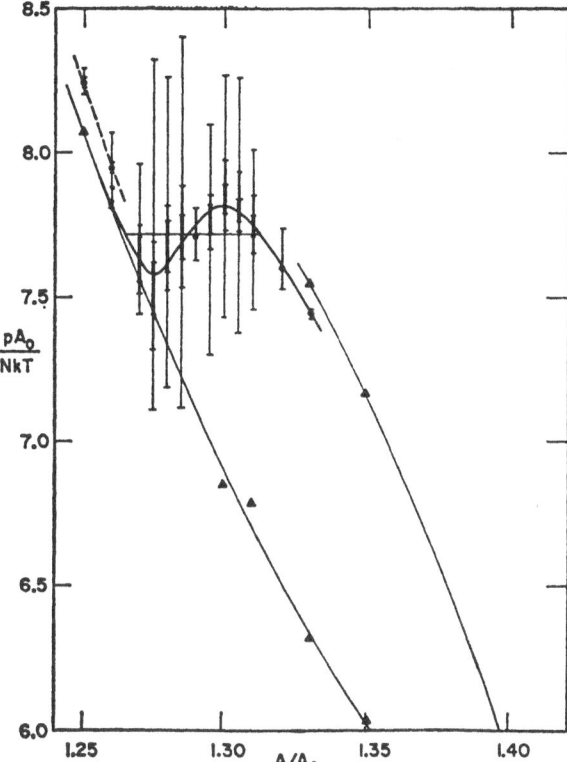

Fig. 6.21. The equation of state of hard disks. The light, middle and vertical lines were discussed in [6.89]

In conclusion, it can be stated that crystallization can be produced in a free system of particles interacting via a simple pair potential; no walls enclosing the system, no periodic boundary conditions, and no artificial crystallization nuclei are needed. Moreover, it is shown by MD in [6.89] that phase transitions can take place in the absence of the *attractive* part of the potential, too. The study in [6.89] dealt with a two-dimensional system consisting of 870 *hard-disk* particles (the form of the potential is shown in Fig. 6.14a). The disks

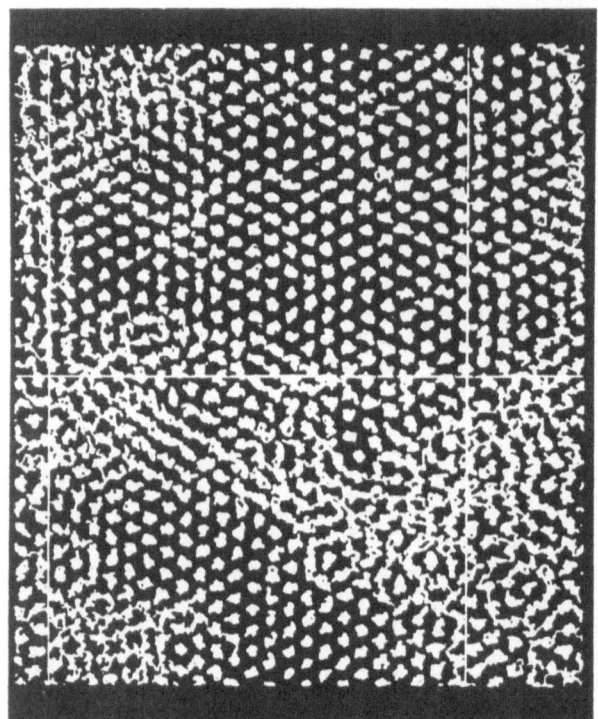

Fig. 6.22. The traces of the centers of particles in the phase transition region showing fluid and crystalline regions

were placed in a periodically repeated rectangular array. Periodic boundary conditions were imposed on the system (otherwise a system of hard-disks would fly apart). In the case of only 72 particles, the particles were either all in the fluid phase or all in the crystalline phase. This is shown in Fig. 6.21 by the two branches drawn through the triangular points. In Fig. 6.21 the reduced pressure $p^* = pA_0/k_BNT$ is shown as a function of the reduced area A/A_0, where A_0 is the area of the system closely packed. It is reported in [6.89] that "in the region of A/A_0 from 1.33 to 1.35 the system fluctuated infrequently between a high-pressure fluid branch and a low-pressure crystalline branch, while at A/A_0 of 1.31 and higher densities the solid phase is always stable".

For a 870-particle system the fluid phase and the crystalline phase exist side by side. This is demonstrated in Fig. 6.22 where the trajectories of the particles (plotted by an oscilloscope) are shown. It can be seen from Fig. 6.22 that there are regions of localized particles (crystallites) between regions of mobile particles (fluid). Another evidence of the coexistence of the two phases are the characteristically large *pressure fluctuations* in the phase transition region; in this region the fluid and the crystalline phases can exist with almost equal probability. It has been reported in [6.89] that "the fluctuations are much larger in the middle of the phase transition region than near the ends". In contrast to the 72-particle system, the system with 870 particles shows in the phase transition region *a van der Waals loop-like* behavior (Fig. 6.21).

6.5.2 Diatomic Molecules Adsorbed on Surfaces

A monolayer consisting on 400 rigid, diatomic molecules was studied by MD in [6.90]. The center of mass of molecules was fixed on a square lattice, and periodic boundary conditions were imposed on the system. The atoms of the different molecules interact through a Lennard-Jones (6,12) potential, see (6.4); the parameters of the potential are $\varepsilon = 1.313 \times 10^{-14}$ erg and $\sigma = 3.708$ Å. The MD calculations were performed with a fixed number density of 5.5397×10^{14} cm^{-2}. The other parameters are

Mass of the molecules: 4.6517×10^{-27} g,
Size of the molecules: 1.2875 Å,
Lattice constant: 4.24875 Å.

The ground-state configuration of this system is shown in Fig. 6.23; the nearest-neighbors are perpendicular while the next-nearest neighbor molecules are parallel to one another.

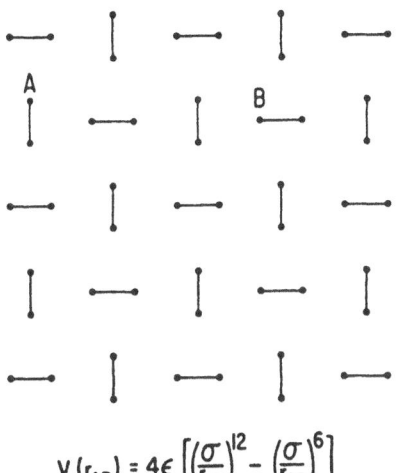

$$V(r_{AB}) = 4\varepsilon \left[\left(\frac{\sigma}{r_{AB}} \right)^{12} - \left(\frac{\sigma}{r_{AB}} \right)^{6} \right]$$

Fig. 6.23. Rigid, diatomic molecules on a square lattice (ground-state configuration)

On the basis of this model the *internal energy* and the *orientational order parameter* have been studied. As regards the internal energy *Kalia* et al. [6.90] stated the following: "In and away from transition region, the temperature dependence of internal energy is found to be independent of the thermal history of the system. The absence of hysteresis and latent heat of transition, and continuous variation of the internal energy with temperature are strong indications of a continuous order-disorder transition". The results for the internal energy as a function of temperature are shown in Fig. 6.24a.

Also the *order parameter*

$$\eta = \left\langle \frac{1}{N} \sum_{i=1}^{N} \cos 2\phi_i \times \exp\left(iQ \cdot R_i\right) \right\rangle \tag{6.74}$$

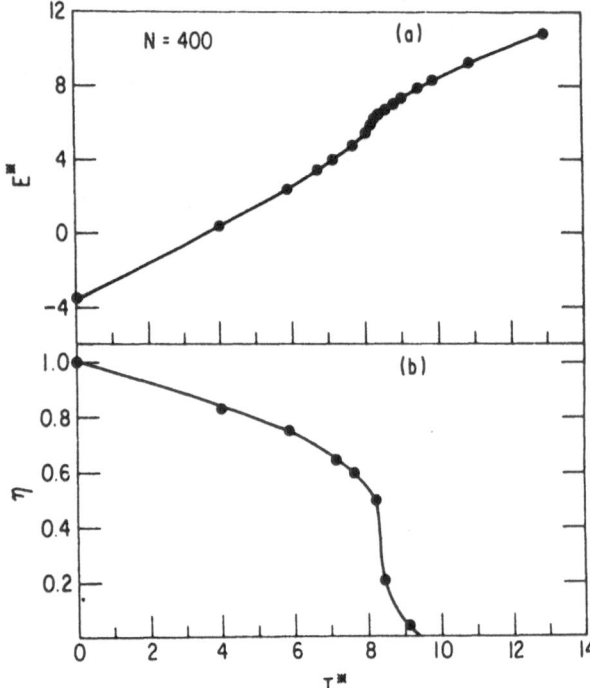

Fig. 6.24. (a) Internal energy per molecule, $E^* = E/\varepsilon$, and (b) order parameter η, as functions of reduces temperature $T^* = k_B T/\varepsilon$

was studied in [6.90] as a function of temperature. ϕ_i is the angle between the molecular axis and the x axis, and R_i is the position vector of the center of mass of the i-th molecule. The phase factor $\exp(iQ \cdot R_i)$ in the case of a square lattice takes the values of ± 1. As can be seen from Fig. 6.24b, η changes continuously in the transition region.

It should be mentioned that diatomic molecules whose atoms are interacting through a *quadrupole-quadrupole* interaction and whose center of mass are fixed on a *triangular* lattice show a *first-order transition* [6.91].

6.5.3 Melting Transition of Near Monolayer Xenon Films on Graphite

Xenon films on graphite were studied as a function of temperature using MD in [6.92]. For the interaction between the various atoms of the xenon-graphite system Lennard-Jones (6,12) pair potentials, see (6.4), has been chosen. The parameters are

xenon-xenon interaction: $\varepsilon/k_B = 225.3 \,\mathrm{K}$, $\sigma = 4.07 \,\text{Å}$;
xenon-carbon interaction: $\varepsilon/k_B = 79.5 \,\mathrm{K}$, $\sigma = 3.74 \,\text{Å}$.

The carbon-carbon interaction was not needed because in the calculations the carbon atoms were not vibrating; they were fixed at their lattice sites. The number of xenon atoms used in the MD calculations was 576. Periodic boundary conditions were imposed on the system with respect to translations parallel to the surface.

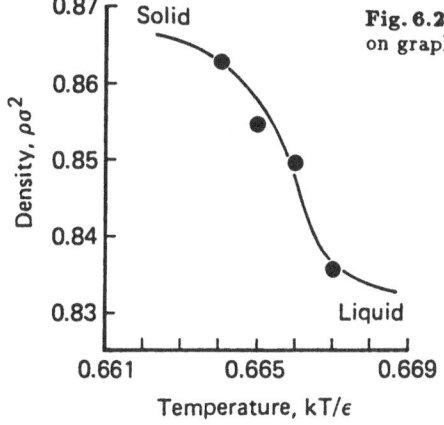

Solid

Liquid

Fig. 6.25. Reduced density $\varrho\sigma^3$ of the first layer (xenon on graphite) as a function of $T^* = k_B T/\varepsilon$

Fig. 6.26. The block-averaged density of the first layer (xenon on graphite), averaged over each 1000 time steps, as a function of time τ for various reduced temperatures $T^* = k_B T/\varepsilon$. The time unit for this figure is 5×10^{-11} s

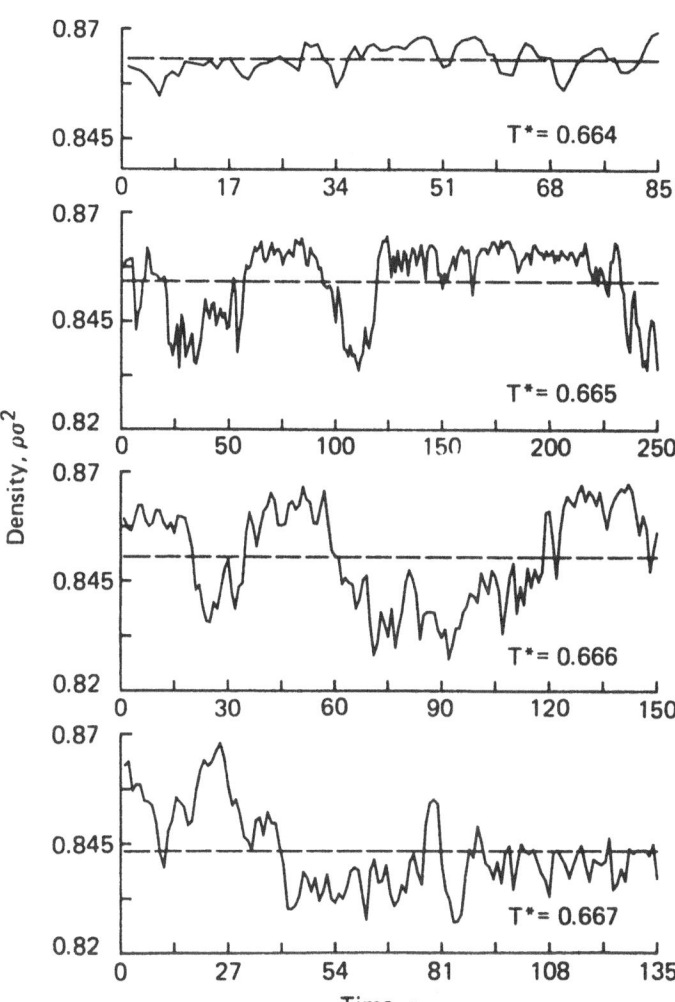

$T^* = 0.664$

$T^* = 0.665$

$T^* = 0.666$

$T^* = 0.667$

Time, τ

MD calculations have been done at reduced temperatures

$$T^* = \frac{k_B T}{\varepsilon} \qquad (6.75)$$

of 0.664, 0.665, 0.666, and 0.667. The mean density of the *first* layer as a function of temperature is shown in Fig. 6.25. It can be seen from Fig. 6.25 that the mean density decreases *continuously* with increasing temperature, and the first layer is in the *solid state* at the temperature 0.664 and in the *liquid state* at the temperature of 0.667; this temperature interval is approximately 7 K. The examination of the *block-averaged* density of the first layer, averaged over 1000 time steps $(5 \times 10^{-11}\text{s})$, provides a deeper insight. In [6.92] the following is reported (see also Fig. 6.26): "We note that at $T^* = 0.664$ the density fluctuates about the mean solid density of 0.863; similarly, at $T^* = 0.667$, the density decays from an initial solid density and fluctuates about the mean liquid density of 0.836. However, for the two intermediate temperatures of 0.665 and 0.666 the first-layer density oscillates between the densities of the solid state and liquid state, the residence time at higher density being significantly longer at the lower temperature".

The results (Fig. 6.26) can be explained as follows [6.92]: For sufficiently high temperatures and xenon densities a second-layer population is likely, and the second layer acts as an atom reservoir for the first layer. In particular, "the liquid monolayer will freeze by adsorbing atoms from the second layer so that its density is representative of the solid state. However, this solid state is not constant in time; over a period of nanoseconds, several atoms of the monolayer xenon solid are promoted by fluctuations to the second layer, thereby lowering the first layers density and driving it back to the liquid phase".

6.5.4 Microscopic Behavior of Krypton Atoms at the Surface

In this subsection we want to study the structure and the dynamics of *free* crystal surfaces. The knowledge of the structural and dynamical behavior of such surfaces is also important for understanding phenomena at the surface. For example, the behavior of adsorbates and chemical reactions at surfaces are influenced by these properties.

We have already pointed out above that a first-principle calculation of surface properties provides an accurate knowledge of the atomic interactions. The only group of materials for which this requirement is nearly satisfied is that of noble-gas solids. In contrast to metals, the potential functions of noble-gas solids do not depend on the density, and they are the same at the surface and in the bulk of the crystal. At free metal surfaces the local background electron density may be changed from its average bulk value and it produces concomitant changes in the potential functions. Such difficulties do not arise at noble-gas surfaces and, therefore, we want to restrict our study to noble-gas (krypton) surfaces.

The mean-square amplitudes of the particles are significantly larger at the surface of the crystal than in the bulk. The harmonic approximation (Chap. 5) should be valid [6.39] for surface vibrations in noble-gas crystals below one sixth of the melting temperature T_m (as compared to about $1/3\,T_m$ for the bulk). The melting temperature of krypton is 116 K. Thus, in the case of krypton usual *lattice dynamics* calculations should only be valid for T<19 K. For T>19 K we have to consider *anharmonic* effects. In MD calculations anharmonicities are treated without approximations and we are not restricted to temperatures lower than 19 K. In this subsection we want to study the structure and dynamics of a krypton surface as a function of temperature [6.93].

As the surface particles are, in general, less bonded than the bulk particles we expect that the structure and the dynamics are more sensitive to variations in temperature than in the bulk of the crystal. Thus, in the study of temperature effects (e.g., phase transitions) of *adsorbed* layers also the crystal surface on which the layer is situated should be treated as temperature dependent and not as a *static* lattice corresponding to a zero temperature (see, for example Sects. 6.4.2,3).

For *krypton* a reliable pair potential is available [6.94]. The krypton potential of *Barker* et al. [6.94] should be more realistic than the Lennard-Jones (6,12) potential because it undoubtedly correlates accurately a much wider range of experimental data. We shall see below that the differences between the potential of Barker et al. and the Lennard-Jones potential give not only rise to quantitative effects in the results of surface properties but also to *qualitative* effects. All the other MD calculations on noble-gas surfaces reported in the literature have been done exclusively with the Lennard-Jones potential. Examples are given in [6.90,92 and 95,96]. Since the potential of Barker et al. for krypton is close to the real pair potential, the results can be considered as a quantitative description.

a) Model and Interaction Potential

Model Parameter. The MD calculations have been performed for the following model: The krypton atoms are arranged as a slab-shaped fcc crystal (Fig. 6.12), the two free surfaces being (100) planes. For this (100) surface, which is an open, square lattice, the effect of premelting should be more pronounced than at the (111) close packed triangular surface. The structure of (100), (111) and (110) surfaces in an fcc crystal are shown in Fig. 6.27. The slab consists of 11 layers of 50 atoms, i.e., the total number of particles used in the calculations was 550. PBC were imposed with respect to translations parallel to the surface. The time step used in the calculation was 10^{-14}s. The calculations have been done for 8500 time steps (0.85×10^{-10}s). The magnitude of the initial velocities of all atoms was chosen to be equal but the Maxwell distribution was reached after 120 time steps (1.2×10^{-12}s).

The box size of the krypton systems is about 28.5 Å (Table 6.1), and the cut-off radius for the potential (the details of the potential are given below) was chosen to be the fourth-nearest neighbor distance (8 Å); the interaction of

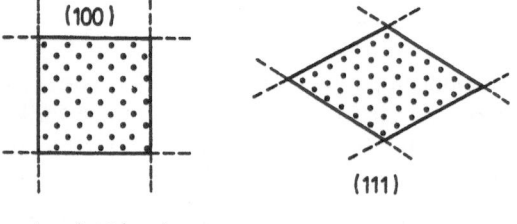

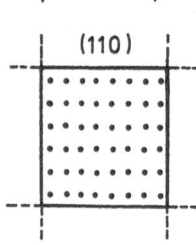

Table 6.1

T [K]	L [Å]	ϱ [atoms/Å³]
7	28.225	0.0222
70	28.625	0.0213
102	29.110	0.0203

Fig. 6.27. Configuration of (100), (111) and (110) surfaces in an fcc crystal

the fifth neighbors has no influence on the results. Thus, we may conclude that the krypton pair potential is short ranged compared to half the box size and, therefore, the effects due to PBC should be small (see also Sect. 6.4). This is confirmed by a more systematic analysis: the particle number has been varied from 108 to 864 and it turned out that in the case of N = 550 no systematic effects arising out of PBC.

All the bulk data which are given in this subsection have been determined also by MD (model without free surfaces, i.e. PBC in all directions) using the same model parameter and the same interaction potential as in the surface calculations.

The calculations have been performed for three temperatures: 7 K, 70 K and 102 K (the melting temperature is 116 K). The density for each temperature has been extracted from the experimental data given in [6.97]. For each temperature the length L of the cubic box and the particle number ϱ are indicated in Table 6.1.

Interaction Potential. The pair potential used in the calculations has the form [6.94]

$$v(r) = \varepsilon[v_0(R) + v_1(R)], \quad \text{where} \tag{6.76}$$

$$v_0(R) = \sum_{i=0}^{5} A_i(R-1)^i \exp\left[\alpha(1-R)\right] - \sum_{i=0}^{2} -\frac{C_{6+2i}}{r^{6+2i} + \delta} \tag{6.77}$$

$$v_1(R) = [P(R-1)^4 + Q(R-1)^5] \exp\left[\alpha'(R-1)\right], \quad R>1$$
$$= 0 \qquad\qquad\qquad\qquad\qquad\qquad\qquad R<1 \tag{6.78}$$

with

$$R = \frac{1}{r_m}, \tag{6.79}$$

r_m being the interatomic distance at the minimum of the potential. The parameters for krypton are [6.94]:

$$\frac{\varepsilon}{k_B}[K] = 201.9 \qquad \qquad C_6 \quad = 1.0632$$

$$r_m\,[\text{Å}] = 4.0067 \qquad \quad C_8 \quad = 0.1701$$

$$\qquad \qquad \qquad \qquad \quad C_{10} \; = 0.0143$$

$$A_0 \quad = 0.23526$$

$$A_1 \quad = -4.78686 \qquad \alpha \quad = 12.5$$

$$A_2 \quad = -9.2 \qquad \qquad \delta \quad = 0.01$$

$$A_3 \quad = -8.0 \qquad \qquad P \quad = -9.0$$

$$A_4 \quad = -30.0 \qquad \qquad Q \quad = 68.67$$

$$A_5 \quad = -205.8 \qquad \quad \alpha' \quad = 12.5.$$

The pair potential based on these parameters is shown in Fig. 6.28.

This potential is consistent with a wide range of experimental data including second virial coefficients, gas transport properties, solid-state data, and measurements of differential scattering cross sections. The Lennard-Jones potential is also shown in Fig. 6.28. Although the shapes of the two potentials are similar, there are considerable quantitative differences. For example, the depth of the minimum of the potential of Barker et al. is approximately 25% greater

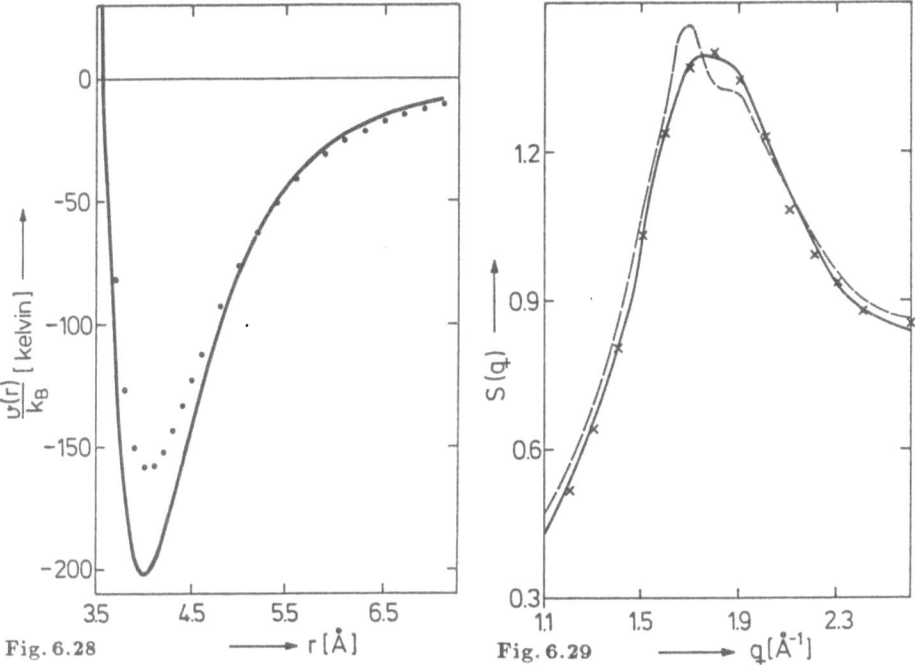

Fig. 6.28.

Fig. 6.29.

Fig. 6.28. Pair potentials for krypton; (*full line*) potential of *Barker* et al. [6.94], (*dotted line*) Lennard-Jones (6,12) potential

Fig. 6.29. Structure factor for gaseous krypton; (*full line*) MD calculation with the potential of Barker et al.; (*dashed line*) MD calculation with the Lennard-Jones (6,12) potential; (*crosses*) experimental results [6.98]

than the depth of the Lennard-Jones potential. The potential of Barker et al. should be more realistic because it undoubtedly correlates accurately a much wider range of experimental data. We did an additional check for the potential of Barker et al. on the basis of very accurate structural data for gaseous krypton [6.98]. For this purpose a MD calculation for gaseous krypton was performed and on the basis of these data we computed the structure factor

$$S(q) = \langle \varrho_{-q} \varrho_q \rangle, \quad \text{where} \tag{6.80}$$

$$\varrho_q = \frac{1}{N^{1/2}} \sum_{i=0}^{N} \exp\left(i\boldsymbol{q} \cdot \boldsymbol{r}_i\right) \tag{6.81}$$

is the microscopic number density. $\langle \ldots \rangle$ denotes again statistical averaging. $\boldsymbol{r}_i$ is the position vector of atom i and $\boldsymbol{q}$ is the wave-vector. $S(q)$ in the Fourier transform of the pair correlation function introduced in Sect. 6.3.3. Figure 6.29 shows the MD results for $S(q)$ obtained from calculations with the potential of Barker et al. and the Lennard-Jones (6,12) potential. As can be seen from Fig. 6.29 the potential of Barker et al. describes undoubtedly better the experimental structure data than the Lennard-Jones (6,12) potential does. We shall see below that the differences between both potentials give rise to relatively large quantitative effects in the results for surface properties – distinctly larger than those given in Fig. 6.29.

In the calculation of the properties of the system the forces acting on the particles are relevant. The force F_i acting on particle i is given by

$$\boldsymbol{F}_i = - \sum_{\substack{j=1 \\ j \neq i}}^{N} \frac{\boldsymbol{r}_{ij}}{r_{ij}} \frac{\partial v(r_{ij})}{\partial r_{ij}}. \tag{6.82}$$

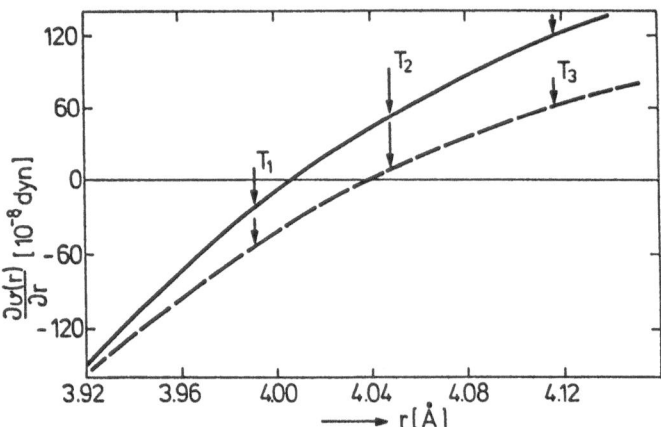

Fig. 6.30. The derivative of pair potentials in the vicinity of the first-nearest neighbor distance. The position of the arrows are the first-nearest neighbor distances at $T_1 = 7\,\mathrm{K}$, $T_2 = 70\,\mathrm{K}$ and $T_3 = 102\,\mathrm{K}$. (*full line*) potential of Barker et al.; (*dashed line*) Lennard-Jones (6,12) potential

Thus, the derivative of different pair potentials should be studied. In Fig. 6.30 $\partial v(r)/\partial r$ are plotted both for the potential of Barker et al. and for the Lennard-Jones potential in the vicinity of the first-nearest neighbor distance; the first-nearest neighbor contribution to the sum in (6.82) should be of considerable relevance. It can be seen from Fig. 6.30 that the differences between the derivatives of the two potentials are large for all temperatures.

b) Results

Structure and Displacements in the Outermost Layer. The positions of the krypton atoms of the outermost layer are represented in Fig. 6.31 for the tempera-

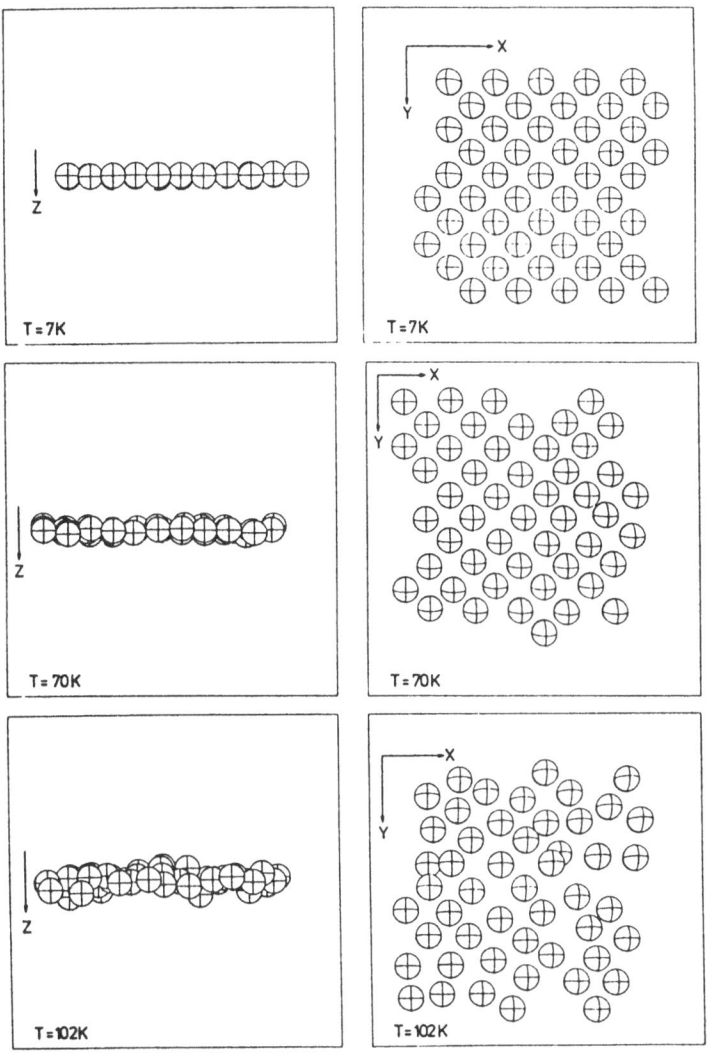

Fig. 6.31. Structure of the outermost layer of 7 K, 70 K and 102 K. Another representation of this configurations is given in Fig. 6.2

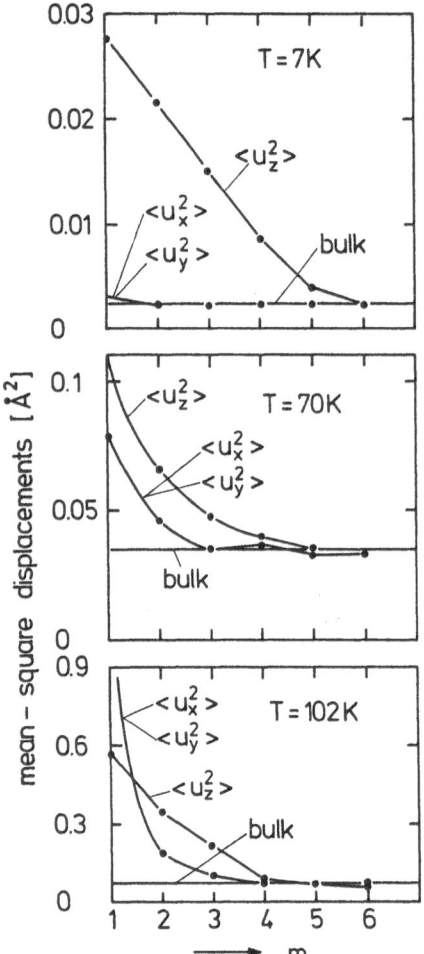

tures of 7 K, 70 K and 102 K. The coordinate z is the direction perpendicular to the surface, and the coordinates x, y are parallel to the surface. The diameter of the atoms is approximately 3.5 Å and the layers consist of 50 atoms. The positions shown in Fig. 6.31 are the configurations after 10^3 time steps in the iteration process of the calculation corresponding to 10^{-11}s.

It can be seen from Fig. 6.31 that for the temperature of 7 K the atoms at the surface form a well-defined lattice parallel to the surface and the structure is identical to that observed for a layer in the bulk of the crystal. The root-mean-square displacement $\sqrt{\langle u^2 \rangle}$ (Fig. 6.32) *parallel* to the surface (0.080 Å) does not differ very much from that in the bulk (0.070 Å). These amplitudes are of the order of 1.5% of the lattice constant (5.645 Å). However, the mean-square amplitude *perpendicular* to the surface is considerably larger in the outermost layer than that in the bulk of the crystal (Fig. 6.32, 7 K).

In the calculation for 70 K we obtain for $\sqrt{\langle u^2 \rangle}$ *parallel* to the surface (Fig. 6.32) a value (0.396 Å) which is approximately 50% larger than in the bulk of the crystal (0.264 Å). Also in the calculation for 70 K (Fig. 6.31) the atoms of the outermost layer, at first glance, form a well-defined periodical lattice which corresponds to that in the bulk. However, there are small lattice distortions superimposed which cannot be explained by the relatively large vibrational amplitudes at the surface alone.

In the calculation for 102 K (Fig. 6.31) a periodical structure parallel to the surface can hardly be recognized. Although the temperature is still 14 K below the melting point, the outermost layer is extensively disordered. In particular, it can be seen from Fig. 6.31 that not all the atoms are arranged side by side but also one upon another, and this situation indicates a *diffusion process*. We have determined the diffusion constant D for the outermost layer. This can be done by studying of the mean-square displacement $\langle r^2(t) \rangle$ as a function of time parallel to the surface. $\langle r^2(t) \rangle$ can be expressed by the velocity autocorrelation function (Sect. 6.4.3) as follows [6.44]

$$\langle r^2(t) \rangle = \frac{4k_BT}{m} \int_0^t (t - s) \frac{\langle v(0) \cdot v(s) \rangle}{\langle v(0)^2 \rangle} ds. \tag{6.83}$$

With the asymptotic form

$$\lim_{t \to \infty} \langle r^2(t) \rangle = 4Dt + \text{const}, \tag{6.84}$$

where

$$\text{const} = \frac{4k_BT}{m} \int_0^\infty s \frac{\langle v(0) \cdot v(s) \rangle}{\langle v(0)^2 \rangle} ds, \tag{6.85}$$

$$D = \frac{k_BT}{m} \int_0^\infty \frac{\langle v(0) \cdot v(s) \rangle}{\langle v(0)^2 \rangle} ds, \tag{6.86}$$

we can estimate the diffusion coefficient D from our MD data. Equation (6.84) is valid for times t when the velocity autocorrelation function is zero. This is approximately fulfilled for the system at T = 102 K for times longer than 2.2×10^{-12} s. Using (6.84) we found from our MD model that D is of the order of $\sim 10^{-5}$ cm^2/s which is also a typical value for liquids.

The occurrence of a diffusion process means that the displacment $\langle u^2 \rangle$ parallel to the surface must be *infinite* (Fig. 6.32, 102 K); clearly, the bulk value remains finite (0.387 Å) below the melting temperature (116 K).

Since D is so large and the structure of the layer is disordered at T = 102 K we may conclude that our system shows the effect of *surface premelting* already investigated for Lennard-Jones systems [6.95,96,99]. In [6.99] the diffusion process has not been studied, and in [6.95] the long-time gradient for $\langle r^2(t) \rangle$ is immeasurably small on the MD scale and, in our opinion, one is hardly able to recognize a diffusive motion. In [6.96] there is a diffusion process parallel and perpendicular to the surface; in our system there is no diffusion perpendicular to the surface at $T = 102\,\text{K}$ indicating that the surface properties are very sensitive to small variations in the interaction potential.

Interplanar Spacings. Let us now investigate the deviations of the mean positions a' of layers near the surface from the mean position a that these layers would have in the bulk of the crystal. In Fig. 6.33 the quantity

$$\Delta b = a'_m - a_m \tag{6.87}$$

is represented as a function of m, where m is the integer labeling the planes of the crystal which are parallel to the surface. We define $m = 1$ for the outermost layer, 2 for the plane just below the upper layer, and so on. It can be seen from Fig. 6.33 that there is a *multi-layer relaxation* for all three temperatures. In the case of $T_1 = 7\,\text{K}$ the relaxation effect is getting small with increasing m. Due to the geometrical symmetry the mean position of the innermost layer ($m = 6$) must be identical to that in the bulk of the crystal; this is fulfilled for 7 K, 70 K and 102 K.

For $m = 1$ (outermost layer) Δb is getting small with increasing temperature. We explain this effect as follows: Let us consider surface particles which are vibrating perpendicular to the surface. When moving inward, they collide with the particles of the second layer ($m = 2$) and the *repulsive* part of the potential is effective, which causes them to reverse their direction of motion. However, when the particles are moving outward, they do not collide with any other particles, but the direction of motion is reversed by the *attractive* part

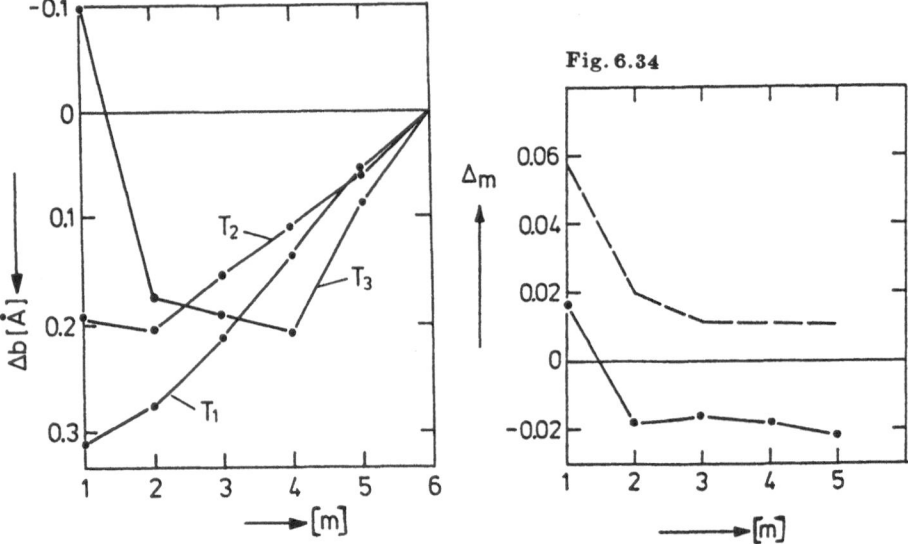

Fig. 6.33. The deviations of the mean positions of layers near the surface from the mean position that these layers would have in the bulk. Δb is defined by (6.87)

Fig. 6.34. Interplanar spacing of surface layers. Δ_m is defined by (6.88). (*Full line*) MD calculation with the potential of Barker et al. (70 K); (*dashed line*) Lennard-Jones system ($T = 57\,\text{K}$)

of the potential in the crystal. Due to the asymmetry of the repulsive and the attractive parts of the pair potential, we observe an additional (additional to the usual thermal expansion) *outward shift* of the mean position of the particle. This *outward shift* of the mean position will be strongly temperature-dependent.

In general, the relaxation as a function of temperature is determined by minimizing the total free energy (including vibrational modes), and the resulting mean displacements a'_m are called the *dynamic displacements*. If the mean displacements are determined by minimizing the static energy we only obtain the *static displacements*. Most of the lattice-dynamical calculations in the literature, [6.29], have been performed on the basis of the *static* displacements. However, it is shown in Fig. 6.33 that Δb varies strongly with temperature, and therefore, the assumption that the mean displacements have their static values breaks down at high temperatures. Thus, in lattice dynamical calculations we have to consider that *both* the harmonic approximation and the validity of the *static* approximation for the mean displacements break down at high temperatures. MD calculations take into account automatically both dynamic displacements and anharmonic effects. The curves in Fig. 6.33 show the *dynamic* displacements. We expect, however, that the curve for $T_1 = 7\,\mathrm{K}$ is close to the *static* case.

Let us compare our results with those obtained from a Lennard-Jones system. In [6.30] the results of the quantity

$$\Delta_m = \frac{a'_{m+1} - a'_m}{a_{m+1} - a_n} - 1 \qquad (6.88)$$

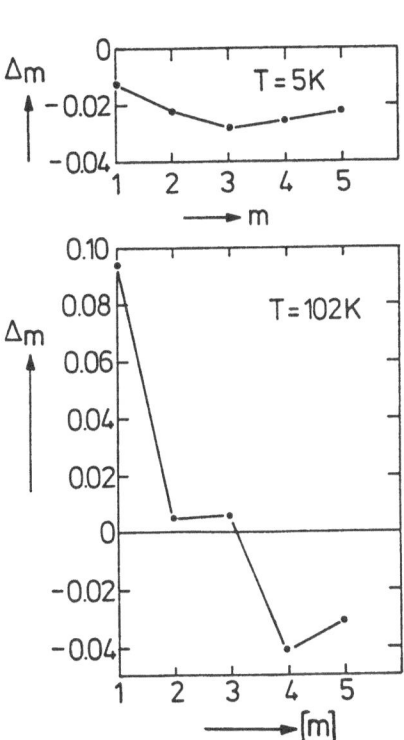

Fig. 6.35. Interplanar spacing of surface layers for the krypton system. Δ_m is defined by (6.88)

have been discussed. Figure 6.34 shows our results for Δ_m at 70 K together with those obtained for the Lennard-Jones system at 57 K [6.30]. As can be seen from Fig. 6.34 there are not only large quantitative differences between our model and the Lennard-Jones system but also *qualitative* differences: for $m \geq 2$ in our system the distances between the layers are *smaller*, and in the Lennard-Jones system they are *larger* than in the bulk of the crystal. The corrections due to the different temperatures for which the calculations have been performed would enhance the differences between the two systems. Furthermore, the results for the lowest temperature (Fig. 6.35) show that the relaxation of our system is *negative* (contraction of the system) and in the case of the Lennard-Jones system it is *positive* (expansion of the system).

Δ_m is approximately independent of m at 7 K (Fig. 6.35). This means that the interplanar spacing is approximately constant for $1 \leq m \leq 6$. This order is obviously disturbed at higher temperatures (Figs. 6.34,35).

c) Conclusions

i) Significant differences are obtained for the structural and dynamical properties at the surface of a krypton crystal if a realistic interaction potential is used instead of the idealized Lennard-Jones potential. In particular, it is shown that surface properties are very sensitive to small variations in the pair potential.

ii) 14 K below the melting temperature we observe the effect of *pre-melting* (Fig. 6.31).

6.6 Final Remarks

The main problem in MD calculations is to find *realistic* interaction potentials. The calculations for krypton (Sect. 6.5.4) have shown that

i)   surface properties are strongly dependent on temperature, and
ii) surface properties seem to be much more sensitive to small variations in the pair potential than bulk properties.

Thus, in the *reliable* description of the structure and the dynamics of surfaces precise knowledge of the interaction is required. However, as we have already stated in Sect. 6.4.3, the *accurate* determination of pair potentials for materials of interest (e.g., metals) is a matter of considerable difficulty even for the bulk, and the surface problem is much more complicated. The only group of materials for which the pair potential is well-known is that of noble-gas solids. In contrast to metals, the potential functions of noble-gas solids do not depend on the temperature, and they are the same at the surface and in the bulk of the crystal. However, there is still an open question in connection with noble-gas surfaces: How relevant are *three-body forces*, etc. at the surface of these materials?

At free *metal* surfaces the local background electron density $n(\mathbf{r})$ may change from its average bulk-value and produces concomitant changes in the potential functions, and because the surface properties are sensitive to variations in temperature, n(r) at the surface should be calculated for *nonzero temperatures*, i.e. instead of the density-functional formalism [6.49] for T = 0, the extended approach (valid for nonzero temperatures) by *Mermin* [6.100] should be used in future developments.

References

6.1   W. Schommers: Phys. Rev. A**16**, 327 (1977)
6.2   C. Hoheisel: Phys. Rev. A**26**, 2998 (1981)
6.3   W. Schommers: Phys. Rev. A**22**, 2855 (1980)
6.4   W. Schommers: Phys. Rev. A**27**, 2241 (1983)
6.5   W. Schommers: Phys. Rev. Lett. **38**, 1536 (1977)
6.6   W. Schommers; Phys. Rev. B**21**, 847 (1980)
6.7   W. Schommers: Phys. Rev. B**22**, 1058 (1980)
6.8   W. Harrison: *Pseudopotentials in the Theory of Metals* (Benjamin, New York 1966)
6.9   O. Krisement: "Vorlesung über Statistische Mechanik und Thermodynamik", Universität Münster (1968)
6.10   L.D. Landau, E.M. Lifshitz: *Statistical Physics* (Pergamon, London 1963)
6.11   I. Prigogine: *From Being to Becoming – Time and Complexity in the Physical Sciences* (Freeman, San Fransisco 1979)
6.12   A. Münster: *Statistical Thermodynamics*, Vol. 1 (Springer, Berlin, Heidelberg 1969)
6.13   A. Münster: *Statistical Thermodynamics*, Vol. 2 (Springer, Berlin, Heidelberg 1974)
6.14   T.L. Hill: *An Introduction to Statistical Thermodynamics* (Addison-Wesley, Reading, MA 1960)
6.15   K. Huang: *Statistische Mechanik*, Bd. I–III (Bibliographisches Institut, Mannheim 1964)
6.16   R. Becker: *Theorie der Wärme*, 3rd (Springer, Berlin, Heidelberg 1985)
6.17   W. Schommers: Z. Physik **257**, 78 (1972)
6.18   S.A. Rice, P. Gray: *The Statistical Mechanics of Simple Liquids* (Wiley, New York 1965)
6.19   J.D. Weeks, D. Chandler, H.C. Andersen: J. Chem. Phys. **55**, 5422 (1971)
6.20   N.K. Ailawadi, D.E. Miller, J. Naghizadzh: Phys. Rev. Lett. **36**, 1494 (1976)
6.21   J.K. Percus, G.J. Yevick: Phys. Rev. **110**, 1 (1950)
6.22   M.J. van Leeuwen, J. Groeneveld, J. de Boer: Physica **25**, 792 (1959)
6.23   N.H. March: In *Theory of Condensed Matter* (IAEA, Vienna 1968)
6.24   W. Abel, R. Block, W. Schommers: Phys. Lett. **58**A, 367 (1976)
6.25   W. Schommers: Phys. Rev. A**28**, 3599 (1983)
6.26   A. Rahman: Phys. Rev. **136**, A405 (1964)
6.27   A.H. Narten: J. Chem. Phys. **56**, 1185 (1972)
6.28   P.A. Nelson: A Molecular Dynamics Study of a Square-Well Fluid, Ph.D. Dissertation, Princeton University (1966)
6.29   R.E. Allen, F.W. de Wette: Phys. Rev. **179**, 873 (1969)
6.30   R.E. Allen, F.W. de Wette, A. Rahman: Phys. Rev. **179**, 887 (1969)
6.31   R.E. Allen, G.P. Alldredge, F.W. de Wette: Phys. Rev. B**4**, 1648 (1971)
6.32   R.E. Allen, G.P. Alldredge, F.W. de Wette: Phys. Rev. B**4**, 1661 (1971)
6.33   G.P. Alldredge, R.E. Allen, F.W. de Wette: Phys. Rev. B**4**, 1682 (1971)
6.34   T.S. Chen, R.E. Allen, G.P. Alldredge, F.W. de Wette: Solid State Commun. **8**, 2105 (1970)
6.35   T.S. Chen, G.P. Alldredge, F.W. de Wette, R.E. Allen: Phys. Rev. Lett. **26**, 1543 (1971)
6.36   T.S. Chen, G.P. Alldredge, F.W. de Wette, R.E. Allen: J. Chem. Phys. **55**, 3121 (1971)
6.37   G. Benedek: In *Dynamics of Gas-Surface Interaction*, ed. by G. Benedek, U. Valbusa, Springer Ser. Chem. Phys., Vol. 21 (Springer, Berlin, Heidelberg 1982) p. 227
6.38   J.E. Black: In *Vibrations at Surfaces*, ed. by R. Caudano, J.-M. Gilles, A.A. Lucas (Plenum, New York 1982)

6.39 F.W. de Wette: In *Interatomic Potentials and Simulation of Lattice Defects*, ed. by P.G. Gehlen, J.R. Beeler, Jr., R.I. Jaffee (Plenum, New York 1972)

6.40 H. Ibach, D.L. Mills: *Electron Energy Loss Spectroscopy and Surfaces Vibrations* (Academic, New York 1982); H. Ibach, T.S. Rahman: In *Chemistry and Physics of Solid Surfaces V*, ed. by R. Vanselow, R. Howe, Springer Ser. Chem. Phys., Vol. 35 (Springer, Berlin, Heidelberg 1984) Chap. 19

6.41 W. Schommers: Solid State Commun. **16**, 45 (1975)

6.42 W. Schommers: Phys. Lett. A**43**, 157 (1973)

6.43 W. Gläser, S. Hagen, U. Löffler, J.-B. Suck, W. Schommers: In *Properties of Liquid Metals* (Taylor and Francis, London 1973)

6.44 P.A. Egelstaff: *Introduction to the Liquid State* (Academic, London 1967)

6.45 R. Zwanzig, N.K. Ailawadi: Phys. Rev. **182**, 260 (1969)

6.46 Diffusion Data **2**, 42 (1968)

6.47 P.G. Gehlen, J.R. Beeler, Jr., R.I. Jaffee: In *Interaction Potentials and Simulation of Lattice Defects* (Plenum, New York 1972)

6.48 M. Rasolt, F. Perrot: Phys. Rev. B**28**, 6749 (1983)

6.49 W. Kohn, L.J. Sham: Phys. Rev. **140**, A1133 (1965)

6.50 F. Manghi, E. Molinari: J. Phys. C**15**, 3627 (1982)

6.51 J.A. Appelbaum, D.R. Haman: Phys. Rev. Lett. **34**, 806 (1975)

6.52 E. Caruthers, L. Kleinmann, G.P. Alldredge: Phys. Rev. B**8**, 4570 (1973)

6.53 J.R. Chelikowsky, M.L. Cohen: Phys. Rev. B**13**, 826 (1976)

6.54 F. Manghi, C.M. Bertoni, C. Calandru, E. Molinari: Phys. Rev. B**24**, 6029 (1981)

6.55 P.M. Morse: Phys. Rev. **34**, 57 (1929)

6.56 K. Mohammed, M.M. Shukla, F. Milstein, J.L. Merz: Phys. Rev. B**29**, 3117 (1984)

6.57 H. Eichler, M. Peyzl: Phys. Stat. Sol. **35**, 333 (1969)

6.58 H.D. Diener, R. Heinrich, W. Schellenberger: Phys. Stat. Sol. B**44**, 403 (1971)

6.59 V. Vitzk, C.R. Perrin, D.K. Bown: Philos. Mag. **21**, 1049 (1970)

6.60 C. Gehlen, A.R. Rosenfield, G.T. Hahn: J. Appl. Phys. **39**, 5246 (1968)

6.61 R. Chang: Philos. Mag. **16**, 1021 (1967)

6.62 R.M.J. Cotterill, M. Doyama: Phys. Rev. **145**, 465 (1966)

6.63 M. Doyama, R.M.J. Cotterill: Phys. Rev. **137**, A994 (1965)

6.64 R.A. Johnson: Phys. Rev. **134**, A1329 (1964)

6.65 L.A. Grififalco, V.G. Weizer: J. Phys. Chem. Solids **12**, 260 (1960)

6.66 W.D. Wilson, C.L. Bisson: Phys. Rev. B**3**, 3979 (1971)

6.67 A. Anderman, W.G. Gehman: Phys. Stat. Sol. **30**, 283 (1968)

6.68 R. Furth: Proc. R. Soc. (London) **183**, 87 (1944)

6.69 L.A. Grififalco, V.G. Weizer: Phys. Rev. **114**, 687 (1959)

6.70 R.G. Lincoln, K.M. Koliwad, P.B. Ghate: Phys. Rev. **157**, 463 (1967)

6.71 F.O. Goodman: Phys. Rev. **164**, 1113 (1967)

6.72 I.M. Torrens: *Interatomic Potentials* (Academic, New York 1972)

6.73 J.L. Lebowitz, J.K. Percus, L. Verlet: Phys. Rev. **153**, 250 (1967)

6.74 R. Block, W. Schommers: J. Phys. C**8**, 1997 (1975)

6.75 R. Abe: Prog. Theor. Phys. **21**, 421 (1959)

6.76 M.C. Abramo, M.P. Tosi: Nuovo Cim. **5**, 1044 (1972)

6.77 B.J. Alder: Phys. Rev. Lett. **12**, 317 (1964)

6.78 P.A. Egelstaff: Ann. Rev. Phys. Chem. **24**, 159 (1973)

6.79 P.A. Egelstaff, D.I. Page, C.R.T. Heard: J. Phys. C**4**, 1453 (1971)

6.80 E. Feenberg: *Theory of Quantum Fluids* (Academic, New York 1969)

6.81 J.S. Kirkwood: J. Chem. Phys. **3**, 300 (1935)

6.82 J.A. Krumhansl, S. Wang: J. Chem. Phys. **56**, 2034 (1972)

6.83 H.J. Raveche, R.D. Mountain, W.B. Street: J. Chem. Phys. **57**, 4999 (1972)

6.84 J.S. Rowlinson: Mod. Phys. **88**, 149 (1966)

6.85 S. Wang, J.A. Krumhansl: J. Chem. Phys. **56**, 4297 (1972)

6.86 Y. Waseda, M. Ohtani, K.J. Suzuki: J. Phys. Chem. Solids **35**, 585 (1972)

6.87 J. Winfield, P.A. Egelstaff: Can. J. Phys. **51**, 1965 (1973)

6.88 F.W. de Wette, R.E. Allen, D.S. Hughes, A. Rahman: Phys. Lett. **29**A, 548 (1969)

6.89 B.J. Alder, T.E. Wainwright: Phys. Rev. **127**, 359 (1962)

6.90 R.K. Kalia, P. Vashishta, S.D. Makanti: Phys. Rev. Lett. **49**, 676 (1982)

6.91   O.G. Mouritsen, A.J. Berlinsky: Phys. Rev. Lett. **48**, 181 (1982)
6.92   F.F. Abraham: Phys. Rev. Lett. **50**, 978 (1983)
6.93   W. Schommers, P. v. Blanckenhagen: Vacuum **33**, 733 (1983)
6.94   J.A. Barker, R.O. Watts, J.K. Lee, T.P. Schafer, Y.T. Lee: J. Chem. Phys. **61**, 3081 (1974);
        M.L. Klein (ed.): *Inert Gases*, Springer Ser. Chem. Phys., Vol. 34 (Springer, Berlin, Heidelberg 1984)
6.95   J.Q. Broughton, L.V. Woodcock: J. Phys. C**11**, 2743 (1978)
6.96   J.Q. Broughton, G.H. Gilmer: J. Chem. Phys. **79**, 5119 (1983)
6.97   G.L. Pollack: Rev. Mod. Phys. **36**, 74B (1964)
6.98   A. Teitsma, P.A. Egelstaff: Phys. Rev. A**21**, 367(1980)
6.99   R.M.J. Cotterill: Phil. Mag. **32**, 1283 (1975)
6.100   N.D. Mermin: Phys. Rev. **137**, A1441 (1965)

Additional References with Titles

Abraham, F.F., J.Q. Broughton: Pulsed melting of silicon(111) and (100) surfaces stimulated by molecular dynamics, Phys. Rev. Lett. **56**, 734 (1986)

Landman, U., et al.: Faceting at the silicon(100) crystal-melt interface: theory and experiment, Phys. Rev. Lett. **56**, 155 (1986)

Rosato, V., G. Cicotti, V. Pontikis: Molecular-dynamics study of surface premelting effects, Phys. Rev. B**33**, 1860 (1986)

Schneider, M., A. Rahman, I.K. Schuller: Role of relaxation in epitaxial growth: a molecular-dynamics study, Phys. Rev. Lett. **55**, 604 (1985)

Schommers, W.: Structural and dynamical behavior of noble-gas surfaces, Phys. Rev. B**32**, 6845 (1985)

Schommers, W., P. v. Blanckenhagen: Study of the surface structure by means of molecular dynamics, Surf. Science **162**, 144 (1985)

7. Surface Phonon Dispersion of Surface and Adsorbate Layers

M. Rocca, H. Ibach, S. Lehwald, and T.S. Rahman
With 22 Figures

The surface phonon spectrum can be investigated by high-resolution electron energy loss spectroscopy in the impact scattering regime. We will describe the general features of electron-phonon interaction. In particular, we will discuss how to explore multiple scattering effects for the measurement of the surface phonons. As examples we present measurements performed for Ni(100) surfaces, both clean and covered with (2×2) overlayers of oxygen and sulfur, and after disordered adsorption of oxygen. While the presence of sulfur or of randomly distributed oxygen adatoms affects the dynamics of the substrate only marginally, a softening of the surface modes is observed when the oxygen adatoms form an ordered c(2×2) structure. We tentatively explain this effect as due to internal strains which build up in the near surface region in presence of the adatoms. This picture is confirmed by the result of lattice-dynamical calculations performed for a model with first and second derivatives of the two-body interaction potentials between first and second nearest neighbours. We show that the inclusion of the stress terms enables us to reproduce the observed phonon frequencies and dispersion.

7.1 Background and Overview

Long time has passed since Lord Rayleigh (J. Strutt) in 1885 forecasted in a theoretical study the existence of surface waves in isotropic elastic continua [7.1]. Surface dynamics has since then gained more and more interest in pure, as well as, in applied research. Nowadays surface waves find an application in many technological devices as, e.g., acoustic delay lines [7.2] and narrow-band-filters [7.3]. Lattice dynamical calculations showed that for short wavelengths a large variety of other surface modes exists [7.4] in addition to Rayleigh waves (Chap. 5). Their frequency and dispersion depends strongly on the coupling between the atoms in the near-surface region. These surface waves play an important role in electron-phonon coupling for the electronic surface states [7.5] and in thin films. It is also believed that in some instances the observed reconstruction of the surface can be explained by the freezing of a low-frequency surface phonon (soft phonon) [7.6], in a manner analogous to that in bulk crystals.

In the long-wavelength limit surface phonons could be investigated by scattering of photons (Brillouin and Raman scattering) [7.7]. No equivalent experimental technique was, on the other hand, available until recently to observe phonon dispersion at shorter wavelengths. For such experiments one needs particles whose wave vector is comparable with the inverse of the surface lattice constant and which do not penetrate deeply into the crystal thus having a high surface sensitivity. Only electrons and thermal atoms (particularly He atoms) fulfill both requirements. Although atom beam scattering had been proposed as a possible technique in a theoretical paper by *Cabrera* et al. [7.8], as early as in 1969, more than ten years passed until a beam with the desidered degree of monocromaticity and intensity could be realized experimentally [7.9]. In the meantime electron energy loss spectroscopy (EELS) has also undergone a rapid development, both on the experimental and on the theoretical side [7.10]. Recently it could be shown by some of the authors [7.11] that the thermal background between the LEED peaks consists mostly of inelastically scattered electrons which have interacted with only one phonon. Although the efficiency of the inelastic scattering mechanism in the large-angle region is much lower than that of the usually investigated small-angle dipole scattering region, which is limited to small momentum transfers, sufficent count rates can nowadays be obtained for high enough energy resolutions (6–7 meV or 50–60 cm^{-1}). Moreover by exploiting multiple-scattering effects [7.12] one is able to resolve peaks whose separation in frequency is much smaller than energy resolution.

The chapter is organized as follows. In the next section we discuss the principal features of the method and some experimental details. We will also give a brief overview of the electron-phonon scattering theory placing particular emphasis on multiple-scattering effects, as they play a major role in the measurements of phonon frequencies and dispersion. In Sect. 7.3 we present the experimental phonon spectra measured on the clean and on the adsorbate covered Ni(100) surface. Interesting effects on phonon dispersion upon adsorption were found to depend also on the order within the adsorbed phase. Finally, we devote Sect. 7.4 to a general discussion of the results. A model is proposed which accounts for most of the features observed in the phonon spectra.

7.2 Description of the Experiment

7.2.1 General Features of Electron-Surface Interaction

In electron energy loss spectroscopy (EELS) primary electrons with an energy between 5 and 300 eV are used. Their wavelength is comparable with the lattice constant of the crystals, and their penetration depth is limited to a few monolayers. Due to the translation symmetry of the crystal along the surface the electrons undergo diffraction in the scattering process and give rise to the well known LEED pattern. In addition to this elastic interaction, electrons can also exchange energy and momentum with the crystal.

Two limiting cases for the inelastic process have been established: dipole scattering and impact scattering. In dipole scattering the electrons scatter from the long-range electric fields which are caused by charge fluctuations near the surface. This scattering mechanism is very effective particularly at low primary energies and for small momentum transfers and has been used frequently to detect vibrational modes of adsorbates. For larger scattering angles the dipole scattering probability vanishes and the majority of the electrons is scattered inelastically via the short-range potential of the ion cores of the crystal.

In inelastic scattering from crystalline surfaces phonon creation and annihilation events obey the following energy and momentum conservation laws

$$E_i = E_f \pm \hbar\omega, \tag{7.1}$$

$$\boldsymbol{k}_{i\parallel} = \boldsymbol{k}_{f\parallel} \pm \boldsymbol{G}_{\parallel} \pm \boldsymbol{Q}_{\parallel}, \tag{7.2}$$

where ω and $\boldsymbol{Q}_\parallel$ are frequency and wavevector of the phonon, $\boldsymbol{G}_\parallel$ is a reciprocal lattice vector of the surface, E and $\boldsymbol{k}_\parallel$ are the energy and the wavevector of the primary particles, and the subscripts i and f refer to the initial and final state of the system, respectively. As phonon energies are in the range of $20\,\mathrm{meV}$ while the energy of the primary electrons is of the order of $100\,\mathrm{eV}$ any particular scattering angle θ_f corresponds, to a very good approximation, to a unique momentum transfer, regardless of the size of the energy loss. This is shown in Fig. 7.1 with the Ewald sphere. For in-plane scattering we have then

$$\theta_f - \sin^{-1}\left[\sin\theta_i \pm \left(\frac{Q_\parallel + G_\parallel}{k}\right)\right], \tag{7.3}$$

where $k = |\boldsymbol{k}_i| = |\boldsymbol{k}_f|$ and θ_i is the angle of incidence. For the interaction with a single phonon and for $\theta_f > \theta_i$ the energy loss peak corresponds to the creation of a phonon with wavevector $+Q$, while the energy gain peak is due to the annihilation of a phonon with wavevector $-Q$. For $\theta_f < \theta_i$ the above holds but

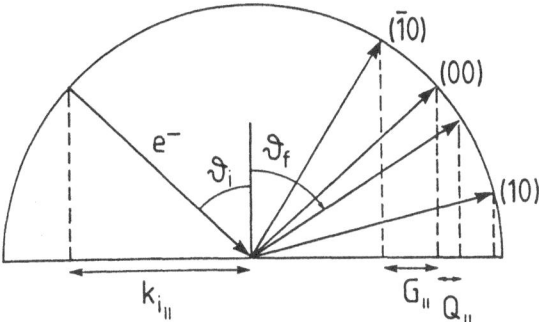

Fig. 7.1. Elastic and inelastic scattering of an electron from a surface. The arrows represent the wavevector of the primary electron. In elastic processes only surface reciprocal lattice vectors are exchanged while for inelastic scattering also the wavevector $\boldsymbol{Q}_\parallel$ has to be considered. As the energy of the phonon, $\hbar\omega$, is much smaller than the primary energy of the electron beam the electrons which have interacted with a phonon of wavevector $\boldsymbol{Q}_\parallel$ are scattered to very good approximation in the direction θ_f independently of the magnitude of the energy loss

the sign of $Q_\parallel$ is inverted. The phonon dispersion curve $\omega(Q_\parallel)$ can therefore be measured by recording the energy losses of the electrons for different scattering angles.

The electron may also engage in inelastic scattering processes involving more than one phonon. In this case the relation between the measured values of ω and $Q_\parallel$ is not unequivocal so that these events contain no valuable information on the phonon dispersion. Their contribution to the inelastic intensity should therefore be reduced as much as possible. An estimate of multiphonon scattering probability can be obtained from the Debye-Waller factor $2W$, which describes the decrease of the elastic intensity I_{el} of the LEED spots with the crystal temperature as:

$$I_{el} = I_0 e^{-2W}, \tag{7.4}$$

where I_0 would be the reflected intensity in the absence of inelastic processes. For the inelastic scattering contribution I_{in} it follows

$$I_{in} = I_0 - I_{el} = I_0 e^{-2W}(e^{2W} - 1) = I_0 e^{-2W}\left[2W + \frac{(2W)^2}{2!} + \ldots\right]. \tag{7.5}$$

One can show that the first term of the expansion represents one-phonon scattering events, the second two-phonon events and so on. In Fig. 7.2 we plot I_{in} versus W using (7.5).

In the high temperature limit, i.e. for $T > \theta_D$, θ_D being the surface Debye temperature, $2W$ reads

$$2W = \frac{3\hbar^2 T}{Mk_B \theta_D^2}|k_f - k_i|^2, \tag{7.6}$$

where k_B is Boltzmann constant, $\hbar$ Planck constant and M is the mass of the crystal atoms. The effective value of θ_D depends on E_i thus reflecting the

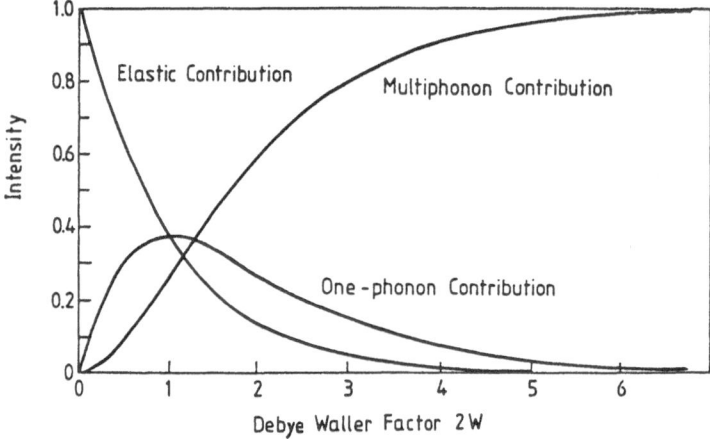

Fig. 7.2. Elastic, one phonon and multiple phonon scattering contributions according to (7.5)

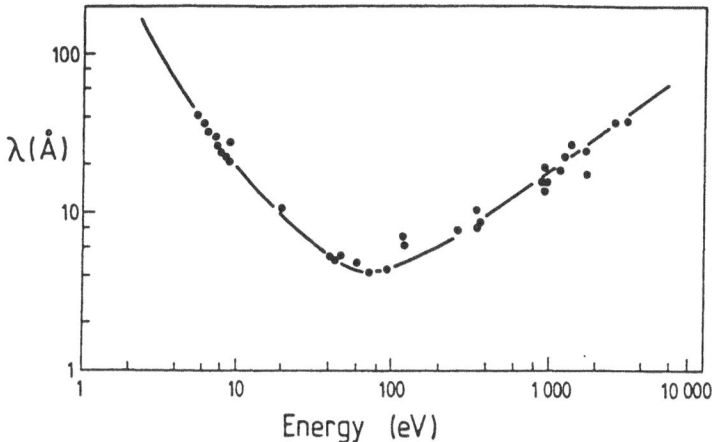

Fig. 7.3. Mean free path λ of electrons in a crystal according to [7.13] and therein cited works. The experimental points were measured on samples of different metals

dependence on energy of the penetration depth of the electrons of the primary beam.

If we introduce in (7.6) $\theta_D = 200\,\text{K}$, i.e. the value reported in the literature for E_i around $100\,\text{eV}$ [7.13] we obtain for typical experimental conditions ($\theta_i \simeq 35°$, $\theta_f \simeq 65°$ and $E_i \simeq 100\,\text{eV}$) $2W = 0.80$ at room temperature and $2W = 0.34$ at $T = 130\,\text{K}$. From Fig. 7.2 we can now see that 37% of the inelastic intensity is due to multiphonon processes at room temperature, while this percentage decreases to 18% at $T = 130\,\text{K}$. Working at low crystal temperature is therefore more desirable.

Another general feature of electron scattering is the small penetration depth of the primary beam into the crystal which makes the technique surface sensitive. Considering that the incident beam scatters from atoms in the layer l_z and that the scattered electron has to emerge to the surface, the attenuation A of the scattering cross section from deeper crystal layers compared to the first layer is roughly

$$A = \exp\left[-(\cos\theta_i + \cos\theta_f)dl_z/\lambda(E)\right] \tag{7.7}$$

with d the vertical distance of the crystal layers and $\lambda(E)$ the mean-free path of electrons. As the attenuation is principally due to the excitation of plasmons and of electron hole pairs $\lambda(E)$ is nearly independent of the substrate material in the case of metals (Fig. 7.3) [7.13–15]. According to (7.7) the probability for phonon scattering from deeper layers is small for grazing angles and $E_i \sim 100\,\text{eV}$, while at steeper angles and larger energies the possibility of exciting vibrations of deeper crystal layers increases.

7.2.2 Experimental Details

The experiments, we want to describe, were performed on a Ni single crystal cleaned by repeated cycles of Ne-ion bombardment and annealing to $1400\,\text{K}$

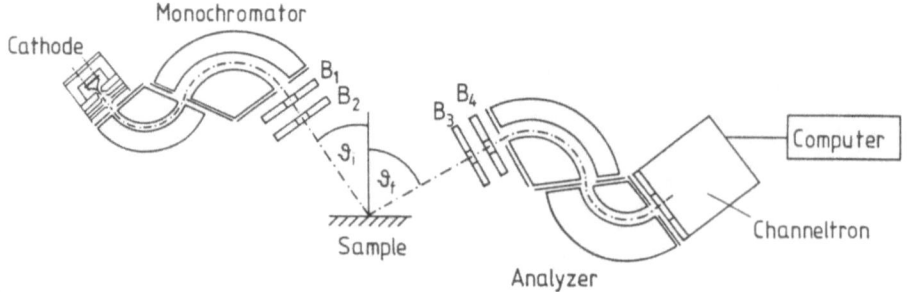

Fig. 7.4. Schematic drawing of the experimental set up for EELS

until the crystal was leached of carbon and sulfur and no traces of either impurity could be detected neither by our cylindrical mirror analyzer (CMA) Auger spectrometer nor by EELS. The measurements were performed under UHV conditions, i.e. in the low 10^{-11} Torr range. The apparatus was also equipped with a LEED system and a dosing system for the gas inlet.

The EEL spectrometer is schematically shown in Fig. 7.4. The electron beam, produced by a cathode, is brought to the desired degree of monocromaticity by an energy dispersive 127°cylindrical capacitor. The electrons are then accelerated to the primary energy which can be chosen in the range between 1 and 500 eV. The focus on the surface is optimized by the $B_1 - B_2$ lens system. The backscattered electrons are decelerated and focused by the $B_3 - B_4$ lenses onto the entrance slit of another 127°cylindrical capacitor and analyzed with respect to energy. Electrons are finally collected by a channeltron. Spectra were recorded with an energy resolution of 60 cm^{-1} and the data were sampled at 5 cm^{-1} intervals with a total sampling time of 10–20 s/channel. A typical number of counts was 50–300 per channel in the loss peaks. The raw spectra were then smoothed and peak positions determined from mathematical properties of Gaussians.

For experimental reasons the monochromator and not the analyzer is adjusted to obtain the required momentum transfer. We therefore specify our scattering conditions by quoting values of the scattering angle θ_f, $Q_{\|}$, and the incident energy E_i which then determines θ_i according to (7.3). In our EELS

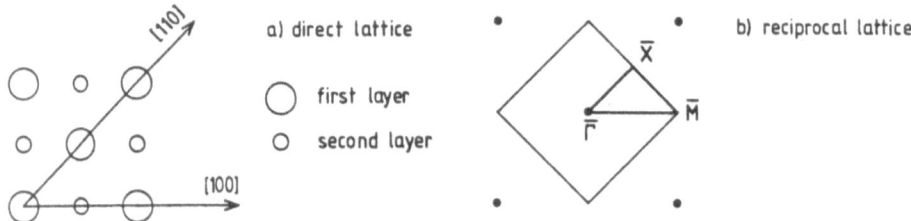

Fig. 7.5. Direct and reciprocal lattice of the (100) surface of Ni. The triangle $\overline{\Gamma}\,\overline{X}\,\overline{M}$ is the irreducible element of the surface Brillouin zone and contains the whole information on the phonon spectrum

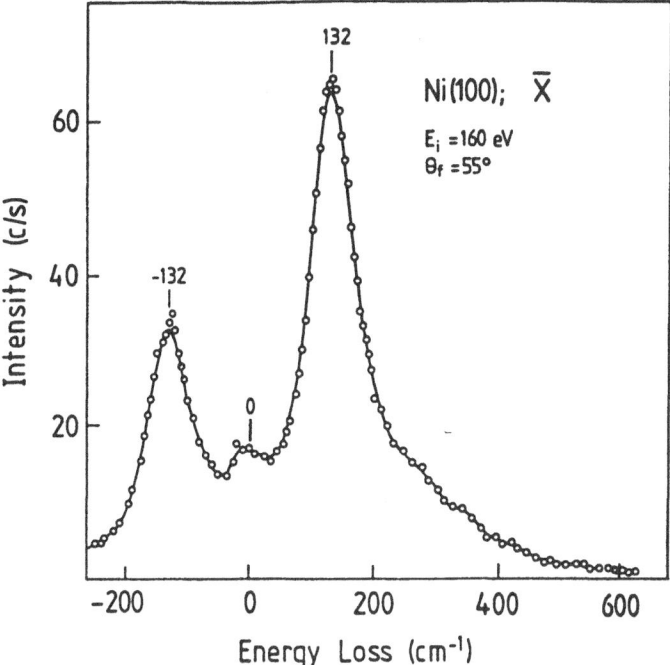

Fig. 7.6. EELS spectrum in $\overline{X}$. The scattering conditions correspond to a maximum of the cross section of S_4

experiments the scattering plane was aligned along a high symmetry direction of the surface as this greatly simplifies the interpretation of the data. In the case of Ni(100) these directions are the $\langle 100 \rangle$ and $\langle 110 \rangle$ directions as shown in Fig. 7.5. They correspond to the directions $\overline{\Gamma}\,\overline{M}$ and $\overline{\Gamma}\,\overline{X}$ of the two-dimensional surface Brillouin zone, respectively.

In Fig. 7.6 we show an unsmoothed sample spectrum recorded at $\overline{X}$ for the clean Ni(100) surface with the crystal at room temperature and for scattering conditions corresponding to a maximum of the cross section of the S_4 mode (Rayleigh mode). The spectrum was sampled at $5\,\mathrm{cm}^{-1}$ intervals with a total sampling time of $15\,\mathrm{s/channel}$. The inelastic peaks correspond to electrons which have annihilated (energy gain) and created (energy loss) one surface phonon in the scattering process, respectively. The peak at $\omega = 0$ is due to diffuse elastic scattering of electrons with defects of the surface such as kinks, steps or randomly distributed adatoms. For fixed scattering parameters its intensity is therefore a measure of the degree of disorder present on the surface. The intensity I of the energy gain peak is proportional to the Bose factor n

$$n = [\exp(\hbar\omega/k_BT) - 1]^{-1} \tag{7.8}$$

while that of the energy loss peak is proportional to $1 + n$ so that

$$\frac{I_{gain}}{I_{loss}} = \frac{n}{1+n} = \exp\left(-\hbar\omega/k_B T\right) \tag{7.9}$$

in accord with the measured ratio.

The broad and unstructured background in Fig. 7.6 extends beyond the highest vibrational frequency of the crystal ($295\,\mathrm{cm}^{-1}$ for Ni) and is caused by multiple-phonon processes. In order to minimize this multi-phonon contribution most of the measurements were performed with the crystal cooled down to 130 K. Working at room temperature, on the other hand, is convenient when one deals with a low-frequency phonon, as in this case the Bose factor n is comparable to or even larger than unity. It may also be convenient to have a significant gain peak in order to determine more precisely the zero on the energy scale of the spectrum in cases where the diffuse elastic intensity is small. Changes in the phonon frequencies, observed between 130 K and room temperature, were generally found to be within the experimental error.

7.2.3 Multiple-Scattering Effects

Electrons are strongly scattered from the ion cores of the crystal so that multiple elastic scattering effects are superimposed on the general features of inelastic scattering described in the preceding paragraphs. Because of multiple scattering the phonon cross section is a strong and sometimes rapidly varying function of the scattering parameters.

The inelastic scattering probability has been evaluated by *Tong* et al. [7.12] with the assumption that the muffin tin ion potentials follow rigidly the instantaneous positions of the atoms $\{R\}$. This assumption is reasonably well satisfied, except in dipole scattering where the potential explicitly depends on the ions being non rigid. Following Tong et al. the differential scattering probability reads

$$\frac{dP}{d\Omega} = \frac{m}{2\pi^2\hbar^2} A \frac{\cos^2\theta_f}{\cos\theta_i} E_f |\langle \{n_{Q_{\|}s}\}_f | f(k_f, k_i; \{R\}) | \{n_{Q_{\|}s}\}_i \rangle|^2, \tag{7.10}$$

where m is the electron mass, $\{n_{Q_{\|}s}\}$ is the occupation number of the mode s at the wavevector $Q_{\|}$, A is the area of the crystal, and $f(k_f, k_i; \{R\})$ is the scattering amplitude. The scattering amplitude is inversely proportional to $\sqrt{A}$ so that the total scattering cross section in (7.10) is independent of the crystal area, as expected. For small displacements u_j of the atoms from the crystal positions $R_j^{(0)}$, the scattering amplitude can be developed to first order in u_j as follows

$$f(k_f, k_i; \{R\}) = f(k_f, k_i; \{R^{(0)}\}) + \sum_\alpha \sum_j \left(\frac{\partial f}{\partial R_{j\alpha}}\right)_0 u_{j\alpha} + \ldots, \tag{7.11}$$

where α denotes the cartesian directions. The displacements $u_{j\alpha}$ can be expressed in terms of phonon annihilation and creation operators a_s and a_s^+ as

$$u_j = \sum_{Q_{\|}s} \frac{e^{(s)}(Q_{\|}, l_z)}{[2M\omega_s(Q_{\|})N\hbar^{-1}]^{1/2}} e^{iQ_{\|}\cdot R_{j\|}}[a_s^+(Q_{\|}) + a_s(Q_{\|})]\qquad(7.12)$$

with N the number of unit cells, M the mass of the substrate atoms and $e^{(s)}(Q_{\|}, l_z)$ the polarization vector of the phonon s at $Q_{\|}$ and in the layer l_z. Inserting (7.12) into (7.11) and then into (7.10) one has

$$\frac{dP}{d\Omega} = \frac{m}{2\pi^2\hbar} A \frac{\cos^2\theta_f}{\cos\theta_i} E_f \frac{1}{2MN} \sum_{l_z, Q_{\|}, s} \sum_\alpha \frac{1}{\omega_s(Q_{\|})}$$

$$\times e_\alpha^{(s)}(Q_{\|}, l_z) \left(\frac{\partial f}{\partial R_{j\alpha}}\right)_0 [n\delta(\omega + \omega_s)\delta(k_{f_{\|}} - k_{i_{\|}} - Q_{\|})$$

$$+ (n + 1)\delta(\omega - \omega_s)\delta(k_{f_{\|}} - k_{i_{\|}} + Q_{\|})].\qquad(7.13)$$

Let us first evaluate the derivative of the scattering potential $(\partial f/\partial R_{j\alpha})_0$ in the kinematic (single-scattering) approximation. As we know, this approximation is too crude for electrons but it can give us some insight into the origin of the selection rules operating in impact scattering without going through a rigorous proof.

In the kinematic approximation the total scattering amplitude f is

$$f = f_0(\Delta k) \sum_j \exp(i\Delta k \cdot R_j),\qquad(7.14)$$

where f_0 is the scattering amplitude at each atom and $\Delta k = k_f - k_i$ is the transfered momentum. Assuming that f_0 does not depend on the displacement of the atoms from their equilibrium positions we obtain for the first derivative of f

$$\left(\frac{\partial f}{\partial R_{i\alpha}}\right)_0 = if_0(\Delta k)\Delta k_\alpha \exp\left(i\Delta k \cdot R_j^{(0)}\right).\qquad(7.15)$$

Equation (7.15) tells us that the inelastic scattering probability is proportional to

$$|\Delta k \cdot e^{(s)}(Q_{\|}, l_z)|^2.\qquad(7.16)$$

For specular scattering $\Delta k_x = \Delta k_y = 0$ so that only vertical vibrations can be excited, while for in-plane scattering only Δk_y is zero so that all modes polarized in the scattering plane xz are allowed for detection. These selection rules have been demonstrated to hold also for the dynamical scattering theory [7.12].

The derivative of the scattering amplitude, with the inclusion of multiple-scattering terms, can be written as [7.12]

$$\left(\frac{\partial f}{\partial R_{j\alpha}}\right)_0 = \left\langle k_f|(G + GT_0G)\left(\frac{\partial V\{R\}}{\partial R_{j\alpha}}\right)_0(1 + GT_0)|k_i\right\rangle\qquad(7.17)$$

with G the propagator of the electron inside the crystal, and T_0 the transfer matrix. The subscript o indicates that the quantity has to be computed with the crystal ions in their equilibrium positions. The different factors in (7.17) describe from right to left: the multiple scattering of the electron as it enters the crystal, the inelastic scattering event, and the transport of the electron back to the vacuum.

For $V\{R\}$ *Tong* et al. assumed a muffin tin potential which was further assumed to move rigidly with the atom positions. This means that no charge transfer inside the unit cell is permitted. The theory therefore fails to reproduce the strong increase of the scattering cross section of dipole active modes for vanishing momentum transfers. On the other hand, it is very difficult with methods presently available to go beyond this approximation. $[\partial V(\{R\})/\partial R_{j\alpha}]_0$ differs for the different directions and particularly for displacements parallel and perpendicular to the surface. Moreover, it will in general be different for surface atoms and for atoms located in deeper crystal layers. As the polarization vector of each mode (contained in u_j) has different amplitudes in different directions and in different layers, multiple scattering will affect the behaviour of the cross section of each mode in a different way as a function of the scattering parameters.

The result of the theory of *Tong* et al. has recently been verified for the cross section of the S_4 and of the S_6 modes at the $\overline{X}$ point of the Brillouin zone of the clean Ni(100) surface [7.16–18]. As it is shown in Fig. 7.7, these modes correspond to vertical and to longitudinal displacement of surface layer atoms, respectively. According to kinematical arguments the cross section of S_6 should be about 25 times smaller than the one of S_4, this factor being independent of the impact energy. The dynamical scattering theory, however, predicted $S_4(\overline{X})$ and $S_6(\overline{X})$ to display strongly oscillating intensities with occasionally both modes having a comparable cross section (at $\theta_f = 65°$ and $E_i = 155\,\text{eV}$, e.g.).

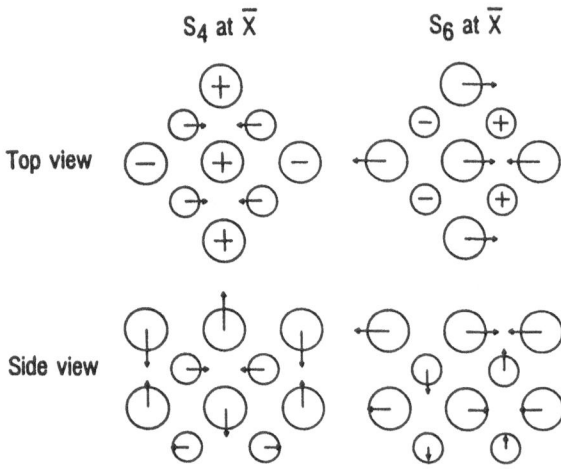

Fig. 7.7. Displacement pattern caused by the phonons S_4 and S_6 at $\overline{X}$

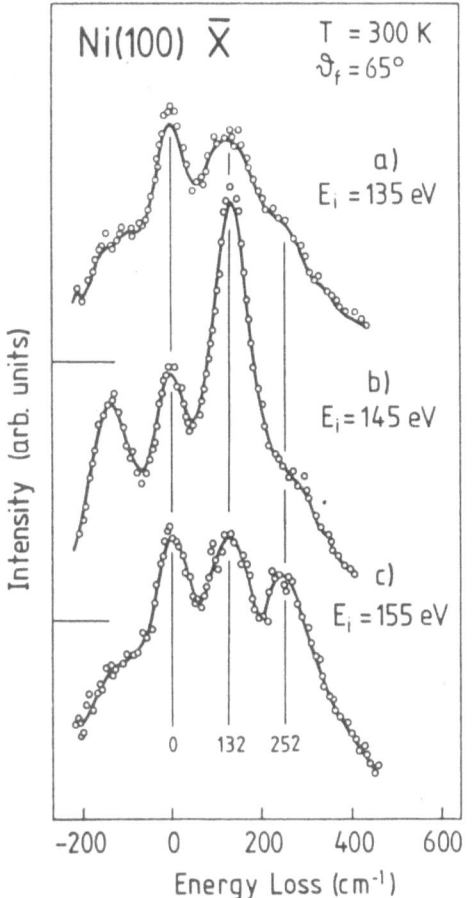

Fig. 7.8. Energy-loss spectra for Ni(100) $\overline{X}$ for different energies of the primary electrons at $\theta_f = 65°$

Inside the figure:

Ni(100) $\overline{X}$ T = 300 K $\vartheta_f = 65°$

a) $E_i = 135$ eV

b) $E_i = 145$ eV

c) $E_i = 155$ eV

0   132   252

Intensity (arb. units)

Energy Loss (cm^{-1})

-200   0   200   400   600

Experimental spectra taken for this geometry with the energy E_i varied in 10 eV steps are shown in Fig. 7.8. In accord with the theoretical prediction we have little inelastic intensity at $E_i = 135$ eV. This energy corresponds to negative interference conditions for both surface modes. On the other hand spectrum (b) recorded at $E_i = 145$ eV corresponds to a relative maximum of the S_4 cross section with still no intensity for S_6. Finally in spectrum (c) at 155 eV two energy loss peaks of comparable intensity are clearly present. Their frequencies lie below the bulk band and in a gap, respectively, so that the peaks can be unambiguously attributed to the S_4 and the S_6 phonons.

The comparison of the measured intensities to the result of the calculations over a wide energy range is shown in Fig. 7.9. As inelastically scattered electrons are distributed over the whole solid angle the acceptance angle α_i of the spectrometer directly affects the experimentally observed intensities. The dependence of α_i on the primary beam energy can be estimated from Abbe's sine law [7.10] which describes the conservation of phase space in the imaging process of the lens system and reads

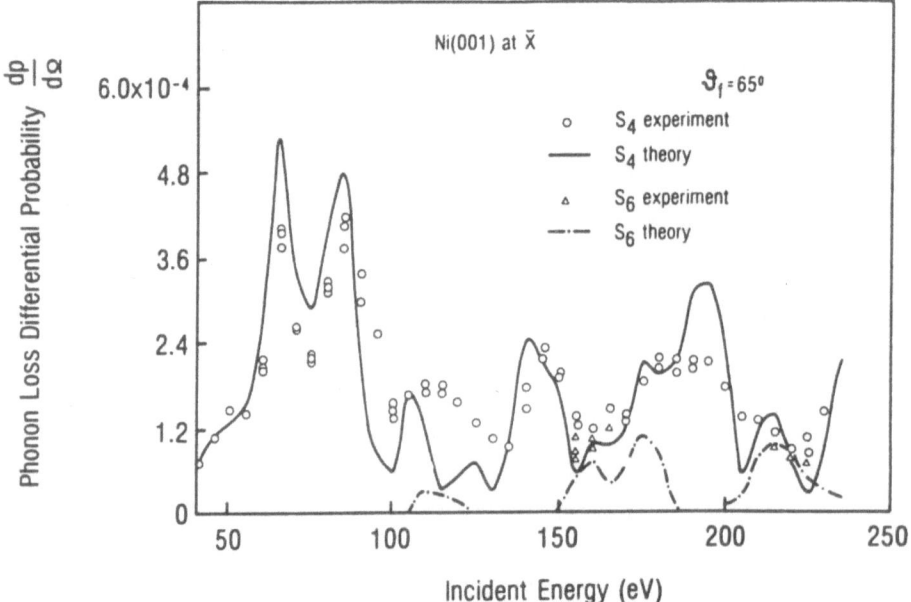

Fig. 7.9. Inelastic scattering intensity as a function of the primary energy. The full and dashed lines are the calculated cross sections for $S_4(\overline{X})$ and $S_6(\overline{X})$, respectively. In the calculation the outer interlayer spacing of the Ni crystal has been assumed reduced by 2.5% in order to match the measured S_4/S_6 intensity ratio

$$M \cdot E_i \alpha_i = \text{constant} \tag{7.18}$$

where M is the magnification factor of the lens system. M can be determined independently from the size of the illuminated area on the sample and was found to be about $M \sim 1$ and nearly constant in the range between 50 and 220 eV. Due to the symmetrical design of the spectrometer the same is assumed to be valid for the area viewed by the analyzer. The experimental intensities reported in Fig. 7.9 were therefore multiplied by E_i, in order to correct for α_i being proportional to E_i^{-1}.

Absolute intensities can be reproduced in EELS only within a factor of 2. Data recorded in different sets of measurement on different days therefore agree only in the displayed structure, not in the absolute values. Except for a necessary overall scaling, the features in the intensity vs. energy curves were reproducible within 10%, as can be estimated from the scatter of the data points. Other sources of error were the positioning of the scattering angle, which can be reproduced only within $\pm 0.5°$, and the determination of the primary energy, for which $\Delta E \sim \pm 0.5$ eV. Moreover for E_i in the range from 180 eV to 210 eV a surface resonance contributed to the measured intensity of S_4 [7.16,17]. With these limitations understood the agreement between theory and experiment can be regarded as fairly good, since the major features of the

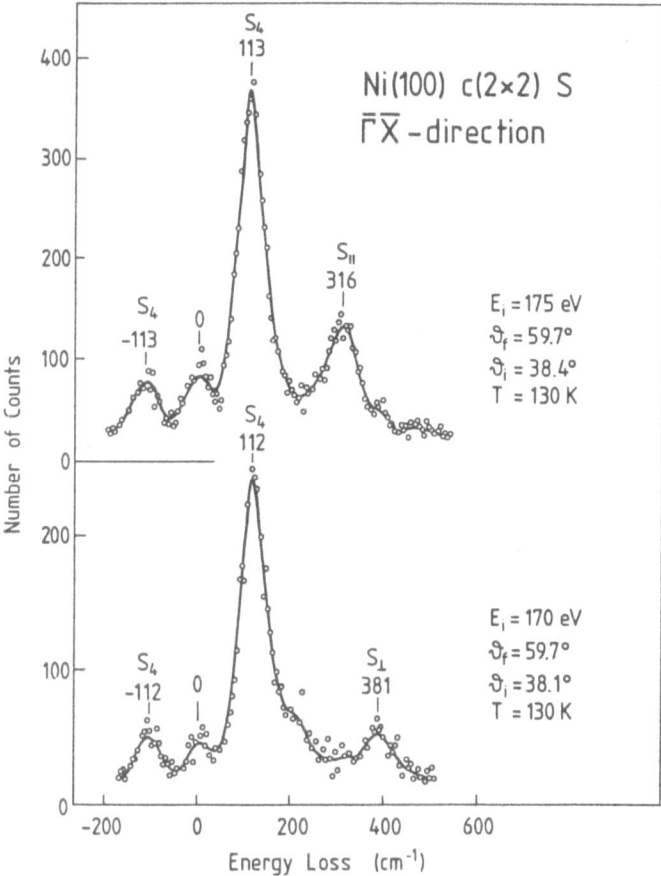

Fig. 7.10. Energy loss spectra taken for Ni(100) c(2×2)S along $\bar{\Gamma}\bar{X}$ for $Q_{\parallel} = 1.64\,\text{Å}^{-1}$ at impact energies differing by only 5 eV. For $E_i = 175\,\text{eV}$ the spectrum displays a loss peak due to the parallel branch of the sulfur modes, while at $E_i = 170\,\text{eV}$ the loss peak corresponds to the vertical branch

intensity vs. energy curve are reproduced by the calculation. In the calculation a 2.5% contraction of the first interlayer spacing of the crystal was assumed. With this contraction also the measured ratio of the intensity of the energy loss peaks corresponding to S_4 and to S_6 could best be reproduced [7.16–18].

The reported intensity-energy curve tells us how multiple scattering effects may be used in order to improve the potential of EELS. The largest cross section for each mode can be found by a systematic search for the best scattering conditions. The search for optimum intensities should be performed on a relatively fine grid since even 5 eV changes in the energy of the primary electrons can give rise to substantial variations in intensity. Once a high cross section for a particular mode is achieved the angle of incidence Θ_i can be varied to probe other values of $Q_{\parallel}$. Particular values of E_i and θ_f can be found for which the cross section for a particular dispersion branch is nearly constant over a good part of the Brillouin zone so that the phonon dispersion can be measured.

We would like to point out that by utilizing the multiple scattering effects modes can be distinguished from each other even when differing in frequency less than the energy resolution. This can be achieved if scattering conditions exist where one has a significant cross section for only one mode. A particularly dramatic example of such a situation is shown in Fig. 7.10 for electron scattering off the $c(2\times2)S$ overlayer on Ni(100) [7.19]. The spectra were taken in the same position of the analyzer for impact energies differing by only 5 eV. As one can see the loss peak due to the Rayleigh wave at 112 cm^{-1} is unaffected by the small change in E_i while the loss peak at 316 cm^{-1} disappears and another loss at 381 cm^{-1} appears. These peaks correspond, as we will see later, to the (essentially) parallel and perpendicularly polarized branches of the adsorbate modes, respectively.

7.3 Phonon Dispersion Curves

7.3.1 The Clean Ni Surface

Phonon dispersion has been measured for the clean Ni(100) surface along the $\overline{\Gamma}\,\overline{X}$ and the $\overline{\Gamma}\,\overline{M}$ directions of the surface Brillouin zone (Fig. 7.1). The energy-loss spectra show a variety of structures which have been assigned to different surface modes as well as to surface resonances inside the bulk band [7.11,20,21]. In all measurements the scattering plane was aligned with the sagittal plane, i.e. the plane spanned by the surface normal and the direction of $Q_{||}$. Since for $\overline{\Gamma}\,\overline{X}$ and $\overline{\Gamma}\,\overline{M}$ the sagittal plane coincides with a mirror plane of the crystal, all normal modes have to be either odd or even with respect to this symmetry operation and the motion of the nickel atoms in the sagittal plane and perpendicular to it are decoupled. The selection rules allow further only even modes to be observed for in-plane scattering. Such surface modes are S_4 and S_6 along $\overline{\Gamma}\,\overline{X}$ and S_1 along $\overline{\Gamma}\,\overline{M}$. In accord with the selection rule there is no evidence for the odd surface mode S_1 in the data on $\overline{\Gamma}\,\overline{X}$ [7.11,21].

The observed energy losses as a function of momentum are shown in Fig. 7.11 for (a) $\overline{\Gamma}\,\overline{M}$ and (b) $\overline{\Gamma}\,\overline{X}$. The projection of the bulk phonons on the surface Brillouin zone (bulk band) is shaded in the figure. Data points which are positioned below the bulk band or in the gap near $\overline{X}$ are due to the interaction of the electrons with pure surface modes. In the figures genuine surface modes are marked as filled symbols. The S_4 mode along the $\overline{\Gamma}\,\overline{X}$ direction is partly imbedded into the bulk band. It retains, however, its character as a pure surface mode since the overlapping bulk states are decoupled from it being odd with respect to the mirror plane. Resonances inside the bulk band are denoted by open symbols. Shoulders or not well resolved peaks are marked as broken symbols. Data points differing in frequency less than experimental energy resolution (60 cm^{-1}) were measured under different scattering conditions.

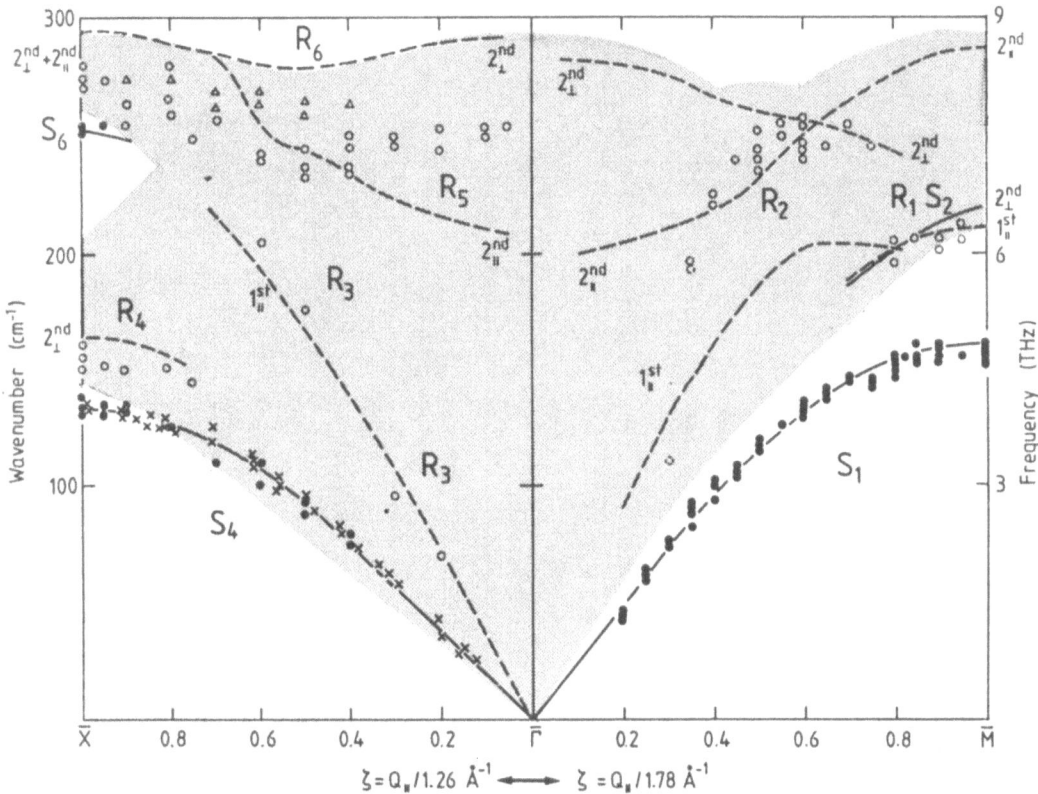

Fig. 7.11. Collection of the data points for Ni(100) reported as a function of the reduced wave vector ς: a) $\overline{\Gamma}\,\overline{X}$, b) $\overline{\Gamma}\,\overline{M}$. Full symbols mark losses caused by surface phonons, while open symbols refer to resonances inside the bulk band. Broken symbols indicate poorly resolved peaks or shoulders. Crosses indicate data taken from [7.11]. Data points for the R_6 resonance are evidenced by triangles. Full and broken lines result from a lattice dynamical calculation for surface modes and surface resonances, respectively. For the resonances we indicate also the layer in which the spectral density reaches its maximum

The energy-loss spectra were taken at regular intervals in reciprocal space. The phonon wavevector in the figure is normalized with respect to the zone boundary and denoted by ς. Here $\varsigma = 1$ corresponds to $1.78\,\text{Å}^{-1}$ at $\overline{M}$ and to $1.26\,\text{Å}^{-1}$ at $\overline{X}$.

Sample spectra with the S_4 and S_6 modes have already been shown in the preceding Sect. 7.2. Similar spectra for the S_1 mode on $\overline{\Gamma}\,\overline{M}$ are presented in Fig. 7.12 for different $Q_\parallel$ vectors [7.19]. The frequency of the Rayleigh waves at $\overline{X}$ and $\overline{M}$ are $132\pm2\,\text{cm}^{-1}$ and $155\pm5\,\text{cm}^{-1}$, respectively, and the frequency of S_6 at $\overline{X}$ is $252\pm5\,\text{cm}^{-1}$ [7.20,21].

The vertical motion of the atoms in the first and the second crystal layers at $\overline{M}$ are decoupled [7.4]. This gives rise to a new vertically polarized phonon called S_2 which is principally localized in the second layer of the crystal. We

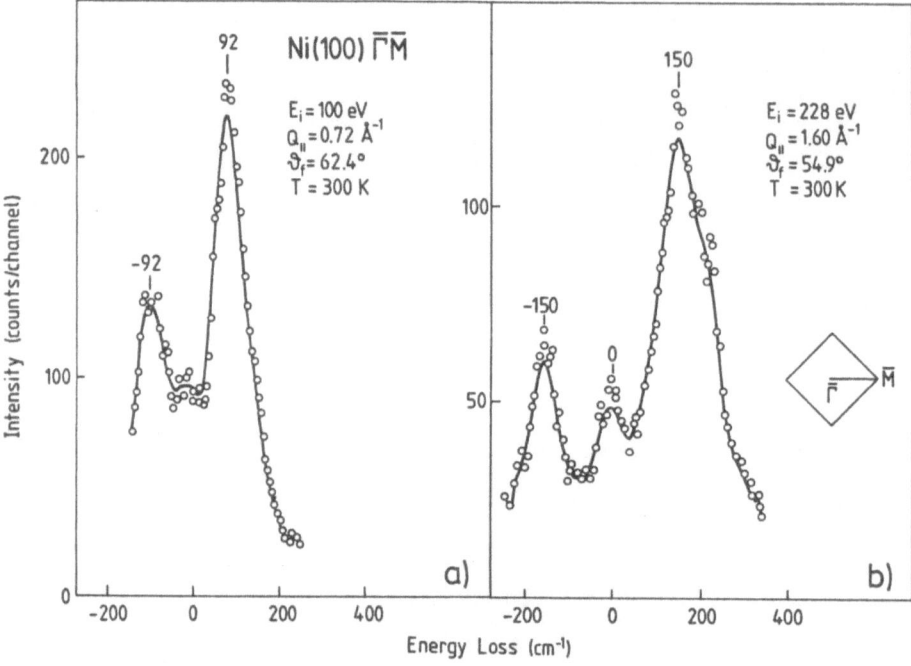

Fig. 7.12. Typical energy-loss spectra for scattering conditions corresponding to a high cross section for the S_1 mode on $\bar{\Gamma}\bar{M}$. The shoulder, which is clearly displayed in spectrum (b) is probably due to the creation of the S_2 phonon localized in the second crystal layer

assume that the shoulder in the energy-loss peak in spectrum (b) of Fig. 7.12, is due to the excitation of this phonon as the spectrum was recorded under the scattering conditions ($E_i = 200\,eV$, $\theta_f = 55°$, $\theta_i = 35°$) where, according to (7.7) the information on the second layer is reduced by only 50%. Unfortunately we were not able to find scattering conditions for which S_2 has a larger cross section than S_1. The data points appearing in this frequency range in Fig. 7.11 were recorded at very glancing angles of incidence, so that it seems unlikely to have excited a mode located in the second layer of the crystal. We will come back to this point in the following.

Other features in the phonon spectrum lie inside the bands of bulk phonons and are due to high spectral densities and/or to surface resonances. These features can also display a high intensity under particular scattering conditions because of multiple-scattering effects. An example of such a peak is shown in Fig. 7.13. The energy-loss spectrum has been recorded at half the way between the $\bar{\Gamma}$ and $\bar{M}$ points. The lower frequency peak is due to the creation of a Rayleigh phonon (S_1) while the higher frequency peak is due to the R_2 resonance. As the spectrum was recorded with the crystal cooled down to $T = 130\,K$ one can rule out the possibility that this high-frequency loss can be due to multiple-phonon processes. Because of the form of the dispersion (Fig. 7.11) we exclude that the loss can be caused by a surface magnon. Resonances inside the

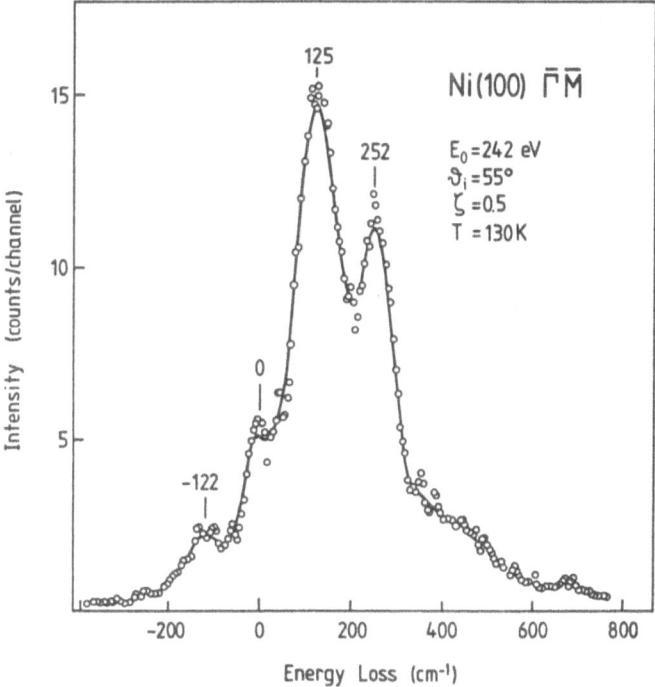

Fig. 7.13. Energy loss spectrum for $\varsigma = 0.5$. The peaks are due to annihilation of S_1, diffuse elastic scattering, creation of S_1 and to a surface resonance in the bulk phonon band. The maximal frequency of the bulk band for this $Q_\parallel$ is $\sim 280\,\mathrm{cm}^{-1}$. The background at higher frequencies is therefore due to multi-phonon excitation

volume band have been observed also by inelastic He scattering on the (111) surfaces of Ag, Au and Cu [7.22,23] and their presence seems to be more the rule than an exception. Most of the high-frequency losses on $\overline{\Gamma}\,\overline{M}$ as well as on $\overline{\Gamma}\,\overline{X}$ were best observed at relatively high energies and steep angles of incidence. As there the penetration depth of the electrons in the crystal is largest we have to consider the possibility that these resonances are mainly localized in deeper crystal layers.

The data in Fig. 7.11 are compared to the results of a lattice dynamical calculation (full lines for surface modes and broken lines for surface resonances). The calculations were performed for a nearest-neighbor force-constant model of the crystal using the Fourier transformed Green's function method [7.24]. In this model the value of the force constant in the bulk is obtained from a one parameter fit to the phonon-dispersion curves, as measured by neutron scattering. In order to reproduce the frequency of the surface branches one has to assume a stiffening of the force constant k_{12} between the atoms of first and second crystal layer by 20% with respect to the bulk value [7.11,20]. Such a stiffening of the force constants indicates that the Ni(100) surface is contracted. This hypothesis is confirmed by the analysis of the ratio of the S_4 to the S_6 cross section near to $\overline{X}$ which indicates that the outer interlayer distance d_{12} is contracted by 2.5±0.8% with respect to the bulk value [7.16–18] and from the results of Rutherford backscattering experiments of *Frenken* et al. [7.25,26] who also find a 3.2±0.5% contraction of d_{12}. No readjustment of the force constant in the surface plane seems necessary for a good fit to the experiment. The calculations also reproduce most of the observed resonances [7.21]. According to these calculations the lower-frequency loss R_3 is due to the threshold of the longitudinal bulk phonons (broken line indicated by $1^{st}_{\parallel}$ in Fig. 7.11), whereas the features R_2, R_4, R_5 and R_6 are due to surface resonances localized principally in the second layer of the crystal ($2^{nd}_{\parallel}$ and $2^{nd}_{\perp}$). In case of R_1 an assignment is possible both to a first layer longitudinal resonance as well as to the second layer mode S_2. This last possibility is however unlikely because most data points are measured for glancing angles of incidence where the information from deeper crystal layers is strongly reduced due to the vanishing penetration of the primary electrons into the crystal. The frequency and the dispersion of these structures can be fairly well reproduced in the lattice dynamical calculations with the 20% stiffening of k_{12} without the need of a more complex force field [7.21].

7.3.2 (2×2) Adsorbate Phases on Ni(100)

The surface phonon spectrum is strongly modified by the presence of an adsorbed phase. We will show this with the data on hand for oxygen [7.27–29] and sulfur [7.19] adsorption on Ni(100). These two elements belong to the same chemical group, that of the chalcogenides, but differ in their mass, their covalent radius, and their binding energy to nickel, with mass and radius larger for sulfur and the binding energy larger for oxygen. On Ni(100) oxygen as well as sulfur form p(2×2)[1] and c(2×2) structures at a coverage of $\theta = 0.25$ and $\theta = 0.50$, respectively. Direct and reciprocal lattice for these structures are shown in Fig. 7.14.

Due to the presence of the overlayer the character of the surface modes of the bare surface may be modified and new modes may come into existence. In case of c(2×2) structure the Rayleigh wave persists as a principally shear vertically polarized phonon, while the S_6 mode of the clean surface now corresponds to a breathing motion of the substrate atoms around the adatoms. From symmetry considerations we will refer to this mode as A_1 mode.

Sulfur and oxygen overlayers have been studied by different techniques, among them LEED [7.,30], EXAFS [7.31–33], photoemission spectroscopy [7.34 –36], thermal helium beam scattering [7.37], SEELFS [7.38] and EELS in the dipole scattering regime [7.39–41]. According to these results the adatoms sit in the fourfold hollow sites. The vertical distance $R_{\perp}$ of the adsorbate from the Ni surface was determined to be around 1.35 Å for sulfur and 0.9 Å for oxygen

---

[1] Recent endeavoures to produce a p(2×2) sulfur overlayer in our laboratory indicate that the p(2×2) LEED pattern remains diffuse on very clean and well ordered nickel surfaces.

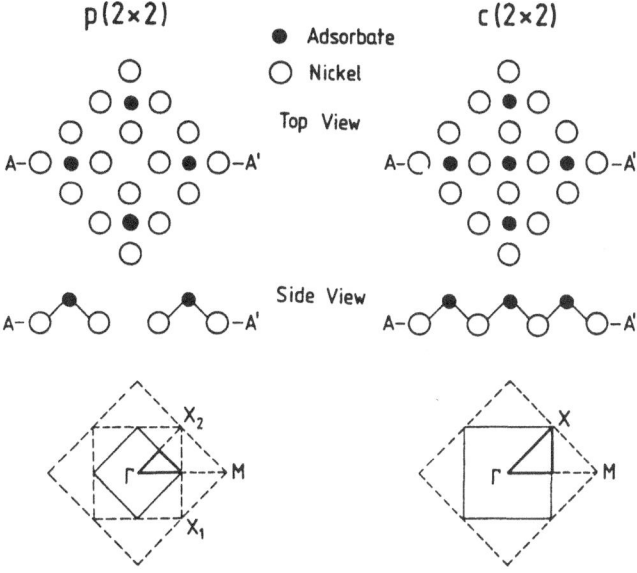

Fig. 7.14a,b. Superstructures observed after ordered S and O adsorption on Ni(100): (a) direct lattice, (b) reciprocal lattice

for both ordered structures. The values obtained for $R_\perp$ by EXAFS, which is believed to date to be the most accurate technique, is 1.37 ± 0.03 Å for sulfur and of 0.86 ± 0.07 Å for oxygen [7.32,33].

Although the adsorption geometry remains the same, the frequency of the vertical oxygen-Ni(100) vibration drops from $430\,\mathrm{cm}^{-1}$, as measured for the p(2×2) structure to $310\,\mathrm{cm}^{-1}$ for the c(2×2) structure, indicating a strong modification of the adsorbate-surface bonding. In case of sulfur, on the other hand, no major shift of the vertical vibrational frequency takes place with coverage.

Many different models have been proposed in the past to explain the different behaviour of the two adsorbates. *Upton* and *Goddard* proposed some years ago that the two phases of oxygen correspond to an oxydic and to a radical state of oxygen [7.42]. In this case, however, the adsorption distance should be of 0.28 Å in case of the c(2×2) structure, in disagreement with the experimental result. More recently *Bauschlicher* and *Bagus* suggested using cluster calculations that the difference in vibrational frequency between the two oxygen structures is due to the lateral coupling between the adsorbate atoms [7.43] and that the changes in the bonding are relatively small. On the other hand, lattice dynamical calculations [7.47] show that such lateral interactions are incapable of explaining this frequency shift at the $\overline{\Gamma}$ point, for the two types of overlayers.

The observed phonon dispersion for the c(2×2) phases of oxygen [7.27] and sulfur [7.19] are shown in Fig. 7.15. In the picture the band of bulk states is

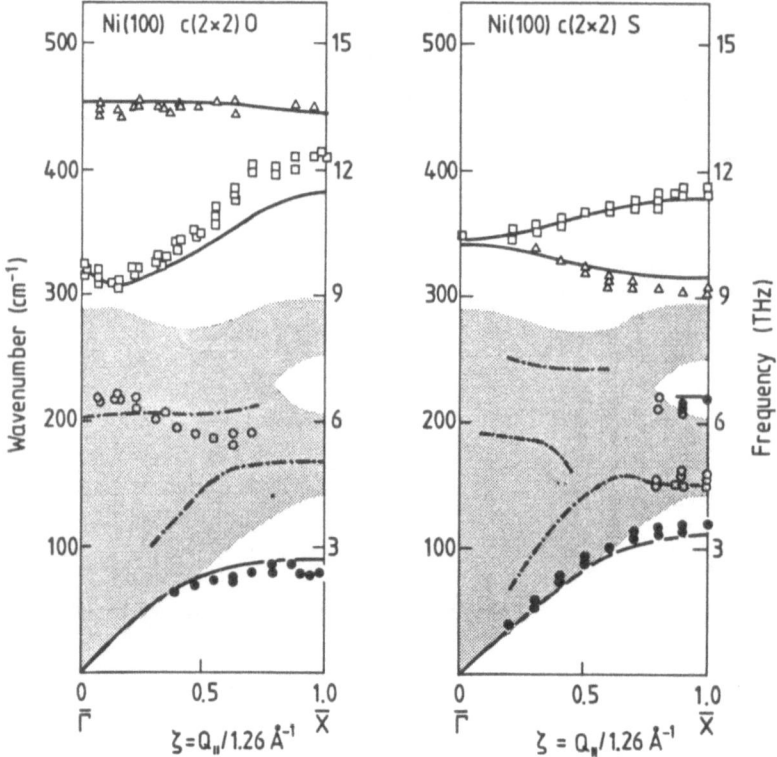

Fig. 7.15. Collection of the data points for the c(2×2) O and S superstructures on Ni(100). Squares and triangles refer to the (principally) vertically and longitudinally polarized branches of the adsorbate modes, while crosses denote the Rayleigh wave and open circles the S_6 mode and surface resonances. Full and broken lines result from a lattice dynamical calculation for surface modes and surface resonances, respectively

again indicated by the shadowed region. Above the bulk bands we have surface modes of the adsorbate. Below the bulk band and inside its gap, surface modes associated with the substrate show up. These are the Rayleigh wave and, in the gap near $\overline{X}$, the A_1 mode. Moreover, we also observed some intense resonances inside the bulk band. Examples of spectra corresponding to scattering conditions favourable to the observation of the Rayleigh wave for the c(2×2)S overlayer are shown in Fig. 7.16. The spectra were taken at room temperature and the peaks correspond as usual to phonon annihilation and creation. In some spectra also the adsorbate modes can be seen. Other measurements taken for this system have already been reproduced in Fig. 7.10 and have been discussed in Sect. 7.2.

The observed phonon dispersion curves could be reproduced by lattice dynamical calculations [7.19,47], based on a nearest-neighbor central force model. The parameters in the calculation are the vertical adsorbate substrate distance $R_\perp$, the second derivative of the pair potentials ϕ_{10}'' between the adatom and

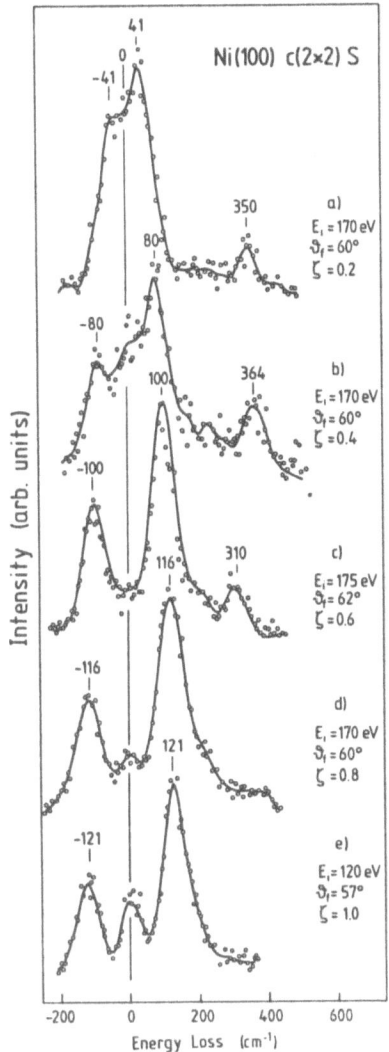

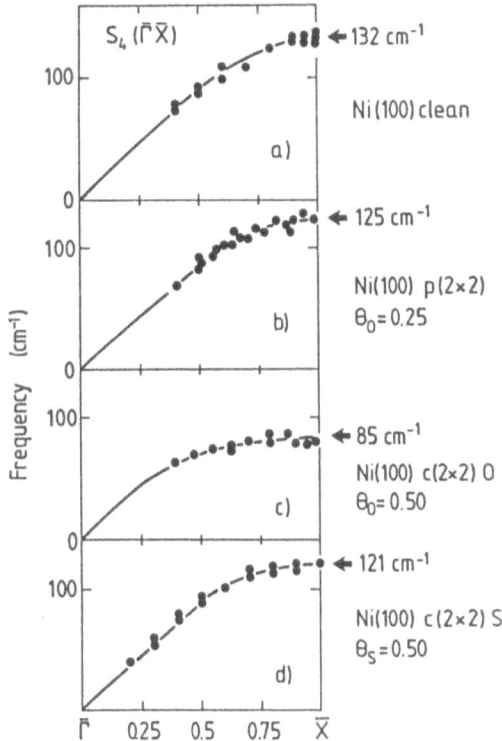

Fig. 7.17a–d. Dispersion curves for the S_4 phonon for the clean surface (**a**) and for the (2×2) structures of O (**b** and **c**) and S (**d**). The data points for the p(2×2) structure of oxygen were taken from [7.28]

◄ **Fig. 7.16.** Set of measurements for the Rayleigh wave on the c(2×2)S structure. In some of the spectra also the vertical branch ($\varsigma = 0.2$ and $\varsigma = 0.4$) or the parallel branch ($\varsigma = 0.6$) of the surface modes show up

the substrate atoms, and ϕ''_{12} between the atoms of first and second substrate layer. In the case of oxygen a similar term ϕ''_{00} representing the direct lateral interaction between the oxygen atoms has also been included. As in earlier calculations, the coupling constant between the atoms in the bulk crystal has been obtained from a one parameter fit to the measured phonon dispersion curves. In Fig. 7.15 the results of the model calculation are shown by full lines for the surface modes while dashed lines denote the position of intense surface resonances inside the volume band. In the case of oxygen a good fit to the phonon dispersion can be reached by placing the adatoms at $R_\perp \sim 0.9\,\text{Å}$, i.e. the value determined by EXAFS. To reproduce the frequency of $S_4\,(\overline{X})$ however one had to reduce the force constant between first and second layer atoms to 30% of its value in the bulk. While the reduction of the force constant is in qualitative agreement with the observed surface relaxation of ~5% [7.26] there is some doubt regarding the magnitude of the effect. We will come back to this point in the next subsection.

In the case of sulfur the vertical distance of the sulfur atom was enlarged to 1.45 Å in order to obtain a fit between the simple force constant model and the experimental data. However, more recent lattice dynamical calculations with non-central forces [7.44] seem to provide a fit to the dispersion curves even for $R_\perp = 1.35$ Å, which is the vertical distance according to EXAFS.

7.3.3 Effect of Random Adsorption of Oxygen on the Rayleigh Wave

A particular interesting aspect of the data on the $c(2\times2)$ oxygen overlayer is the shift of the S_4 phonon frequency. In Fig. 7.17 we compare the S_4 dispersion curve along $\overline{\Gamma}\,\overline{X}$ for the clean Ni(100) surface [7.11] and for the $p(2\times2)O$ [7.28], $c(2\times2)O$ [7.27] and $c(2\times2)S$ [7.19] structures. As one can see, $\omega_{S_4}(\overline{X})$ shifts from the clean surface value of 132 to 120 cm^{-1} for $c(2\times2)S$, and to 125 cm^{-1} for $p(2\times2)O$ and to only 80 cm^{-1} for $c(2\times2)O$.

In case of the sulfur overlayer the magnitude of the shift compares with what is expected from a mass loading effect. This effect stems from the fact that the adatom, due to their small mass, follow rigidly the motion of the surface atoms causing practically an increase of their effective mass.

In case of $p(2\times2)O$ the shift in the frequency of S_4 is too large to be accounted for by mass loading, and even more so for $c(2\times2)O$. A modification of the force constant coupling the first and the second Ni layer was therefore assumed in the lattice-dynamical calculations in order to reproduce the data. The physical reasoning was that the establishing the oxygen-nickel bonds reduces the number of electrons available for the backbonding of the nickel surface atoms and that the reduced force constant reflects the weakening of the bonding. The different behaviour of oxygen and sulfur could then be tentatively explained as being due to the weaker binding of sulfur to Ni(100).

However, to our own surprise we discovered that randomly adsorbed oxyen atoms produce only a small shift of the frequency of S_4, much smaller than when the adatoms are allowed to order on the surface [7.29]. Some experimental spectra showing this effect are reproduced in Fig. 7.18. The disordered overlayer was obtained by adsorbing oxygen at 130 K which inhibits the diffusion of the adsorbed atoms. Even at coverages as high as 50% no LEED superstructure could be observed.

The collection of the measured frequencies as a function of coverage is given in Fig. 7.19. The crosses denote the frequency of S_4 at $\overline{X}$ in the ordered surface structures while circles and triangles denote different sets of measurements obtained for random oxygen adsorption. The small frequency shift for random adsorption is by and large consistent with what is expected from a mass-loading effect. An estimate of the effect of a random distribution of atoms on the phonon frequencies has been given in a theoretical work by *Benedek* et al.

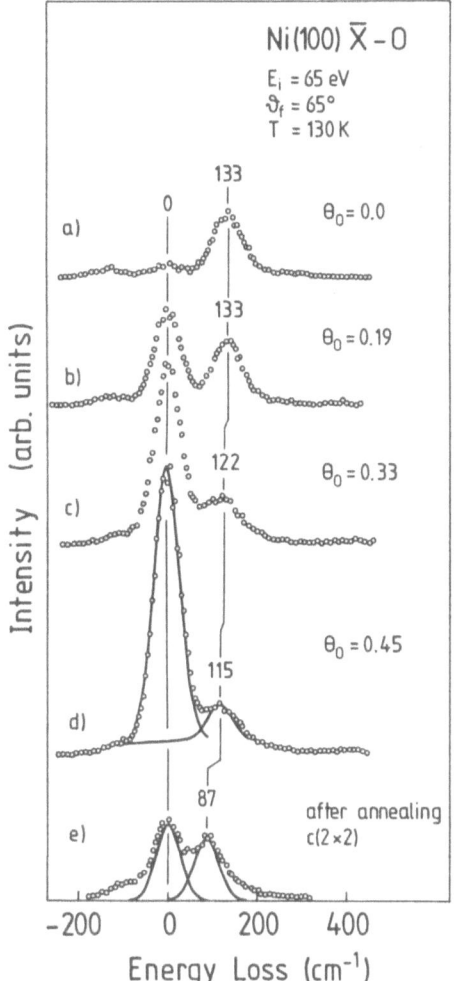

Ni(100) $\overline{\text{X}}$ – O

$E_i = 65\ eV$
$\vartheta_f = 65°$
$T = 130\ K$

a)   0   133 $\theta_0 = 0.0$

b)   133 $\theta_0 = 0.19$

c)   122 $\theta_0 = 0.33$

         $\theta_0 = 0.45$

d)   115

e)   87   after annealing
         c(2×2)

Intensity (arb. units)

-200   0   200   400

Energy Loss (cm^{-1})

Fig. 7.18a–e. Effect of random oxygen adsorption on the frequency of $S_4(\overline{X})$. Spectrum (a) was recorded before exposing the crystal to oxygen. Spectra (b, c and d) correspond to increasing oxygen coverages. Afterwards we annealed the crystal for 5′ at 420 K in order to allow the adlayer to order. The frequency of S_4 falls then to 85 cm^{-1} (spectrum e)

[7.45]. The presence of adatoms causes a renormalization of the dynamical matrix. If the adsorbate mass M_A is much smaller than the substrate mass M_S and the adsorbate induces no changes in the coupling between substrate atoms then the eigenfrequencies are modified as a function of coverage θ

$$\omega^2 = \omega_0^2 \left(1 - \frac{M_A}{M_S} \theta \right), \tag{7.19}$$

where ω_0 and ω are the phonon frequencies on the clean and adsorbate covered surface, respectively. Equation (7.19) is strictly valid only for a phonon whose frequency is given by $\omega = \sqrt{\phi''/M}$, ϕ'' being the second derivative of the pair potential, and M the mass of the surface ions. This is not the case for ω_{S_4} at $\overline{X}$ so we do not expect a perfect agreement of (7.19) with experiment. The result is reported in Fig. 7.19 by a dashed line and shows that only a minor

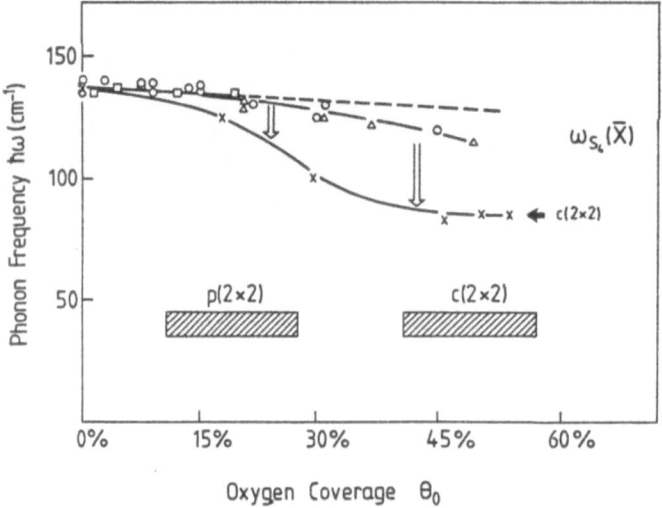

Fig. 7.19. Collection of experimental data for $S_4(\overline{X})$ for ordered (crosses) and disordered (triangles and squares) oxygen adsorption on Ni(100). Typical errors on the data points are of $\pm 5\,cm^{-1}$ for the frequency and $\pm 1\%$ for the coverage. The coverage ranges for which LEED structures were observed for ordered overlayers are marked by hatched areas

adjustment of the force constants is now required to describe the experiment. As it is difficult to see how long-range order should affect nearest-neighbour force constants one is inclined to search for alternative explanations as to why the frequency of the S_4 mode at $\overline{X}$ is reduced with the ordered $c(2\times 2)$ oxygen overlayer as much as it is.

7.4 Internal Strains, Phonon Anomalies and Reconstruction

7.4.1 A Lattice-Dynamical Model with Internal Strains

In this subsection we wish to present a lattice-dynamical model which goes beyond the simple nearest-neighbor model and is, on the other hand, simple enough to not require too many unknown parameters. The central postulate of the model is that of an attractive interaction between the oxygen atom and the second layer nickel atom. Such an additional force is suggested by total-energy calculations [7.46] and also by the experimental observation that oxygen has a tendency to slip into the crystal on surfaces containing defect sites. It therefore appears that on an ordered surface the oxygen is sterically hindered from moving into the surface by the geometric arrangement of the surface atoms. Our lattice-dynamical model includes therefore central force interactions between the oxygen atoms and the second nearest-neighbor nickel atoms underneath the fourfold hollow site. We further assume that this force is attractive, i.e., that the pair potential $\phi_{02}(r_0 - r_2)$ between the oxygen atom placed at r_0 and the nickel atom placed at r_2 is not at its minimum, i.e. $\phi'_{02} \neq 0$. Equilibrium conditions then require ϕ'_{01} (the first derivative of the pair potentials between the oxygen atom and the first layer nickel atoms) as well as ϕ'_{12} (the first derivative of the pair potential between the first-layer nickel atom and the nickel atom underneath the oxygen atom) also be nonvanishing at the equilibrium positions. The equations for equilibrium then read

$$\phi'_{02} + 4\phi'_{01}\hat{n}_z(10) = 0, \tag{7.20}$$

$$2\phi'_{01}\hat{n}_z(10) + 2\phi'_{12}\hat{n}_z(12) = 0, \tag{7.21}$$

$$\phi'_{02} + 4\phi'_{12}\hat{n}_z(21) = 0 \tag{7.22}$$

with $\hat{n}_z(ij)$ the z-component of a unit vector directed along the line connecting atoms placed at positions r_j and r_i

$$\hat{n}(ij) = \frac{r_j - r_i}{\langle r_j - r_i \rangle}. \tag{7.23}$$

The three equilibrium conditions are met by having

$$\phi'_{01} = -\frac{1}{4\hat{n}_z(10)}\phi'_{02} \quad \text{and} \tag{7.24}$$

$$\phi'_{12} = -\frac{1}{4\hat{n}_z(21)}\phi'_{02}. \tag{7.25}$$

No further ϕ' are needed to compensate the forces on the nickel atoms. Thus our choice of forces is rather convenient, although not unique. We note that the model assumes that the first derivative of the pair potential between the first layer nickel atoms and those second layer nickel atoms which do not have an oxygen atom above in the fourfold hollow site to be zero. It would therefore be consistent with the model to have the second derivative, ϕ''_{12}, different, depending on whether the bond connects first and second layer nickel atoms with, or without an oxygen atom above. In the interest of having a minimum set of unknow parameters we take these two ϕ''_{12} as being equal. We also assume that all nickel atoms remain at their ideal crystal positions, although a buckling of the second layer would be consistent with the assumption that oxygen induces internal stresses. Allowing for a buckled second layer, however, would again require non zero ϕ' all the way into the bulk, which makes the model unnecessarily clumsy. We have also carried out calculations where ϕ'_{12} is not zero for those nickel atoms which do not have an oxygen atom above, rather it can assume a range of values. In this model the second layer does get buckled and the effect decays all the way into the bulk. However, we find that such additional effects do not bring about a significant change in the calculated dispersion curves.

We have evaluated the lattice dynamics of the model described above by noting that for non-zero first derivatives the effective force constants are

$$K_{\alpha\beta}(ij) = \frac{\phi'_{ij}}{|r_j - r_i|}\delta_{\alpha\beta} + \left(\phi''_{ij} - \frac{\phi'_{ij}}{|r_j - r_i|}\right)n_\alpha(ij)n_\beta(ij). \qquad (7.26)$$

Although the inclusion of the non-zero first derivative and the coupling of oxygen to the second-neighbor nickel atom directly below brings in several new terms, the basic form of the equations of motion for the atomic displacement remains the same as in earlier calculations [7.24,47]. We have proceeded here with the calculations along the same lines, as discussed. In short, the technique is based on solving exactly a set of coupled equations for the motion of the atoms in the surface layers (in this case the adsorbate layer and three nickel layers), using a Fourier transformed Green's function method. The equations describing the motion of the atoms in the bulk layers are solved by invoking an exponential ansatz. The bulk solutions are then matched on to the equations related to atomic displacements in the surface layers. The features in the phonon spectral

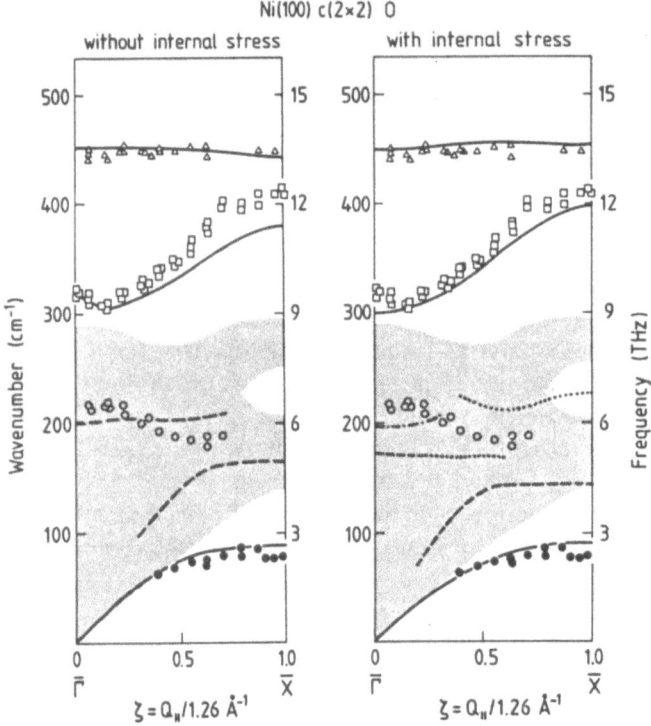

Ni(100) c(2×2) O

Fig. 7.20a,b. Surface modes for the c(2×2) oxygen covered Ni(100) surface calculated in two models: (a) Force constant model as used previously [7.47] with $\phi''_{01} = 2.50$, $\phi''_{00} = 0.22$, and $\phi''_{12} = 0.3$ (in units of the bulk force constant of nickel. (b) Results for a model with an additional attraction force between oxygen and the second layer nickel atom underneath which causes an internal stress. Parameters are $\phi''_{01} = 2.5$, $\phi''_{00} = 0.28$, $\phi''_{02} = 0.7$, $\phi''_{12} = 0.8$, and $\phi'_{02} = 0.35\,a_0$

density plots, which are obtained directly from the Green's function solutions of the equations of motion, can then be identified with the various adsorbate and substrate surface modes and resonances and their dispersion obtained.

In Fig. 7.20 we compare the results of such calculations for a $c(2\times2)$ over-layer of oxygen on Ni(100) with those that were previously obtained [7.47] for a model which did not include coupling of the adsorbate to the second-layer nickel atoms directly below. The parameters chosen were $\phi_{01}'' = 2.5$, $\phi_{00}'' = 0.28$, $\phi_{12}'' = 0.8$ and $\phi_{02}'/a_0 = 0.35$, where the lattice constant $a_0 = 2.49$ Å and ϕ_{ij}'' are the second derivatives of the pair potential with $i = 0, 1, 2$ denoting the adsor-bate, first and second layer nickel atoms, respectively. Our choice of $\phi_{12}'' = 0.8$ is in keeping with the moderate expansion of the interlayer distance [7.25]. Once this parameter is fixed the stress parameter ϕ_{02}' follows from the required match to the S_4 mode at $\overline{X}$. Comparison of the two models shows a slightly better fit for the model with the internal stresses. The main advantage of the model with internal stresses is, however, that it provides a natural explanation for a number of other experimental observations which will be discussed in the following.

7.4.2 The Stress-Model and Other Experimental Observations

In the preceding subsection we have identified the internal stresses caused by the attractive interaction between the oxygen adatom and the second-layer nickel atom as the reason for the reduced frequency of the S_4-surface phonon near $\overline{X}$, rather than an extraordinarily small force constant k_{12} between the first and second layer nickel atoms, as had previously been assumed [7.47]. It is a logical consequence of this new model, that the S_4-anomaly should vanish once the stress is reduced. This is believed to occur for the disordered overlayer. Once the long-range order is not retained the stress may be released by shifting the nickel atoms sideways out of their ideal positions (Fig. 7.21). This, according to (7.26), enlarges the force constant effective for the Rayleigh wave at the $\overline{X}$ point and the frequency of the mode returns to a higher value. An estimate of the effect may be obtained by letting ϕ_{02}' become zero for the $c(2\times2)$ overlayer with all other parameters retained. One obtains $\omega_{S_4}(\overline{X}) = 124 \text{ cm}^{-1}$, in reasonable agreement with the observed value for the disordered overlayer.

It is furthermore interesting to study the effect of removing every second oxygen atom in order to obtain the $p(2\times2)$ surface (Fig. 7.21b). The unbal-anced forces on the first layer nickel atoms would then give rise to a sideways shift. We have calculated the distortions of the lattice within a slab of 8 nickel layers which reproduces reasonably well the dispersion of the adsorbate modes and the Rayleigh wave shown in Fig. 7.20. The resulting equations for the dis-placements $u_\alpha(l_z\kappa)$ of the atom κ' in layer l_z are linear in the displacements save for the $u_\alpha(l_z\kappa)$ appearing in the projections of the bond vectors $\hat{n}_\alpha(ij)$, which enter the equations. Assuming that the displacements $u_\alpha(l_z)$ are small the resulting equations may be solved easily using standard methods. The cal-culated displacements were then used to solve the set of equations again but

a)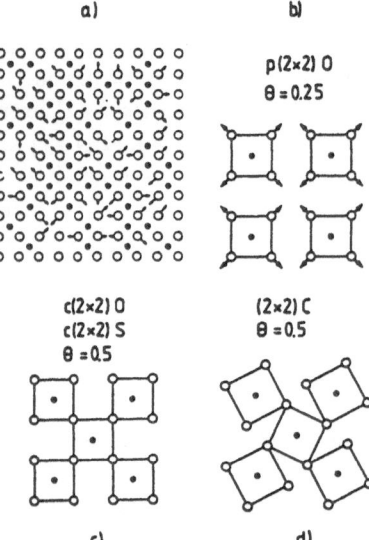

p(2×2) O

θ = 0.25

Fig. 7.21a–d. Effect of adsorption on Ni(100): in case (a) and (b) the Ni atoms can accomodate to the strain induced by the presence of the adsorbate by displacing themselves from the lattice sites. In case (c) no displacement is possible. In (d) we show the reconstruction of Ni(100) after carbon adsorption

b)

c(2×2) O
c(2×2) S
θ = 0.5

(2×2) C
θ = 0.5

c)   d)

with bond vectors $n_\alpha(ij)$ calculated with the atoms in the displaced positions. This was repeated until self-consistency was achieved. For the parameters used for the oxygen overlayer only two cycles were found to be necessary. We have found the static displacements $u_\alpha(l_z\kappa)$ shown in Table 7.1.

As can be seen from Table 7.1 the main displacements are an inwards motion of the oxygen atom and a sideways motion of the first-layer nickel atom, i.e. a reconstruction of the surface. The possibility of such a reconstruction so far has not been explored with structural methods such as LEED or EXAFS. In fact, EXAFS is not very sensitive to the type of reconstruction proposed here, since the nearest-neighbour oxygen nickel distance changes only by 0.03 Å (using the displacements of Table 7.1). Some indirect evidence that the proposed reconstruction might occur is provided by the frequency ratio of the parallel ($640\,\mathrm{cm}^{-1}$) and perpendicular oxygen vibration ($430\,\mathrm{cm}^{-1}$) for the p(2×2) surface. *Lloyd* and *Hemminger* [7.48] noticed that this ratio can be fitted by invoking a force field with three-body forces whereas only $570\,\mathrm{cm}^{-1}$ would be obtained for the frequency of the parallel vibration in a central-force model fitted to the frequency of perpendicular motion. While one cannot a priori ex-

Table 7.1. Static displacements [Å] in case of a p(2×2) overlayer with $\phi_{01}'' = 2.5$; $\phi_{12}'' = 0.8$; $\phi_{02}'' = 0.9$; $\phi_{02}' = 0.35a_0$

$u_z(01) = -0.17$	vertical displacement of oxygen atom
$u_x(11) = +0.10$	sideways displacement of first layer nickel atoms
$u_z(11) = -0.01$	vertical displacement of first layer nickel atoms
$u_z(21) = +0.00$	vertical displacement of second layer nickel atoms (underneath oxygen)
$u_z(22) = -0.06$	vertical displacement of second layer nickel atoms (without oxygen above)
$u_\alpha(l_z\kappa) < 0.02$	all other

clude that three-body forces are important, it would remain unexplained why the p(2×2) oxygen overlayer is unique in that. The proposed reconstruction would lead automatically to a higher frequency of the parallel vibration.

Finally, we comment on the relation of the model of internal strains to a reconstruction of the surface which occurs with a c(2×2) overlayer of carbon atoms [7.49,50] (Fig. 7.21d). It was noticed recently by some of the authors [7.51] that the reconstruction of the carbon-covered nickel surface is identical to the displacement pattern of an odd mode at the $\overline{X}$ point of the unreconstructed c(2×2) surface. This mode belongs to the A_2 representation of the C_4 point group relevant at $\overline{X}$ [7.28,52]. Using the nearest-neighbor force constant model it was argued that the A_2 mode at $\overline{X}$ becomes soft with the carbon overlayer. It was suggested there that one possible way this could happen was if the force constant k_{12} became vanishingly small, in a continuing trend as was then perceived from the observation on similar overlayers of oxygen and sulfur on Ni(100). While it was shown that the A_2 mode at $\overline{X}$ was the only mode to become soft under this condition it remained difficult to perceive how the coupling constant ϕ''_{12} should manage to become zero without completely removing the first nickel layer with the carbon atoms from the substrate. The model with adsorbate induced internal stresses provides a much more elegant explanation. As with the nearest-neighbor central force model the particular mode of A_2 symmetry remains localized entirely within the first nickel layer and the eigenfrequency of the mode is simply given by

$$M_s\omega^2_{A_2}(\overline{X}) = \phi''_{12} + 2\phi'_{12}\frac{1}{a_0} + 2\phi'_{01}\frac{1}{(a_0^2/2 + R_\perp^2)^{1/2}}$$

$$= \phi''_{12} - \frac{1}{2a_0}\left(\sqrt{2} + \frac{a_0}{R_\perp}\right)\phi'_{02} \qquad (7.27)$$

with a_0 the surface lattice constant of the clean surface and $R_\perp$ the vertical distance of oxygen above the surface. In deriving this equation we have used (7.24–26) and assumed that the distance between the first and second nickel layer is the same as in the bulk. It is easily seen that with the set of parameters used in Fig. 7.20 for describing the c(2×2) oxygen overlayer the A_2 mode remains at finite frequency and the c(2×2) structure is stable. If ϕ'_{02}, the attractive interaction between the adsorbate and the second-layer substrate atom, increases the mode becomes soft and the surface reconstructs into the rotated phase, as depicted in Fig. 7.21d. [Because of its symmetry elements this structure is designated as p4g(2×2)]. Once the surface has been reconstructed, the internal stress is reduced. If one takes the anomalously low frequency of the S_4 mode at $\overline{X}$ as an indication of the presence of internal stresses, as our lattice-dynamical calculations indicate, the frequency of the S_4 mode should be high again on the reconstructed carbon surface, which is in fact the case [7.17,53]. If, on the other hand, ϕ''_{12} were to become zero for the carbon overlayer one would expect the S_4 mode to be also of low frequency in the reconstructed case.

Clearly, the strength of the interaction between the adsorbate and the second-layer substrate atom is likely to increase with smaller distances $R_\perp$ of the adatom. The stability of the $c(2\times2)$ overlayer depends also on the vertical distance of the adatom directly for pure geometrical reason. The smaller $R_\perp$ is, the larger are the internal stresses ϕ'_{12} and ϕ'_{01} necessary to keep the adatom from moving inside. This is illustrated in Fig. 7.22 where we have plotted the critical ratio of ϕ'_{02}/a_0 and ϕ''_{12} which separates the unreconstructed $c(2\times2)$ phase from the reconstructed $p4g(2\times2)$ phase as a function of the vertical distance of the adsorbate. One immediately sees that the $c(2\times2)$ overlayer is less likely to be stable for small $R_\perp$. Carbon has a vertical distance of only 0.1 Å [7.49,50], so that even with the parameters used for fitting the experimental data of oxygen the carbon surface would reconstruct. On the other hand, for sulfur with an adatom distance of 1.35 Å the $c(2\times2)$ is even more stable in keeping with the experimental result.

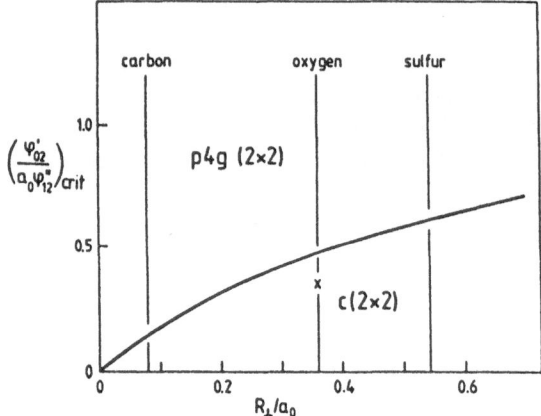

Fig. 7.22. Zero temperature phase diagram for the $c(2\times2)$ and $p4g(2\times2)$ phases with the vertical distance of the adsorbate as parameter. One notices that for any non zero attractive interaction between the adsorbate and the second layer substrate atom ϕ'_{02}, the surface is more likely to be in the reconstructed $p4g(2\times2)$ phase when $R_\perp$ is small. In addition ϕ'_{02} is expected to be larger when $R_\perp$ is small

We can therefore conclude that the model of internal stresses induced by the adsorbate atoms provides an elegant and unifying description of a large variety of experimental observations concerning the structure and dynamics of (100) surfaces. The model may well pertain to other surfaces and adsorbates.

References

7.1   Lord Rayleigh: London Math. Soc. Proc. **17**, 4 (1885), reprinted in Scientific Papers by Lord Rayleigh, Vol. 2 (Dover, New York 1964)
7.2   J.L. Bluestein: Appl. Phys. Lett. **13**, 412 (1969)
7.3   H. Mattews: *Surface Wave Filters* (Wiley, New York 1977)

7.4	R.E. Allen, G.P. Alldredge, F.W. De Wette: Phys. Rev. B4, 1661 (1971); E.A. Ash, E.G.S. Parge (eds.): *Rayleigh-Wave Theory and Application*, Springer Ser. Wave Phen., Vol. 2 (Springer, Berlin, Heidelberg 1985)
7.5	M. Strongin, O.F. Kammener, J.E. Crow, R.D. Parks, D.H. Douglass, J.R. Jensen, M.A. Jensen: Phys. Rev. Lett. **2**, 1320 (1968)
7.6	E. Tosatti: In *Festkörperprobleme* **15**, 113 (Vieweg, Braunschweig 1975) A. Fasolino, G. Santoro, E. Tosatti; Phys. Rev. Lett. **44**, 1684 (1980)
7.7	See, for example, for Brillouin scattering: *Low Dimensional Cooperative Phenomena*, ed. by M.J. Keller (Plenum, New York 1975) and for enhanced Raman scattering see review papers in *Surface Enhanced Raman Scattering*, ed. by R.K. Chang, T.E. Furtak (Plenum, New York 1982)
7.8	N. Cabrera, V. Celli, R. Manson: Phys. Rev. Lett. **22**, 346 (1969)
7.9	G.B. Brusdeylins, R.B. Doak, J.P. Toennies: Phys. Rev. B**27**, 3662 (1983)
7.10	H. Ibach, D.L. Mills: *Electron Energy Loss Spectroscopy and Surface Vibrations* (Academic, New York 1982)
7.11	S. Lehwald, J.M. Szeftel, H. Ibach, T.S. Rahman, D.L. Mills: Phys. Rev. Lett. **50**, 518 (1983)
7.12	S.Y. Tong, C.H. Li, D.L. Mills: Phys. Rev. Lett. **44**, 407 (1980) C.H. Li, S.Y. Tong, D.L. Mills: Phys. Rev. B**21**, 3057 (1980)
7.13	G. Ertl, J. Küppers: *Low Energy Electrons and Surface Chemistry* (Verlag Chemie Weinheim 1974) p. 196
7.14	H. Ibach (ed.): *Electron Spectroscopy for Surface Anlaysis*, Topics Current Phys., Vol. 4 (Springer, Berlin, Heidelberg 1977) p 1
7.15	J.C. Shelton: Surf. Sci. **44**, 305 (1974)
7.16	Mu-Liang Xu, B.M. Hall, S.Y. Tong, M. Rocca, S. Lehwald, H. Ibach, J. Black: Phys. Rev. Lett. **54**, 1171 (1985)
7.17	M. Rocca: Ph.D. thesis, KFA-Report Jül-2000 (June 1985)
7.18	M. Rocca, S. Lehwald, H. Ibach, Mu-Liang Xu, B.M. Hall, S.Y. Tong: In *The Struc- ture of Surfaces*, ed. by M.A. Van Hove, S.Y. Tong, Springer Ser. Surf. Sci., Vol. 2 (Springer, Berlin, Heidelberg 1985)
7.19	S. Lehwald, M. Rocca, H. Ibach, T.S. Rahman: Phys. Rev. B**31**, 3477 (1985)
7.20	M. Rocca, S. Lehwald, H. Ibach, T.S. Rahman: Surf. Sci. Lett. **138**, L123 (1984)
7.21	M. Rocca, S. Lehwald, H. Ibach, T.S. Rahman: Surf. Sci. **171** (1986) published
7.22	R.B. Doak, U. Harten, J.P. Toennies: Phys. Rev. Lett. **51**, 578 (1983)
7.23	U. Harten, J.P. Toennies, Ch. Wöll: In *Dynamical Phenomena at Surfaces, Interfaces and Superlattices*, ed. by F. Nizzoli, K.H. Rieder, R.F. Willis, Springer Ser. Surf. Sci., Vol. 3 (Springer, Berlin, Heidelberg 1985)
7.24	T.S. Rahman, J.E. Black, D.L. Mills: Phys. Rev. B**25**, 883 (1982)
7.25	J.W.M. Frenken, R.G. Smeenk, S.F. Van der Veen: Surf. Sci. **135**, 147 (1983)
7.26	J.W.M. Frenken, J.F. Van der Veen, G. Allan: Phys. Rev. Lett. **51**, 1876 (1983)
7.27	J. Szeftel, S. Lehwald, H. Ibach, T.S. Rahman, D.L. Mills, J. Black: Phys. Rev. Lett. **51**, 268 (1983)
7.28	J. Szeftel, S. Lehwald: Surf. Sci. **143**, 11 (1984)
7.29	M. Rocca, S. Lehwald, H. Ibach: Surf. Sci. Lett. **163**, L738 (1985)
7.30	J.E. Demuth, D.W. Jepsen, P.M. Marcus: Phys. Rev. Lett. **31**, 54 (1973)
7.31	S. Brennan, J. Stöhr, R. Jäger: Phys. Rev. B**24**, 4871 (1981)
7.32	P.H. Citrin, P. Eisenberger, R.C. Hewitt: Phys. Rev. Lett. **41**, 309 (1978)
7.33	J. Stöhr, R. Jäger, T. Kendelewicz: Phys. Rev. Lett. **49**, 142 (1982)
7.34	R.J. Orders, B. Sinkovic, C.S. Fadley, R. Thehan, Z. Hussain, J. Lecante: Phys. Rev. B**30**, 1838 (1984)
7.35	J.J. Bartou, C.C. Bahr, Z. Hussain, S.W. Robey, J.G. Tobin, L.E. Kelbranoff, D.A. Shirley: Phys. Rev. Lett. **51**, 272 (1983)
7.36	D.H. Rosenblatt, J.G. Tobin, M.G. Mason, R.F. Davis, D.A. Shirley, C.H. Li, S.Y. Tay: Phys. Rev. B**23**, 3828 (1981)
7.37	K.H. Rieder: Surf. Sci. **128**, 325 (1983)
7.38	M. De Crescenzi, F. Antonangeli, C. Bellini, R. Rosei: Phys. Rev. Lett. **50**, 1949 (1983)
7.39	S. Andersson: Surf. Sci. **79**, 385 (1979)
7.40	S. Andersson, P.-A. Karlsson, M. Persson: Phys. Rev. Lett. **51**, 2378 (1983)

7.41   S. Lehwald, H. Ibach: In *Vibrations at Surfaces*, ed. by R. Caudano, J.M. Gilles, A.A. Lucas (Plenum, New York 1982)
7.42   T.H. Upton, W.A. Goddard: Phys. Rev. Lett. **46**, 1635 (1981)
7.43   C.W. Bauschlicher, Jr., P.S., Bagus: Phys. Rev. Lett. **52**, 200 (1984)
7.44   J. Black: Private communication
7.45   G. Benedek, G. Nardelli: Phys. Rev. **155**, 1004 (1967)
7.46   H. Ibach, J.E. Müller, T.S. Rahman: Phys. Trans. Royal Soc. (to be published)
7.47   T.S. Rahman, D.L. Mills, J.E. Black, J. Szeftel, S. Lehwald, H. Ibach: Phys. Rev. B**30**, 589 (1984)
7.48   K.G. Lloyd, J.H. Hemminger: Surf. Sci. **143**, 509 (1984)
7.49   J.H. Onuferko, D.P. Woodruff, B.W. Holland: Surf. Sci. **131**, 245 (1983)
7.50   K.H. Rieder, H. Wilsch: Surf. Sci. **131**, 245 (1983)
7.51   T.S. Rahman, H. Ibach: Phys. Rev. Lett. **54**, 1933 (1985)
7.52   H. Ibach: In Advances in Physics (in press)
7.53   M. Rocca, S. Lehwald, H. Ibach, T.S. Rahman: To be published

Subject Index

Adsorbate structures 21
  adsorbate mode frequencies 180
  C 272
  CH_3O 44
  CO 35, 42, 44
  D 60
  H 26, 29, 60, 61
  HCO_2 45
  H_2O 62
  location of adsorbed atoms 111, 130
  metals with adsorbates 179
  $NiSi_2$ 60
  O 21, 37, 39, 40, 44, 157, 180, 186, 262, 272
  O_2 45
  organic molecules 35
  S 42, 262
  Si 47
  W 48
Adsorbate-induced substrate reconstruction 29
Ag(111) 176
Ag(110)+Ar, Kr, Xe 190
Al(111)+O 186
Ammonia synthesis 5
Angle-resolved photoelectron spectroscopy 2
Angle-resolved photoemission extended fine structure (ARPEFS) 42
  Ni(100)+S 42
Angular-dependent photoelectron diffraction (ADPD) 42
Anharmonic surface effects 199–243
Ar 11, 190
Atomic beam scattering 25
  He-diffraction 26
  Ne-diffraction 30
Atom-surface potential 25
Au(110)(1×2) 23, 33, 48, 52, 55
Au(111) 73, 176, 177
Au islands on Si 51
Auger electrons (ion induced) 120, 130
Average values 205
Axilrod-Teller forces 192

Barker potential 231
Blocking 56
  monte carlo trajectory calculations 58

Bravais lattices 155
Brillouin scattering 245
Brillouin zone types 163
Bulk modes 165

Canonical ensemble 207
Catalysis 4
Catalytically active surfaces 45
  active sites 45
Channeling 56, 111
  angular yield profile 120, 137
  Auger-electrons (ion induced) 120, 130
  Computer simulations 125
  continuum potential 111
  dechanneling 119
  flux density contours 118, 134
  flux distribution 116
  surface channeling 111
  thermal displacements 113
  Thomas-Fermi potential 113
Charge-density contours 4
Charge-density relaxation 31
Cluster growth 37
Commensurate overlayers 183
Continued fraction method 168
Continuum methods 164
Continuum potential 111
Contour of surface electron density 25
Correlation functions 204
  pair correlation function 210
  triplet correlation function 224
  velocity autocorrelation function 215
Corrugation 25
  function 25
Cut-off radius (interaction) 215
Cu(100) 175
Cu(110) 21, 31, 178
Cu(111) 176
Cu(100)+CH_3O 44
Cu(100)+CO 44, 183, 189
Cu(100)+c(2×1)O 39
Cu(100)+HCO_2 45
Cu(110)+(2×1)O 21
Cu(100)+O_2 45
Cu(100)+O, S, CH 189
Cu(100)+Xe 190
Cylindrical mirror analyzer 40

Debye-Waller factor 248
Dechanneling 119
Density in phase space 205
Density-functional formalism 217
Diffusion 48, 237
  coefficient 48, 216, 237
  Kr 237
  W on Ir(110) 48
Dipole scattering 62
Dispersion curves for electron states 3
Distortion 17
  displacive 17
  reconstructive 17
Dynamical theory 30
Dynamics of adsorbate layers 179, 245
  Ag(110)+Ar, Kr, Xe 190
  Ag(111)+Ar, Kr, Xe 190
  Al(111)+O 186
  commensurate overlayers 183
  Cu(100)+CH, O and S 189
  Cu(100)+CO 183, 189
  Cu(100)+Xe 190
  fcc(100) substrate 186
  fcc(110) substrate 190
  fcc(111) substrate 183
  incommensurate overlayers 190
  metals with adsorbates 179
  Ni(100)+CO 181
  Ni(100)+O 180, 181, 186
  Ni(111)+O 183
  Ru(001)+O 190

Eikonal approximation 26
Electron density 217
  density-functional formalism 217
  static electronic response function 217
Electron energy loss spectroscopy (EELS)
  246
  Debye-Waller factor 248
  elastic contribution 248
  electron-surface interaction 246
  experimental set up 250
  multiphonon contribution 248, 252
  multiple-scattering effects 252
  Ni(100) 251, 255, 258
  Ni(100)+c(2×2)C 272
  Ni(100)+c(2×2)O 262, 270
  Ni(100)+c(2×2)S 257, 262
  Ni(100)+p(2×2)O 262
  Ni(100)+p(2×2)S 262
  one-phonon contribution 248, 252
Electron induced desorption 62
Electron scattering 76
Electron stimulated desorption ion angular
  distribution 62
Electron-phonon coupling 13, 245
Energy dependent photoelectron
  diffraction (EDPD) 42

Ensembles (statistical) 206
Ewald condition 20

Facets 89
Fe(310) 31
FIM 46
  Au(110)(1×2) 48
  W(100)c(2×2) 48
  W(110)+Si 47
Fourier backtransformation 39
Frequency spectrum 215

GaAs(111)(2×2) 33
Glancing incidence x-ray diffraction 22
  InSb(111)(2×2) 35
Glide lines 37
Gran canonical ensemble 207, 213
Graphite 37, 228
Green's function method 167, 182, 183, 270

Hamilton equations 204
Hard corrugated wall model 26
Hard disks 225
He-diffraction 26
He-scattering 173, 261
Heterogeneous catalysis 4
High-resolution electron microscopy
  (HREM) 71
  Ag-particles 88
  Au(100) 84
  Au(110) 97
  Au(111) 73
  dispersion surface 77
  facets 89
  image formation 74
  image simulation 84
  lens aberration 80
  MgO 96
  reflection geometry 106
  resolution 80
  steps 89
  transmission geometry 103
  WO$_3$ 91
HREELS 62
Hydrogen 26
  Ni(110)+H 26
  Pd(110)+H 29

Image formation 74
Image simulation 84
Impact collision scattering spectroscopy 61
Impact scattering 63
Incommensurate overlayers 35, 190
InP(100) 1
InSb(111)(2×2) 35
Interaction Potentials 201, 213, 214
  at surfaces 216
  attractive part 216
  cut-off radius 215

Morse potential 218
  repulsive part 216
Interfaces 213
Interference conditions 20
Internal strains 268
Ion scattering 31
  Cu(110) 31
  Si(111)(2×1) 33
Ion scattering spectroscopy 56
  Ni(100)+H, D 60
  Si(111)(2×1) 56
  Si(111)+epitaxial NiSi$_2$ 60
  steering effect 61, 111
Ir(100)(1×5) 53
Ir(110) 48
  W on Ir(110) 53

Kinematical theory 22
Kinetic oscillations 7
Kr 11, 190, 230

Lateral resolution 50
Laue condition 20
Layer-by-layer growth 37
LEED 30
  Au(110)(1×2) 33
  Cu(110) 31
  dynamical theory 30
  Fe(310) 31
  GaAs(111)(2×2) 33
  InP(110) 1
  intensity versus voltage 30
  multiple scattering 30
  Ni(311) 31
  Ni(100)+(2×2)C 37
  Ni(100)+c(2×2)CO 35
  retarding field analyzer 30
  Si(111)(2×1) 33
  Video-LEED 37
Lennard-Jones model 11, 201, 224, 227
Lens aberration 80
Liouville operator 204
Liouville theorem 215
Location of adsorbed atoms 135

Magnetic structures 25
Maxwell distribution 208, 221
Mean-square displacement 236
Mean-square velocity 208
Melting 58
  Kr 237
  Pb(110) 58
  premelting 237
Microcanonical ensemble 206
Miller indices 156
  fcc surfaces 156
  bcc surfaces 153
Missing row model 25
Molecular beam epitaxy 38

Molecular dynamics 199–243
  correlation functions 204
  cut-off radius 215
  diffusion 237
  frequency spectrum 215
  Hamiltons equations 204
  hard disks 225
  initial values (coordinates, velocities) 219
  interaction potentials 213
  Kr 230
  Lennard-Jones systems 224, 227, 228
  Liouville operator 204
  Maxwell distribution 208, 221
  mean-square displacements 236
  molecules adsorbed on surfaces 227
  models for the bulk 211
  models with surfaces and interfaces 213
  pair correlation function 211
  pair potential approximation 200, 210
  periodic boundary conditions 211
  phase space 204
  phase transitions 224
  premelting 237
  statistical ensembles 205
  statistical error 215
  stochastic boundary conditions 212
  temperature 223
  thermodynamic limit 207
  time evolution 219
  time step 204
  total information 204
  trajectory in phase space 204
  velocity autocorrelation function 215, 237
  Xe on graphite 228
Morse potential 218
Multi-layer relaxation 238
Multiple scattering 30, 252

Ne 11
Ne-diffraction 30
Neutrons 25
NEXAFS 43
  Cu(100)+CH$_3$O 44
  Cu(100)+CO 44
  Cu(100)+HCO$_2$ 45
  Cu(100)+O$_2$ 45
  Ni(100)+O 44
Nitrogen on graphite 37
Ni(100) 171, 175, 251, 255, 258
Ni(110) 31, 114, 130, 171, 178
Ni(111) 165, 177
Ni(311) 31
Ni(100)+(2×2)C 37
Ni(100)+c(2×2)CO 35, 42
Ni(100)+CO 181
Ni(100)+c(2×2)O 40, 157, 181, 186, 262, 263, 270
Ni(100)+c(2×2)S 257, 262, 264
Ni(110)+H 26

Ni(111)+(2×2)H 61
Ni(100)+H, D 60
Ni(110)+H₂O 62
Ni(100)+O 44, 180
Ni(110)+O 130
Ni(111)+O 183
Ni(110), oxigen adsorption and oxidation 37
Ni(100)+p(2×2)O 154, 181, 186, 262
Ni(100)+p(2×2)S 262
Ni(100)+S 43
NMR 64
Noble gases 11, 201
Noble gases on graphite 37, 228
Nozzle sources 25

Optical methods for electronic surface
  transitions 63
Oxidation of CO 7

Pair potential approximation 200
  cut-off radius 215
  metals 202, 210
Patterson synthesis 23, 35
Pb(110) 58
Pd(110)+H 29
Periodic boundary conditions 211
Periodic overlayers 163
Phase space 205
  density 205
  trajectory 205
Phase transitions 224
  crystallization 225
  hard disks 225
  two-dimensional systems 224, 225
Phonon density of states 10, 13
Phonons 62, 146, 157, 245
  adsorbate mode frequencies 180
  Ag(111) 176
  Ag(110)+Ar, Kr, Xe 190
  Ag(111)+Ar, Kr, Xe 190
  Al(111)+O 186
  anomalies 268
  Au(111) 176, 177
  bare metals 164
  bulk bands 165
  case of a bare periodic surface 161
  case of a periodic overlayer 163
  case of no periodicity 157
  commensurate overlayers 186
  continued fraction method 168, 181
  continuum methods 164
  Cu(100) 175
  Cu(110) 178
  Cu(111) 176
  Cu(100)+CO 183, 189
  Cu(100)+O, S, CH 189
  Cu(100)+Xe 190
  dynamical matrix 160
  fcc(100) substrate 183

fcc(110) substrate 190
fcc(111) substrate 183
fcc(100) surfaces 174
fcc(110) surfaces 177
fcc(111) surfaces 176
frequency spectrum 215
Green's function calculations 167, 182, 183,
  270
He-scattering 173, 261
incommensurate overlayers 190
internal strains 268
metal with adsorbates 179
methods of mode frequency calculation
  164
Ni(100) 171, 175, 258
Ni(110) 171, 178
Ni(111) 165, 177
Ni(100)+c(2×2)C 272
Ni(100)+CO 181
Ni(100)+c(2×2)O 181, 262, 270
Ni(100)+c(2×2)S 257, 262
Ni(100)+O 183
Ni(111)+O 183
Ni(100)+p(2×2)O 181, 262
Ni(100)+p(2×2)S 262
Pt(100) 176
Pt(332) 179
Pt(111), stepped 179
Rayleigh wave 176, 258, 264, 266, 271
resonances 166, 258, 260, 256
Ru(001)+O 190
slab methods 165
soft phonons 172, 176
spectral density methods 167
surface bands 165
T-matrix method 167
Photoelectrons 38
  diffraction 41, 42
Photon induced desorption 62
Premelting 237
Pseudo surface modes (resonances) 166,
  258, 260, 265
Pt 7
Pt(100) 176
Pt(100)(5×20) 53
Pt(111) 179

Raman scattering 245
Rayleigh wave 176, 258, 264, 266, 271
Reciprocal lattice 19, 162
Reconstruction 1, 17, 130, 154, 156, 272
  adsorbate induced 29, 272
  Au(110)(1×2) 23, 33, 48, 52, 55
  GaAs(111)(2×2) 33
  InSb(111)(2×2) 34
  Ir(100)(1×5) 53
  missing row reconstruction 133, 156
  Ni(100)+c(2×2)C 272
  Ni(110)+O 130

Pt(100)(5×20) 53
Si(111)(2×1) 56, 63
Si(111)(7×7) 54
W(100)c(2×2) 48
Reflection geometry 106
Reflection-adsorption infrared
   spectroscopy 62
Relaxation 1, 17, 30, 79, 90, 154, 238
   effect on electronic surface properties 1
   InP(110) 1
   Kr 238
   multi-layer relaxation 238
   Ni(110) 31
Resonance scattering 63
Resonances 166, 258, 260, 265
Rh(111), coadsorption of benzene and CO
   35
RHEED 37
   molecular nitrogen on graphite 37
   Ni(110), oxygen adsorption and oxidation
   37
   noble gases on graphite 37
Ru(001)+O 190
Rutherford backscattering 61, 262

Scanning tunneling microscopy 49
   Au islands on Si 51
   Au(110)(1×2) 52
   lateral resolution 50
   Si(111)(7×7) 54
Search-light effect 39
SEELFS 40
   Ni(100)+c(2×2)O 40
SEM (STEM) 55, 72
SEXAFS 38
   x-ray core absorption edge 38
   Cu(110)+c(2×2)O 39
Shadowing 56
Si(111)(2×1) 33, 56, 63
Si(111)(7×7) 54
Si(111) + epitaxial NiSi$_2$ 60
Slab methods 165
Soft phonons 172, 176
Spectral density methods 167
Statistical ensembles 206
   canonical ensemble 207
   grand canonical ensemble 207
   microcanonical ensemble 206
Statistical error 215
Steering effect 61, 111

Stepped surfaces 20, 89
   influence on surface channeling 127
Stochastic boundary conditions 212
Structural phase transitions 8, 224, 226, 227,
   228, 230
Superconducting transition temperature 14
Surface unit cell 18, 154
Synchrotron sources 22, 39

Theory of vibrations 158–198
Thermal displacements 113
Thermal equilibrium 221
Thermodynamic functions 9
   entropy 9
   free energy 9
   harmonic approximation 9
   specific heat 9
Thermodynamic limit 207
Thomas-Fermi potential 113
Time evolution (molecular dynamics) 219
Time step 204
T-matrix method 167
Trajectory in phase space 205
Trajectory of scattered projectiles 113
Transmission electron microscopy (TEM)
   55, 72
   Au(110)(1×2) 55
Transmission geometry 103

UPS 63
   angle-resolved UPS(ARUPS) 56, 63
   Si(111)(2×1) 63

Vacancy-buckling model 33
Velocity autocorrelation function 215
   frequency spectrum 215
Video-LEED 37

W(110)+Si 47
W(100)c(2×2) 48
Wetting 37
   complete, incomplete 37
Work-function 8

Xe 11, 190, 228
X-rays 22
   core absorption 38
   glancing incidence 22
XPS 63